FROM PARTICLES
TO PLASMAS

Marshall N. Rosenbluth

Lectures honoring Marshall N. Rosenbluth

FROM PARTICLES
TO PLASMAS

Edited by James W. Van Dam

Institute for Fusion Studies,
The University of Texas at Austin

Addison-Wesley Publishing Company, Inc.

The Advanced Book Program

Redwood City, California • Menlo Park, California • Reading, Massachusetts
New York • Amsterdam • Don Mills, Ontario • Sydney
Bonn • Madrid • Singapore • Tokyo
San Juan • Wokingham, United Kingdom

Publisher: *Allan M. Wylde*
Production Administrator: *Karen L. Garrison*
Editorial Coordinator: *Aida Adams*
Electronic Production Consultant: *Mona Zeftel*
Promotions Manager: *Celina Gonzales*

ISBN 0-201-15680-6

ABCDEFGHIJ-MA-898

Foreword

Scientists from around the world gathered at the University of Texas of Austin on February 5 and 6, 1987, to participate in an international symposium on Dynamics of Particles and Plasmas and to honor Professor Marshall N. Rosenbluth on his sixtieth birthday. A few months later, a special session was held for the same purpose during the Annual Controlled Fusion Theory Conference (or "Sherwood Meeting"), April 15, 1987, in San Diego. The invited lectures that were presented at both occasions are contained in the present volume.

The title of this book, *From Particles to Plasmas*, has more than one meaning. First, it reflects how the scientific career of Marshall Rosenbluth has evolved, beginning in the field of elementary particle physics and extending into his major area of plasma physics. Secondly, it is meant to suggest the wide spectrum of subject matters addressed in the individual lectures, ranging from numerical simulation and space physics and accelerators to various subfields in the physics of plasmas. In the third place, the title is a reference to the way in which the theoretical description of plasmas is often constructed, namely, starting from the motion of single particles and then incorporating collective effects.

Most of the contributions in this book do concern various aspects of fusion plasma physics, which is the field in which most of Marshall Rosenbluth's scientific contributions have been and are being made. In this field his eminence and authority are indicated by the sobriquet "pope of plasma physics" that is often applied to him.

A brief account of the life and work of Marshall Rosenbluth is in order. He was born in Albany, New York, on February 5, 1927. He did his undergraduate study at Harvard University, graduating in 1946, and then three years later, at the age of 22, obtained a Ph.D. from the University of Chicago. He was briefly an instructor at Stanford University and then joined the thermonuclear program at Los Alamos Scientific Laboratory in 1950. In 1956 he moved to the John Jay Hopkins Laboratory of the General Atomic Division of General Dynamics Corporation in La Jolla, California, and in 1960 also became a professor in the department of physics of the University of California at San Diego. He moved cross-country in 1967, joining the Institute for Advanced Study in Princeton as a professor in the school of natural sciences. At the same time he was a senior research physicist at the Plasma Physics Laboratory of Princeton University. In 1980 he was attracted to the University of Texas at Austin by the opportunity to establish the new Institute for Fusion Studies. For seven years he was its director, as well as Fondren Professor of Plasma Physics. In late 1987 he returned to California where he is currently a professor again at the University of California at San Diego and also a senior scientist at GA Technologies.

His work has spanned a wide range. It is appropriate to quote from the citation that accompanied the Enrico Fermi Award, which he received in 1986:

> "Marshall Rosenbluth's theoretical contributions have been central to the development of controlled thermonuclear fusion. He is widely considered to be the leading plasma theorist of this generation, including work of central significance for fusion research and development.

> "Dr. Rosenbluth was one of the first to recognize the reason for the instabilities that plagued the attempts in the 1950's to generate stable plasmas. He suggested a way to cure these instabilities which, together with Russian contributions, led to the design of the Tokamak, which is the principal configuration used for current fusion experiments.

> "In particular, in the 1950's his work provided a physical understanding of plasma behavior from the individual particle point of view. This was followed by a mathematical formulation of the kinetic theory which provided a rigorous basis for modern plasma physics. In the early 1960s, his work on the theory of inhomogeneous plasmas and the mathematical descriptions he developed laid the foundation for much of the subsequent work on plasma stability.

> "Marshall Rosenbluth's work on the theory of plasma stability and plasma confinement spans virtually the entire range of fusion research. His work on resistive instabilities again was the pioneering work in this area. In addition, he has made major contributions to the understanding of neoclassical diffusion, trapped particle instabilities, parametric instabilities, and the nonlinear development of magnetohydrodynamic instabilities. Understanding such instabilities is critical to the development of plasma confinement approaches that could lead to useful fusion energy systems.

"Dr. Rosenbluth has been an important contributor to two other areas of energy research. His analysis of the scattering of relativistic electrons by nuclei provided the theoretical underpinning for experimental work by Hofstadter and others on the structure and charge distribution within nuclei. He also made significant contributions at Los Alamos to the design of the first thermonuclear weapon.

"Throughout his career, Dr. Rosenbluth has made important contributions as a scientific advisor in the fields of energy and national defense through service on JASON, National Academy of Science committees, and many other bodies. Dr. Rosenbluth has been the recipient of the E. O. Lawrence Memorial Award (1964), the Albert Einstein Award (1967), and the James Clerk Maxwell Prize in Plasma Physics (1976). He is also a member of the American Academy of Arts and Sciences and of the National Academy of Sciences."

The last chapter of this book includes a list of the publications of this distinguished scientist. This list was brought up to date to cover work published through 1987. It is not a complete list, however, since some of his Los Alamos and JASON work is classified and since he is still actively involved in research.

The considerable efforts of many persons contributed to the organization of the Austin symposium and of the San Diego special session and to the publication of these lectures. The International "Rosenbluth Symposium" on Dynamics of Particles and Plasmas was arranged under the auspices of the Institute for Fusion Studies of the University of Texas at Austin by an organizing committee that consisted of Herbert L. Berk, Patrick H. Diamond, Richard D. Hazeltine, C. Wendell Horton, David W. Ross, and James W. Van Dam (chairman). The symposium was opened by Hans Mark, chancellor of the University of Texas System. The chairmen for the various sessions were Tihiro Ohkawa, Albert Simon, Paul H. Rutherford, Kyoji Nishikawa, Norman Rostoker (who also emceed the banquet), Ravi N. Sudan, Steven L. Weinberg, Alec A. Galeev, and John Killeen. The coordinator for the symposium was Carolyn Valentine, and other secretarial and administrative assistance was provided by Dawn East, Saralyn Stewart, and Joyce Patton.

The special session at the Annual Controlled Fusion Theory Conference was organized by Chuan S. Liu (chairman), with partial assistance from a committee composed of Herbert L. Berk, Bruno Coppi, Russell Kulsrud, L. Donald Pearlstein, Paul H. Rutherford, Ravi N. Sudan, and James W. Van Dam.

The publication of this book was supported by a grant from Maxwell Laboratories, Inc., which was generously arranged by Dr. Alan C. Kolb, its chairman and chief executive office and a long-time friend of the honoree. The expenses of many of the scientists who attended the symposium and the special session were supported by the Office of Fusion Energy of the U. S. Department of Energy. The work of the Institute for Fusion Studies itself is jointly funded by the U. S. Department of Energy and by the State of Texas.

The encouragement and expert assistance of Allan M. Wylde, vice president for advanced books, Carl Hesler, Jr., general manager of the Austin office, and their colleagues at Addison-Wesley Publishing Company have been greatly appreciated. Technical typesetting of the manuscript was done in *TEXtures* by Suzy Crumley, Lisa Hall, and Laura Patterson. Many helpful stylistic comments and other suggestions were received from Richard Hazeltine.

Last, but not least, I thank the authors for their excellent contributions to this volume, which is intended as a tribute to an outstanding career and a remarkable person.

James W. Van Dam

Institute for Fusion Studies
The University of Texas at Austin
Austin, Texas

Contents

x   Contents

Lectures at a Special Session of the 1987 Sherwood Controlled Fusion Theory Conference

FROM PARTICLES
TO PLASMAS

Richard F. Post
Lawrence Livermore National Laboratory
University of California
Livermore, California 94550

Fusion Research and Plasma Physics: A Story of Paradigms

I was pleased and honored and, I confess, somewhat unnerved by the prospect of presenting the keynote talk on this occasion. Pleased and honored, because I have such warm feelings toward our honoree; unnerved, by the task of doing justice to the topics I would like to discuss.

A paradigm — Webster defines the word as meaning "an outstandingly clear example." Thus: "a story of paradigms" — because I believe fusion research and modern plasma physics together contain a remarkable spectrum of paradigms, one of them being our honoree himself.

As the title of my talk suggests, I intend to talk about more than one topic and about more than one outstanding example, but all are interrelated by a common denominator.

The common denominator is, of course, the goal of fusion power, a goal no less desirable today than when it was first being seriously discussed — at least four decades ago.

One thing I share with Marshall Rosenbluth is that both of us have been involved in fusion research for more than three of those four decades. Looking back, I am not sure that any one of us who was involved in the research at that time realized at the outset just how formidable a problem the achievement of fusion power represented — and still represents. A broad hint as to the collective naiveté at the time can be derived from a quote taken from the conclusion of an article on fusion research published in 1956.[1] The quote reads: "It is the firm belief of many of the physicists actively engaged in controlled fusion research in this country that all of the technological problems of controlled fusion will be mastered, perhaps in the next few years." So much for prophecy!

In my defense (since I was the one who wrote that article), I did in a later paragraph in the same article introduce a caveat and a more plausible prediction. First, the caveat: "That an early success in achieving a self-sustaining controlled fusion reaction would lead to economically competitive power in the near future is highly unlikely." And the prediction: "... in the fusion reaction are implicit new dimensions — those of power obtained, possibly by direct electrical conversion, from an inexpensive, safe, and virtually inexhaustible fuel. These possibilities will surely someday play a dominant role in shaping the world of the future."

Since those early days the motivations for pursuing fusion research to a practical result have grown even stronger than they were then. Perceived problems (then) of the progressive depletion of fossil fuels, of air pollution from their use, and of potential hazards from fission power plants have become real problems today, and fusion is increasingly recognized as one of the best, perhaps the only viable, long-range response to these concerns.

My first paradigm, then, within which all later ones will be contained, is fusion research itself. By all measures it is an outstanding example of a long-range applied science effort aimed at achieving a specific and vitally important objective for the benefit of all of mankind. If I try to think of another comparably outstanding example, I can think of only one: research toward the elimination of cancer.

This brings me to the second paradigm I will discuss — namely, the emergence and the growth of the research field of high temperature plasma physics, a research field that flourished in response to the problems posed by the fusion goal. It is a true paradigm in that it represents a classic example of the development of a whole new field of scientific endeavor stimulated by a perceived important human need.

It has become increasingly apparent to us all, that as the field of fusion power research has broadened and deepened during the past three decades, the central issue has always been and still remains: to understand the plasma state of matter. All roads lead to Rome, and all roads to fusion must have their surfaces laid on a roadbed (which is no bed of roses!) of plasma physics. As long as there remain stretches of the roadbed not firmly in place, passage to the goal is still not assured.

It is in this area of laying the plasma physics roadbed where we can from the outset clearly discern the dominant role played by Marshall Rosenbluth, and I will later on refer specifically to some of his contributions to the field. First, however, I would like to define what I mean by "the modern field of plasma physics." I am referring to the study of the plasma state of matter under conditions where there exist strong mutual interactions between the plasma and the electromagnetic fields — of external and/or internal origin — in which it is immersed. In the fusion context it is also typically implied that inter-particle collisions and atomic phenomena enter only as weak perturbations, rather than as dominant processes. This circumstance is in marked contrast to the way these processes enter into most of the older disciplines of plasma physics.

In the context of the research toward fusion power, our imperative is, therefore, that we must achieve a thorough understanding of the physics of plasma in the so-called collisionless regime. An historian might, therefore, examine the growth of our knowledge of plasma physics over the past three or four decades and tag those periods when landmark advances were made. For my purposes I prefer to take another approach. Namely, since it is the properties of an unfamiliar form of matter that must be understood, I will begin by asking: What properties must we understand in order to utilize this form of matter — in particular, for fusion? Armed with such a list we could then begin to measure our ability to predict and to control the behavior of this state of matter. Having that ability is obviously an essential step on the way to achievement of fusion power.

My hierarchy of properties that need to be understood goes as follows:

- At the first level, understanding the intricacies of the motions of charged particles in strong magnetic and electric fields.
- Next, having in hand the formalism by which to predict the consequences of collisional interactions among an ensemble of charged particles when they are immersed in strong magnetic and/or electric fields — i.e., the Fokker-Planck problem.
- Next, understanding the nature of the pressure equilibrium state between a plasma and a magnetic field, including the stability of this state — i.e., the magnetohydrodynamic (MHD) problem in its most basic form.
- Next, understanding the stimulation and the propagation of waves in plasmas, and the conditions required for the onset — and for the control — of unstable modes of these waves.
- Finally, getting it all together by developing theoretical techniques for predicting, and effective means for controlling, the transport of particles and of energy within plasmas under the spectrum of conditions foreseen to be encountered in fusion power systems.

It should come as no surprise that Marshall Rosenbluth has made major contributions to the theory undergirding every single one of the items in the properties checklist that I have given. I do not have the intention, nor would there be time in a talk ten times as long as this one, to discuss all of these contributions, but along the way I will single out a few of them for special mention. In fact, most of those examples will be taken from mirror research. I have chosen them because they are the ones that are most familiar to me. Another speaker might well have chosen a different set to illustrate these same points.

In fact, now is a good time to recall some early work by Rosenbluth that has been like a two-edged sword in my own — I'll be generous and say "persistent" instead of "dogged" — pursuit of the mirror idea. This work was Rosenbluth's contribution to converting the Fokker-Planck equation into forms that made it eminently useful for magnetic fusion research.[2] This contribution comes under the rubric of the second item in my hierarchy of plasma knowledge — namely, the unraveling of the role of collisions in the velocity space of fusion plasmas. Out of this early work came the "Rosenbluth potentials" and their subsequent employment in the development of increasingly sophisticated computer codes. These codes have played, and continue to play, a crucially important role in both the theoretical analysis of mirror confinement and stability and in the design and interpretation of experimental data from mirror systems. These same computer codes are also being increasingly applied in the analysis of other fusion systems, including the tokamak.

A moment ago I said "a two-edged sword" in connection with the applications of the Fokker-Planck equation to mirror systems. I said that, because the Fokker-Planck equation provides a mercilessly rigorous standard by which to judge the fusion power balance potential of proposed mirror fusion systems. The existence of this standard, with its predictions of marginal Q values for simple mirror systems, led — or probably I should say "forced" — me to invent a direct conversion system for mirrors, and later led Dimov and also Fowler and Logan to the invention of the tandem mirror.

Reduction of the Fokker-Planck equation to tractability is only one of many instances in which Rosenbluth's work has contributed in a fundamental way to magnetic mirror research. In fact, my third paradigm will be drawn from one of those contributions. But before I discuss that example I will mention two earlier works that have a special place, not only in mirror physics, but also in the wider field of fusion plasma research.

The first of these contributions comes under the third category of my hierarchy of plasma understanding, namely, the basic problem of MHD equilibrium and stability in magnetically confined plasmas. Today it is difficult to recall that at the time this work was performed — 1957 — the ideas concerning MHD stability were in a very primitive state, and in the mirror approach, experimental results were either incomplete or actually misleading on that issue. It was at this point that Rosenbluth and Longmire[3] published a landmark paper, setting forth in the clearest of terms an analysis — based on orbit theory rather than on fluid equations — of the MHD stability of mirror machines in the axially symmetric forms that were then in use. Though a dismay at the time to mirror researchers, that paper was a harbinger of the solution to the mirror MHD stability problem, and I am sure it also provided a stimulus to later theoretical work dealing with MHD stability in other systems.

Now for the second notable contribution: Those of us in the business (at that time) of exploring the stability of axially symmetric mirror systems were given a partial reprieve from the concern raised by the Rosenbluth-Longmire paper by another paper five years later. This 1962 paper was the major work by Rosenbluth, Krall, and Rostoker[4] on "finite-orbit stabilization." In this paper Rosenbluth and his co-workers employed the now familiar "method of characteristics" to solve an otherwise very difficult theoretical problem. The paper not only helped explain earlier experimental results on mirrors, but its concepts and methods of analysis fed into an ever-widening circle of later important analytical works in which the use of the particle kinetic equation approach to stability theory would be essential.

Rosenbluth's intimate involvement at the time in helping mirror researchers to sort out some very puzzling data can be deduced from the fact that he was a co-author with Ellis, Ford, and Post in a publication[5] based on results from our mirror experiment at Livermore named "Table Top." In this experiment (which we reported in 1960, two years before the Rosenbluth-Krall-Rostoker "finite-orbit stabilization" paper), plasma injection and magnetic compression were used in an axially symmetric mirror field to produce a spindle-shaped column of dense and apparently stable plasma composed of 20 kilovolt temperature electrons and much colder ions. The puzzles posed by the Table Top data were twofold: First, in view of the Rosenbluth-Longmire theory, why was the column grossly stable at all? And second, in the light of the then-prevalent appearance of Bohm diffusion in stellarators, why did this plasma have a transverse diffusion rate that was five orders of magnitude slower than the Bohm rate?

As to the first question, Rosenbluth's contribution was to introduce a preview of the idea of finite orbit stabilization. Taken together with the idea of the so-called "line tying" stabilization effect, we had at least a partial answer to the first puzzle.

As to the second puzzle, we had to wait for many years in order to begin to explain that one, in this case within a much broader set of plasma issues.

In fact, those same issues are the ones that are involved in what I will submit as my next paradigm. The issues involved here are the ones contained in the fourth item of the list of plasma physics issues that I alluded to earlier — namely, microinstabilities. Early work by Harris[6] and others in this country and by Vedenov and Sagdeev[7] and others in the Soviet Union gave clear warning that departures from velocity-space isotropy, or distortions away from a Maxwellian speed distribution, could give rise to unstable growing waves — microinstabilities — in collisionless plasmas. One of these early works, describing an instability that is of particular interest recently to mirror researchers, was presented by Rosenbluth in one of his lectures at the 1960 plasma school at Risö in Denmark.[8] It concerned what is now known as the "Alfvén ion cyclotron instability," an Alfvén-wave mode driven unstable by velocity-space anisotropy of the ions.

Now for the solution of the microinstability problem in mirror systems. This one took over ten years to unfold. I have highlighted it here because, not only is it an example with which I am very familiar, but also for the reason that it personifies a particular approach to the solution of a major problem in plasma physics. It is also the one that I remember with the greatest pleasure, since in its early history it involved a close and fruitful collaborative effort with Rosenbluth in developing a theoretical attack on the problem.

By 1964, having been shown the way by the now-classic "Ioffe experiment" (which was a paradigm in its own right), at Livermore we had succeeded in suppressing all MHD-like plasma activity in our experiments by the use of mirror fields of the magnetic well type. It was therefore becoming increasingly apparent from the experimental data that the new circumstance limiting particle confinement was the presence of high frequency microinstabilities. These one could blame, in a qualitative way, on the known "loss-cone" nature of the ion and/or electron distribution functions of mirror-confined plasmas, with their velocity-space anisotropy and their non-Maxwellian speed distributions. It seemed that the time was ripe for making a directed theoretical attack on the problem, addressing the problem in terms that were specific to the mirror machine, rather than in general terms. The hope was that one would thereby learn how to suppress the microinstabilities, once having deduced the conditions required for their stimulation. As it turned out, that was not merely a pious hope. It was therefore in 1964 that I appeared on Marshall Rosenbluth's doorstep at General Atomic in San Diego. I came equipped with some experimental results and some fuzzy ideas as to how the problem might be formulated theoretically. In what seemed to be (and may have actually been) overnight, Rosenbluth came back with an elegant analysis that yielded criteria for the stabilization of what has come to be called the "high frequency convective loss-cone" mode.[9]

Out of the further evolving of this collaboration, including some time together at Culham, came the work on the "drift cyclotron loss cone" (DCLC) mode,[10] in which Marshall contributed the analytical work and I contributed the generation of needed computer codes. This collaboration was so stimulating to me that I was able to come up with an analysis of the idea of "warm plasma stabilization"[11] of the two modes for a meeting at Gatlinburg about a year later.

Though we were by no means the only ones who were working on the theory of microinstabilities at the time, I feel that Rosenbluth's clear formulation and analysis of the problem, in the specific context of mirror physics, played a decisive role in what would eventually become a major success story in mirror research.

Following those two early papers there elapsed ten years before the happy ending. During that interim period, ever more realistic theoretical models were analyzed and increasingly sophisticated experiments were performed in an attempt to verify the theoretical predictions. Though there were successes, for example, in making contact with the convective loss cone mode and its suppression through experimental checks and through theoretical work by Baldwin and Callen,[12] there remained a major mystery. The mystery was that under plasma conditions where theory predicted that the DCLC mode should be strongly driven, activity at the cyclotron frequency was weak and/or end losses were not noticeably enhanced. Work on Ioffe's mirror machine in Moscow[13] seemed to demonstrate the reality of the mode and indicated the positive effects of warm plasma stabilization, but left unexplained the behavior of the 2XII experiment at Livermore. In 2XII, a dense plasma at about 1 kilovolt ion temperature was created that seemed to decay at about classical collisional rates (that is, in this case, within about a half a millisecond) without evidencing the presence of the DCLC mode.

But any complacency was shattered and the "mystery" was exploded when 2XII, rebuilt as 2XIIB, was brought into operation with its powerful array of neutral beams. Now, as the plasma density and temperature built up, there appeared a virulent instability that destroyed the confinement and that had all the hallmarks of the DCLC mode. The happy ending was provided by Fred Coensgen and his coworkers, who overnight tried out warm plasma stabilization on 2XIIB and showed that it really worked.[14] On the theory side it was the paper of Baldwin, Berk, and Pearlstein[15] in 1976 that employed quasi-linear theory to explain the puzzle by showing how the DCLC mode drove itself to marginal stability through generating a warm plasma stream by its own activity. This theory thereby explained both the weakness of the mode at lower temperatures and its increasing effects at high temperatures. When the 2XIIB experiment was followed by the use of sloshing ions in the TMX-Upgrade experiment to further weaken the DCLC mode and to suppress the Alfvén ion cyclotron mode, the story was essentially complete. A decade and a half of cooperative effort between theory and experiment solved what had once been the problem that was most threatening to the future of mirror systems.

Among the several reasons that I chose the mirror microinstability story as the third paradigm is that I intend to use it as a model, at the conclusion of my talk, to argue for another hoped-for, future paradigm. But, before going into that, I would like to address, briefly, two other topics.

The first of these topics, one to which I cannot hope to do justice, concerns the wide scope of Marshall Rosenbluth's contributions to the entire field of fusion plasma physics. He was, for example, the author of a landmark paper in 1972 on the role of parametric instabilities in laser-irradiated plasmas.[16] Recently, he has been author or co-author on numerous papers on the theory of high power free-electron lasers. I am sure that stories of Rosenbluthian contributions similar to the one I have given from the mirror research field could be given by colleagues in many other plasma research fields.

From those other fields I recall some especially notable works, including the paper on average-minimum-B systems he co-authored in 1964 with Harold Furth,[17] the paper on neoclassical transport in 1972 with Hazeltine and Hinton,[18] and the paper on the tilt instability of the spheromak with Bussac[19] in 1979.

During the years that I surveyed through a literature search, namely, 1960 through 1985, I found that Rosenbluth was author or co-author of some 195 published papers on plasma physics topics (corresponding to an average rate of 7.5 papers per year and a peak rate which reached 14 per year). This does not even count the numberless unpublished reports that he authored or co-authored. He also has served on countless blue ribbon panels or committees, given myriad talks, and helped who knows how many other researchers with physical insights or helpful hints on analysis. More specifically, we in the mirror business have continued to benefit from his advice and counsel and from his contributions to tandem mirror confinement and stability theory.

There is something more that I want to say in all this. From my own experience and from that of others, there emerges the picture of a gentleman in the best of sense of the word and of a physicist who is willing to listen (which is a rare breed in our kind!). These attributes are coupled with an uncanny ability to analyze difficult problems by methods exactly tuned to the problem at hand — being neither more nor less complicated than is required to do the job. I guess you could call it "laziness with style" — and what remarkable style!

An additional observation about Marshall Rosenbluth that I would make is to emphasize that his dedication to the fusion goal runs much deeper than being an accomplished professional in the scientific side of fusion. He also has consistently been a staunch advocate of fusion research in all of his contacts both inside and outside the scientific community. For example, in his remarks made on the occasion of his receiving the Fermi Award in February, 1986, he called on the nation's leaders to take the long view with respect to their support for fusion research. He asked all of us the disturbing question: "Can we be a proud and successful nation twenty years from now if we abandon the struggle?" Then he answered his own question with the upbeat response: "I am not pessimistic. I have a great faith in the wisdom of American's people, and in the workings of the American system of government. In often mysterious and sometimes tortuous ways the right decisions are made, the path to greatness is followed."

A phrase in that last sentence resonates with what I would like to submit as my candidate for a future chapter of this "story of paradigms": "mysterious and sometimes tortuous ways" is not a bad description both of how fusion plasmas behave and of the steps in the growth of our understanding of that state of matter. Though we have made enormous progress in ascending the steps of the ladder of understanding that I earlier enumerated, we have not yet reached the top of the ladder. The last step that I listed has not yet been climbed. If you recall, it was "... developing theoretical techniques for predicting, and effective means for controlling, the transport of particles and of energy within plasmas under the spectrum of conditions foreseen to be encountered in fusion power systems." What I am suggesting is that the same philosophy should now be adopted in tackling that major remaining area of ignorance in plasma physics that was perforce adopted by the mirror community in solving the microinstability problem for mirrors. There the paradigmatic sequence was: theory to expose the problem, experiment to lend it reality, theory to propose the solution, followed by experiment to confirm the predictions. Could not the same sequence work for the last remaining puzzle? That is, to elucidate those processes that cause anomalous transport (drift waves, for example), to use theory to define geometries, and/or to suggest techniques to suppress or to minimize that transport (for example, feedback methods), then to test the theory in the laboratory. The end-result device might or might not look like a tokamak, a reversed field pinch, a mirror, or whatever — but it would be a winner, and it would be predictable. What is not predictable is whether such a program would indeed be successful. On the other hand, what is predictable is that the problem it would addresses will not solve itself.

I am sure that everyone, and especially our honoree, recognizes that the issue that I am now discussing is by no means a new one, since it underlies almost everything that we do in magnetic fusion research. But just possibly we may have, in the pursuit of specific devices, lost sight of the larger aspects of this last step. I believe there is a real opportunity here to generate a new paradigm in the fusion story.

So at the end of my talk — in summing it all up, what is the most striking paradigm in my story of paradigms, the most remarkable example? Isn't it obvious? And I am honored to have participated in this occasion that has been laid on for him.

ACKNOWLEDGMENT

I acknowledge, with thanks, discussions with H.L. Berk, B.I. Cohen, T.K. Fowler, and L.D. Pearlstein during the preparation of this paper. This work was performed under the auspices of the U.S. Department of Energy by the Lawrence Livermore National Laboratory under contract #W-7405-ENG-48.

REFERENCES

1. R.F. Post, Rev. Mod. Phys. **28**, 338 (1956).
2. M.N. Rosenbluth, W.M. MacDonald, and D.C. Judd, *Phys. Rev.* **107**, 1 (1957).
3. M.N. Rosenbluth and C.L. Longmire, *Ann. of Physics (N.Y.)* **1**, 120 (1957).
4. M.N. Rosenbluth, N.A. Krall, and N. Rostoker, *Nucl. Fusion Suppl.*, Part **1**, 143 (1962).
5. R.F. Post, R.E. Ellis, F.C. Ford, and M.N. Rosenbluth, *Phys. Rev. Lett.* **4**, 166 (1960).
6. E. Harris, *Phys. Rev. Lett.* **2**, 34 (1959).
7. A.A. Vedenov and R.Z. Sagdeev, in *Plasma Physics and the Problem of Controlled Thermonuclear Reactions* (Pergamon Press, London, 1959), Vol. III, p. 332.
8. M.N. Rosenbluth, *Risö Report* No. 18, 189 (1960).
9. M.N. Rosenbluth and R.F. Post, *Phys. Fluids* **8**, 547 (1965).
10. R.F. Post and M.N. Rosenbluth, *Phys. Fluids* **9**, 730 (1966).
11. R.F. Post, in *Proc. of International Conf. on Plasma Confinement in Open-ended Geometry*, ORNL Report Conf-671126, p. 309 (1967).
12. D.E. Baldwin and J.D. Callen, *Phys. Rev. Lett.* **28**, 1686 (1972).
13. M.S. Ioffe, B.I. Kanaev, V.P. Pastukhov, and E.E. Yushmanov, *Sov. Phys. JETP* **40**, 1064 (1975).
14. F.H. Coensgen *et al.*, *Phys. Rev. Lett.* **35**, 1501 (1975).
15. D.E. Baldwin, H.L. Berk, and L.D. Pearlstein, *Phys. Rev. Lett.* **36**, 1051 (1976).
16. M.N. Rosenbluth, *Phys. Rev. Lett.* **29**, 565 (1972).
17. H.P. Furth and M.N. Rosenbluth, *Phys. Fluids* **7**, 764 (1964).
18. M.N. Rosenbluth, R.D. Hazeltine, and F.L. Hinton, *Phys. Fluids* **15**, 116 (1972).
19. M.N. Rosenbluth and M.N. Bussac, *Nucl. Fusion* **19**, 489 (1979).

Bruno Coppi
Department of Physics
Massachusetts Institute of Technology
Cambridge, Massachusetts 02139

Plasma Collective Modes and Transport

Two decades of scientific collaboration with Marshall Rosenbluth, including work on ballooning instabilities while at La Jolla, the ion-temperature-gradient mode while at Trieste, and the resistive internal kink mode while at Princeton, are described, with application to the current status of plasma transport theory.

INTRODUCTION

One day, exactly nineteen years ago, Marshall Rosenbluth walked into my office at the Institute for Advanced Study at Princeton and, with a mixture of sadness and anxiety, remarked that he was already forty-one years old! Today, I am sure that he, along with most of us, has come to accept his age with a feeling of inner peace—a feeling that was beautifully expressed by Giovanni Pascoli, a poet well known in Italy a century ago, in these words: "Il giorno fu pieno di lampi...che pace la sera [The day was full of lightnings; how peaceful is the evening]."

Marshall Rosenbluth has shared with many of us a deep commitment to fusion research, even when it would have been advantageous for him to work in other fields of science. Let me suggest some reasons for this commitment. One reason is that dealing with fusion energy forces us to deal with questions about the primordial origins of the elements that we attempt to use as fuels. Deuterium is a case in point. Why is the abundance of deuterium on earth ten times greater than its cosmic abundance? And to what extent are the models of big bang evolution that justify this circumstance believable? Another reason is that the physics of plasmas allows a description of an incredibly large variety of objects and range of physical regimes. For example, last year I was fortunate to participate in the scientific team that witnessed and reported the encounter of the Voyager 2 spacecraft with the planet Uranus. (Incidentally, this encounter had not been in the original plans for the Voyager mission, but was one of those events that, similar to the proof of scientific feasibility for fusion, had been predicted to be impossible before the next century.) During that encounter, the particle density that was observed to precede the planet bowshock was approximately 2.5×10^{-2} particles per cubic centimeter. This value may be compared with the record high density of 2×10^{15} particles per cubic centimeter that had been obtained earlier in the Alcator tokamak experiment at MIT. Between these two values there is a difference of almost seventeen orders of magnitude and yet plasma physics can span this range of parameter regimes.

Truly, it is a privilege to witness these phenomena and then, with rudimentary strokes, to try to paint a theoretical picture of what we have observed. To express my feelings in this matter, let me borrow some phrases by a mutual friend from the Princeton days, Freeman Dyson. In his well-known book *Disturbing the Universe*, there occurs this fascinating passage:[1] "I do not feel like an alien in this universe. The more I examine the universe and study the details of its architecture, the more evidence I find that the universe in some sense must have known that we were coming."

I find Dyson's thoughts to be more comforting than those of another scientist, Pascal, who took time to reflect upon his life's experiences with different feelings. In his famous book, *La Pensées*,[2] which is a collection of random thoughts, there recur phrases like this one: "The eternal silence of these infinite spaces frightens me" (§206, p. 61). Elsewhere, he writes thus:

> "When I consider the short duration of my life, swallowed up in the eternity before and after, the little space which I fill, and even can see, engulfed in the infinite immensity of spaces of which I am ignorant, and which know me not, I am frightened, and am astonished at being here rather than there; for there is no reason why here rather than there, why now rather than then" (§205, p. 61).

Then, too, there is this famous passage that European high school students are often given to read:

> "Man is but a reed, the most feeble thing in nature; but he is a thinking reed. The entire universe need not arm itself to crush him. A vapour, a drop of water suffices to kill him. But, if the universe were to crush him, man would still be more noble than that which killed him, because he

knows that he dies and the advantage which the universe has over him; the universe knows nothing of this" (§347, p. 97).

In other words, Pascal had a terrifying view of an alien universe, in which man has no place.

LA JOLLA: BALLOONING INSTABILITIES, ETC.

My collaboration with Marshall Rosenbluth began in La Jolla and continued in Trieste and in Princeton. (It was, in fact, this same book, *La Pensées* of Pascal, that I read the night my son was being born in a hospital near the University of California at San Diego and General Atomic, where I was working with Marshall. Coincidentally, Freeman Dyson was there at that time, giving stimulating lectures at the university on such extravagant subjects as extraterrestrial technology, in which he anticipated his later thoughts on the "greening of the universe.")

In the fall of 1964, one of the issues that concerned us was the problem of what later came to be known as ballooning instabilities. In order to illustrate what a ballooning mode is, let me refer to the example of the magnetosphere of the planet Jupiter[3] (see Fig. 1). Its magnetosphere is filled with plasma, mainly coming from the satellite Io, and this plasma is composed mostly of sulphur. A key question, still to be analyzed properly, is how this plasma is unloaded from the magnetic field that tends to confine it. In this case the plasma pressure is of the same order as the magnetic pressure, i.e., $\beta \equiv 8\pi p/B^2 \sim 1$. Therefore, it is possible that some sort of "aneurism," so called because it stretches the magnetic field lines, can develop in the region where the radial pressure gradient is maximum. This aneurism is what we refer to as a ballooning mode.

A similar situation existed in the quadrupole and octopole plasma experiments that were being performed by Tihiro Ohkawa at General Atomic in those days. These devices had internal conductors, for which there were regions of favorable and unfavorable curvature. The question, then, concerned where and when the confined plasma would be ballooning unstable. When we began to study the problem theoretically, we first looked to see if this sort of ballooning mode could exist when the ideal magnetohydrodynamic approximation, viz., $\mathbf{E} + \mathbf{v} \times \mathbf{B}/c = 0$ is satisfied, since these modes would be the most violent. (Strangely, no one had ever examined this problem before.) We found that, in fact, new macroscopic modes are possible even when the ideal MHD frozen-in law is satisfied. Our theory basically applied to multipole experiments, since tokamaks were not yet popular. The theory for ballooning modes was extended to include the stabilizing effects of finite Larmor radius. We also felt that new ballooning modes could be found if the frozen-in law was abandoned.

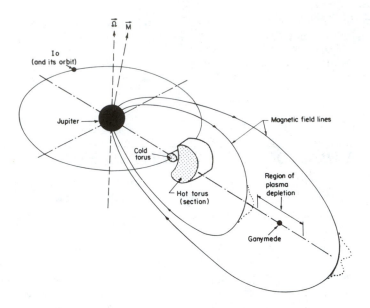

FIGURE 1 Schematic of the magnetic field lines of the planet Jupiter, together with the orbits of its moons Io and Ganymede. [From Ref. 3]

In order to remind ourselves of the theory of the ballooning mode, I may say that it can be regarded as a mixture of a shear Alfvén wave and a Rayleigh-Taylor instability, as can be seen from the following heuristic dispersion relation:

$$\omega^2 = k_\parallel^2 v_A^2 - \frac{g_{\text{eff}}}{r_p}. \tag{1}$$

Here, $k_\parallel$ is the wavenumber in the direction of the magnetic field, v_A is the Alfvén velocity, and $r_p = -p/\left(dp/dr_\perp\right)$ is the pressure gradient scale distance. Also, $g_{\text{eff}} = v_s^2/R_c$, where v_s is the sound velocity and R_c is the local radius of curvature of the magnetic field lines. In reality, $k_\parallel$ is an operator; it simulates the derivative $i(\partial/\partial s)$, where s measures the distance along a field line. Let us identify the length L as the characteristic spatial scale for variation of the magnetic field along its equilibrium direction. In most cases, L is related to a periodicity length. In the case of a multipole plasma device, this length is related to the distance between adjacent conductors. Then, writing $k_\parallel \cong 2\pi/L$, we can express Eq. (1) as

$$\omega^2 = k_\parallel^2 v_A^2 (1 - G), \tag{2}$$

where

$$G = \frac{v_s^2 L^2}{4\pi^2 v_A^2 R_c r_p} \propto \beta \left(\frac{L^2}{R_c r_p} \right). \tag{3}$$

Equation (2) indicates that the key factor that controls the onset of ballooning instability (i.e., when ω^2 becomes negative) is not beta, but rather the parameter G. The existence of these ballooning modes is nowadays well known, but such was not the situation some twenty years ago. Furthermore, at present many of us think primarily about axisymmetric toroidal configurations with a strong current, i.e., tokamaks, for which G becomes $G = \beta \left(q^2 R / r_p \right)$, even though multipole and mirror configurations have played a very important role in advancing our understanding of this instability. Nevertheless, when our findings[4,5] were announced in 1965 at the IAEA conference held in Culham, U. K., they were well received by the fusion community—so well, in fact, that eventually the conclusions from our rudimentary and approximate theory were believed to imply that only low-beta toroidal configurations were possible. (There were some lonely voices—John Clarke's, for one—who thought otherwise.) Large-scale conceptual experiments and fusion power stations were even being designed by the major laboratories on the basis of this conclusion.

By the late 1970's, the situation had gotten to the point where Marshall Rosenbluth and I were once called to Washington to debate whether there was any hope of ever achieving finite values of beta. At the time, I was careful to remind the audience that our 1965 theory, although correct, was based on a simple-minded model and had limitations, which needed to be reconsidered. This is exactly what we then undertook in 1977–78: a critical review of the basis of the initial theory—which in the meantime had been extended to apply to current-carrying, axisymmetric toroidal configurations. In so doing, we discovered that some important phenomena had been left out of the theory and that, when these effects were included, a new stable regime could be found, which we referred to as the "second stability region."[6,7] The reason for this name can be illustrated by means of another heuristic dispersion relation,

$$\omega^2 = k_{\|0}^2 v_A^2 (1 + \varepsilon G)^2 - G k_{\|0}^2 v_A^2 \tag{4}$$

which can be derived from Eq. (2) by letting $k_\| \rightarrow k_{\|0}(1 + \varepsilon G)$. Recall that $k_\|$ is actually a somewhat complicated differential operator. Let $k_{\|0}$ represent the parallel wavenumber in a plasma for which the value of G is small. When the parameter G increases in value so as to be larger than unity, the magnetic field lines actually tend to "stiffen," or become tied together, which has the effect of reducing the connection length L, defined earlier. In other words, $k_\|$ increases when G becomes finite. The result, as can be seen from Eq. (4), is that the ballooning mode is stable $(\omega^2 > 0)$ when $G = 0$, goes unstable as G increases to order unity, but again becomes stable for $G \sim O\left(\varepsilon^{-2}\right)$, which is greater than unity if $\varepsilon < 1$ is assumed. Figure 2 shows a schematic representation of the stable, unstable, and second stability regimes in the two-dimensional parameter space of G and $\hat{S}$, the latter being the magnetic shear, $\hat{S} = d(\ln q)/d(\ln r)$, with q the safety factor. The low-beta "first" stable region would have been the operational domain of the "Gigators," a name that I used in the middle 1970's to refer to proposed toroidal fusion devices that would

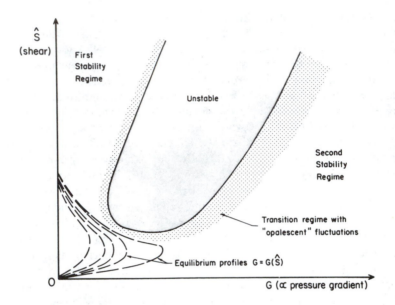

FIGURE 2 Schematic representation of the stability regions for ballooning modes, as a function of the shear S and the parameter G, which is proportional to the plasma pressure gradient.

cost a billion dollars or more. In Fig. 2, the shaded zone near the unstable boundary approximately represents a transition region with "opalescent" fluctuations. In the second stability regime, it may be said that "self healing" of the ideal MHD instability occurs, due to finite beta. Stable, high-beta equilibrium solutions have been obtained in this regime, both analytically[8,9] and numerically.[10,11] Non-ideal phenomena, such as tearing and kinetic effects, are not included in this model. Also shown in Fig. 2 are a family of dashed curves, $G = G(\hat{S})$, each of which corresponds to a given equilibrium plasma-magnetic field configuration. We now believe that it is, in fact, possible to go to finite beta values, either by extension of the first stability region or by methods (e.g., control of the plasma current) that open a "channel" of access to the second stability region.[12]

It may be that the concepts associated with second stability are relevant to understanding the dynamics of Jupiter's magnetosphere. However, in my opinion, an even more important consequence of the discovery of the second stability regime is the ability to think of burning a plasma mixture consisting of deuterium and Helium 3, in the following reaction,

$$D + \text{He}^3 \rightarrow \text{He}^4(3.6\,\text{MeV}) + p(14.7\,\text{MeV}), \tag{5}$$

and thereby achieving a "clean" fusion reactor, i.e., one in which the fraction of energy produced in the form of neutrons is minimized.[13] For many years, we were locked into thinking that only a fusion reactor burning deuterium and tritium could be feasible. But D-T reactors have a number of drawbacks. Society today wants an energy source that is drastically better than fission. The discovery of second stability implies that a reactor could operate at finite beta conditions. This, combined with the knowledge that has been gained about the approach to fusion with strong magnetic fields and the technology that has been developed, especially in Japan, for high-field superconducting devices, lead us to believe that with current-day technology it is possible to burn a D-He^3 plasma fuel. In other words, it would be possible, for the first time, to work on the problem of clean fusion. Long ago, Tamm and Sakharov[14] had the same dream, which they proposed to do by burning D-D. Their parameters were quite inadequate: exorbitant beta values, unsufficient current to confine the 15 MeV protons, and so on. Now, however, this old dream may have been realized, at least on paper.

Other topics that engaged us at San Diego were the theory for the resistive type of ballooning mode, as well as the description of certain resistive modes caused by gravity in the presence of shear,[4] which would tend to remain unstable even in the limit of high temperature. The latter caused us some concern, since we tried to apply to the resistive mode the sophisticated theory that had been developed in a famous paper by Rosenbluth and collaborators on the finite Larmor radius stabilization of the usual interchange mode, but we were unable to obtain any stability. Eventually, after not only finite gyroradius effects, but also the effects of ion–ion collisions, were included, the mode was found to become stable.

TRIESTE: ION-TEMPERATURE-GRADIENT MODE, ETC.

About 1964, I was invited by Marshall Rosenbluth to move to Trieste, to work together with a sizable contingent of other plasma theorists, some of whom are pictured in the photograph of Fig. 3, at the International Center for Theoretical Physics. Among the many projects that were undertaken there, one in particular[15]— which engaged Rosenbluth, Sagdeev, and myself—has had an impact that continues to the present. This problem consisted of assessing the potential effects of modes that can be excited in a magnetically confined plasma in the presence of a significant temperature gradient, i.e., when the ratio of the temperature gradient to the density gradient exceeds a certain value, e.g., $\eta_i = [d(\ln T_i)/dr] / [d(\ln n)/dr] > 2$.

The existence of these η_i modes (also called ion mixing modes in a different connection) was known at the time. In rectangular geometry (x, y, z), with straight field lines $\mathbf{B} = B_0 \mathbf{e}_z$, the normal mode solution for the perturbed ion temperature has the form

$$\tilde{T}_{i\parallel} = \hat{T}_{i\parallel}(x) \exp\left(ik_y y + ik_{\parallel} z - i\omega t\right), \tag{6}$$

and one finds that the thermal energy transport due to this mode is frighteningly large, even though the particle transport is, in general, quite small. What concerned us, then, in 1966 were two questions. First, does this η_i instability continue to exist as a normal mode solution in a realistic magnetic field configuration? This question was an issue because in another study,[16] we had proved that the electron drift wave would not exist as a normal mode in slab geometry with a sheared magnetic field, of the type $\mathbf{B} = B_0 \left(\mathbf{e}_z + \mathbf{e}_y x/L_s\right)$. The second question was: If the η_i modes survive, what is the effective rate of thermal energy transport that can be expected from them? In order to answer the first question, a rather complicated integral equation was derived, for which only an asymptotic solution, valid in the limit of $\eta_i \gg 1$, could be obtained. (Subsequent attempts by others to obtain a more complete solution of that integral equation have still been unsuccessful.) Then, from quasilinear theory, the following thermal diffusion coefficient was estimated:

$$D \cong \left(\rho_i L_s/r_T^2\right) D_{\text{Bohm}}^{(i)}. \tag{7}$$

Here, ρ_i is the ion gyroradius, L_s is the scale length for the shear, r_T is the temperature gradient scale length, and $D_{\text{Bohm}}^{(i)} \cong cT_i/eB_0$ is the coefficient for Bohm diffusion of ions. (It turned out later that more sophisticated theories also arrive at the same result.) Equation (7) indicates that large thermal diffusion can be expected. Mass diffusion by particles, however, is low, smaller by approximately the square root of the electron-to-ion mass ratio. However, since the heat loss is proportional to the density, we speculated that a good divertor would be able to reduce the density at the wall, where the value of η_i is large, and thus keep the actual heat loss small. In this way, the effect of the η_i mode would not be too severe. In those days, we thought that this mode would never be realized, but in more recent years it has become a critical issue with the advanced experiments on the Alcator and TFTR tokamaks.

One area in which consideration of the η_i mode is important is that of inward particle transport. Three examples of experimental evidence for such inward particle transport can be cited. One is the observation of steady-state density profiles with finite curvature (i.e., that are not flat) at the center of the plasma column, in regimes where the particle source in this region is negligible. In fact, the particle balance equation, in the absence of sources and assuming that a diffusion process is present, is given by

$$\frac{\partial n}{\partial t} + \frac{1}{r} \frac{\partial}{\partial r} \left(r D_p \frac{\partial n}{\partial r}\right) = 0. \tag{8}$$

FIGURE 3 Group photograph taken at the International Center for Theoretical Physics, Trieste, Italy, in May, 1966.

Then, in steady state (i.e., $\partial n/\partial t = 0$), since the particle diffusion coefficient D_p is non-zero, the density profile must be absolutely flat. Instead, there is strong evidence that the density profile is not flat, but has finite curvature $\left(\partial^2 n/\partial r^2 \neq 0\right)$, which implies that there must be an inward flow in order to satisfy the conservation of particles. A second piece of clear evidence for the phenomenon of inward particle transport is the observation of increased density due to gas injection. This technique of gas puffing was imported to MIT from our colleagues at Princeton and was strongly exploited in the Alcator experiments. Again, to be able to explain why the density rises when gas is injected at a certain rate, one must assume that there is inward particle transport. The third piece of evidence came from the direct experiments by Mazzucato at Princeton. In these experiments, deuterium was injected into the plasma chamber of the PLT machine. Then, at the same time that neutron production at the center of the plasma was observed to increase, measurements of the excited modes showed that their phase velocity was in the direction of the ion diamagnetic velocity. This observation appeared to be consistent with a theoretical model[17,18] that had been developed for the ion mixing mode (related to the η_i mode). Indeed, the second and third of these three experimental evidences that have been cited here indicate that the proper equation for particle transport is not a diffusion equation, but rather an equation with inflow. Without this effect, one would derive an incorrect estimate for the diffusion coefficient: for the JET tokamak, for instance, it would predict an equivalent particle confinement time of up to five minutes! The ion mixing theory for inward particle transport assumes that the η_i mode is excited at the edge of the plasma column, but because the collisionality is high in this region, the particle inflow driven by the temperature gradient tends to be significantly faster than the transport rate that was calculated from collisionless theory in the original η_i paper.

Another application of η_i mode ideas occurred in the problem that I will refer to as "transport recovery." A little history is necessary to explain this problem. In 1974–75, the Alcator-A experiment at MIT achieved record values for the confinement parameter $n_0\tau_E$, which is the product of the density n_0 and the energy confinement time τ_E. A similar machine (FT) at the Frascati laboratory in Italy, with which I was involved, exhibited confinement properties that were nearly identical. With the success of Alcator-A, it was fairly easy to get federal support for another device (Alcator-C), this one a bit larger than Alcator-A, but without a copper shell. When the density was increased in the new machine, surprisingly the confinement time did not go up as we had expected on the basis of the Alcator-A and the FT experiments. Steve Wolfe, a former student, showed me the density profiles that he had begun to measure. Using his profiles, I estimated the value of η_i to be greater or approximately equal to 5, which was quite unfavorable and pointed to these modes as culprits. The situation looked bad. About this time Ed Kintner, then head of the fusion office of the Department of Energy, in a telephone call reminded me of all the money that had been spent on this new machine and said, "You owe me some confinement." I could find no excuse. In those days it was not yet widely known that the Alcator-A plasmas had been "born" with good density profiles (i.e., relatively peaked), but that the plasmas in Alcator-C were born with unfavorable profiles (i.e., too flat). At the same time the idea of "profile consistency" was being considered.[19] On the basis of such profile consistency, i.e., that in general the temperature profiles are maintained by a macroscopic type of constraint, it could be argued that one cannot tamper with the temperature profile. Therefore, the only profile that could be modified in an attempt to cure the Alcator-C confinement malady was that of the density.[20,21] This was the kind of reasoning that led to the program of pellet injection experiments in Alcator-C, whose eventual happy outcome was the recovery of transport and the achievement of values for $n_0 \tau_E$ that were close to 10^{14}sec/cm^3. Experimental evidence for these record results[22] is shown in Fig. 4. After the Alcator-C had achieved these results, the same experiments were repeated on a much larger machine, the TFTR tokamak at Princeton, in very advanced parameter regimes. Slightly lower temperatures, but record high values for the density, were obtained, along with a value for $n_0 \tau_E$ above 10^{14}sec/cm^3. Again, the interpretation for these results is consistent with the idea that the η_i mode tends not to be excited when the density profile is modified by means of injected pellets.

Another line of work that was started at Trieste, which involved Furth, Rosenbluth, Sagdeev, and myself, was the theory of new modes that can be excited by the presence of "impurity" populations within mostly hydrogenic plasmas.[23] Nowadays, a relatively large variety of experiments on magnetically confined plasmas has shown that the transport of impurities cannot be described by the familiar collisional transport theory. Hence it is reasonable to expect that the role of modes of this kind will receive increased attention in order to explain the experimental observations.

Some interesting work in space physics was also done at Trieste. While Rosenbluth was off on a cruise to Yugoslavia, Laval, Pellat, and I thought together about reconnection in the earth's magnetotail and wrote the first paper on this subject.[24] In a subsequent paper[25] (with Rosenbluth) on the same subject, a simple analysis was given to show that a small amount of temperature anisotropy, i.e., different temperatures along and across the magnetic field lines, can complicate the original prediction of instability and even provide a probable justification for the observed macroscopic stability of the magnetotail. In fact, the earth's magnetotail persists beyond the moon.

Just to mention a few other papers written at Trieste: One concerned new wave-particle resonance interactions that could be expected in complex (i.e., non-slab) geometries.[26] Also, certain kinetic collisionless modes of instability, not derivable from ideal MHD theory, were found and applied to multipole magnetic field configurations.[27-30] We also developed theory for the nonlinear interactions of positive and negative energy modes in plasmas.[31,32] Interestingly, when we arrived at Trieste, Rosenbluth felt that Sagdeev and Galeev, who had been working extensively on nonlinear phenomena, should do some toroidal plasma theory, while we should do some nonlinear theory to learn their methods. The result was that we ended up working on the nonlinear theory of positive and negative energy modes,

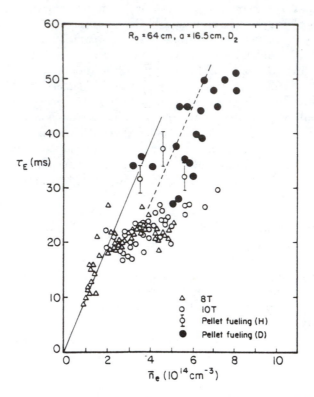

FIGURE 4 Record results from the Alcator-C experiments. [From Ref. 22]

whereas Sagdeev and Galeev, after returning from Trieste to the Soviet Union, went on to develop the neoclassical theory of tokamak transport.

PRINCETON: RESISTIVE INTERNAL KINK

Our joint work on the theory of resistive modes in San Diego and the collaboration begun in Trieste with two new French friends, Laval and Pellat, made it easy for us about ten years later in Princeton to join forces and develop the theory of $m^0 = 1$ modes that produce magnetic reconnection.[33,34] These are of special importance for the interpretation of the so-called "sawtooth oscillation" phenomenon of the electron temperature in the central part of current-carrying toroidal plasma columns.

The crash phase of the central temperature is often attributed to the excitation of $m^0 = 1$ modes that depend on finite plasma resistivity.

CONCLUSION: STATUS OF TRANSPORT THEORY

My concluding remarks concern the current status of plasma transport theory. There is now overwhelming evidence that both energy and particle transport in plasmas are to be described by equations that are intrinsically different from the familiar equations of classical (i.e., collisional) transport theory. An instance was earlier cited in connection with particle transport and the need to include the effect of inflow. For energy transport, perhaps an equation of the following sort[35] for the heat flux q,

$$q = -K\left(\kappa_s, \kappa_f\right)\frac{\partial T}{\partial r}$$

should be considered, involving a diffusion coefficient K that is a function of two different coefficients, one for the slow behavior (κ_s) and one for the fast behavior (κ_f), with some type of relaxation term. Moreover, the inherently different nature of these equations would derive not only from their being of different analytical form, compared to the equations of classical transport theory, but also because the transport coefficients are non-local in nature. Even though the theoretical formulation of the relevant transport equations is not yet complete, the experimental observations do seem to be consistent with the basic characterization of transport processes as induced by the collective modes that are known theoretically. (This point was made earlier in connection with the η_i modes.) Therefore, it would seem that the time has come to propose and to undertake a broadly based experimental program to investigate transport.

Even with all the uncertainties that we face, at the present not only does it appear that breakeven can be attained in fusion reacting plasmas, but experiments are even being designed that, on the basis of existing theory, can in fact achieve ignition.[36] If we wisely concentrate on several key experiments, probably we could witness the accomplishment of ignition in our own lifetime. My hope is that when we meet with Marshall Rosenbluth again in twenty years, we will see such experiments in operation.

ACKNOWLEDGEMENT

It is a pleasure to thank J. W. Van Dam for his skilled and generous editorial work.

REFERENCES

1. Freeman Dyson, *Disturbing the Universe* (Harper and Row, New York, 1979), p. 250.
2. Blaise Pascal, *Pensées* (E. P. Dutton, New York, 1958).
3. R. L. McNutt, P. S. Coppi, R. S. Selesnick, and B. Coppi, "Plasma Depletions in the Jovian Magnetosphere: Evidence of Transport and Solar Wind Interaction," *Journal of Geophysical Research* **92**, 4377 (1987).
4. B. Coppi and M. N. Rosenbluth, "Collisional Interchange Instabilities in Shear and $\int dl/B$ Stabilized Systems," in *Plasma Physics and Controlled Nuclear Fusion Research 1965* (International Atomic Energy Agency, Vienna, 1966), Vol. I, pp. 617–641.
5. H. P. Furth, J. Killeen, M. N. Rosenbluth, and B. Coppi, "Stabilization by Shear and Negative V''," in *Plasma Physics and Controlled Nuclear Fusion Research 1965* (International Atomic Energy Agency, Vienna, 1966), Vol. I, pp. 103–125.
6. B. Coppi, A. Ferreira, J. W.-K. Mark, and J. J. Ramos, "Ideal-MHD Stability of Finite-Beta Plasmas," *Nuclear Fusion* **19**, 715–725 (1979).
7. B. Coppi, A. Ferreira, J. W.-K. Mark, and J. J. Ramos, "A 'Second' Stability Region of Finite-β Plasmas," *Comments on Plasma Physics and Controlled Fusion* **5**, 1–8 (1979).
8. B. Coppi, A. Ferreira, and J. J. Ramos, "Self-Healing of Confined Plasmas with Finite Pressure," *Physical Review Letters* **44**, 990–993 (1980).
9. D. Lortz and J. Nuhrenberg, "Ballooning-Instability Boundaries for the Circular Tokamak," *Nuclear Fusion* **19**, 1207–1213 (1979).
10. H. R. Strauss, W. Park, D. A. Monticello, R. B. White, S. C. Jardin, M. S. Chance, A. M. M. Todd, and A. H. Glasser, "Stability of High-Beta Tokamaks to Ballooning Modes," *Nuclear Fusion* **20**, 638–642 (1980).
11. J. M. Greene and M. S. Chance, "The Second Region of Stability against Ballooning Modes," *Nuclear Fusion* **21**, 453–464 (1981).
12. B. Coppi, G. B. Crew, and J. J. Ramos, "Search for the Beta Limit," *Comments on Plasma Physics and Controlled Fusion* **6**, 109–117 (1981).
13. S. Atzeni and B. Coppi, "Ignition Experiments for Neutronless Fusion Reactions," *Comments on Plasma Physics and Controlled Fusion* **6**, 77–90 (1980).
14. I. E. Tamm and A. D. Sakharov, in *Plasma Physics and the Problem of Controlled Thermonuclear Reactions*, edited by M. A. Leontovich (Pergamon, Oxford, 1961), Vol. 1, pp. 1–47.
15. B. Coppi, M. N. Rosenbluth, and R. Z. Sagdeev, "Instabilities due to Temperature Gradients in Complex Magnetic Field Configurations," *Physics of Fluids* **10**, 582–587 (1967).
16. B. Coppi, G. Laval, R. Pellat, and M. N. Rosenbluth, "Convective Modes Driven by Density Gradients," *Nuclear Fusion* **6**, 261–267 (1966).
17. B. Coppi and C. Spight, "Ion-Mixing Mode and Model for Density Rise in Confined Plasmas," *Physical Review Letters* **41**, 551–554 (1978).

18. T. Antonsen, Jr., B. Coppi, and R. Englade, "Inward Particle Transport by Plasma Collective Modes," *Nuclear Fusion* **19**, 641–658 (1979).

19. B. Coppi, "Nonclassical Transport and the Principle of Profile Consistency," *Comments on Plasma Physics and Controlled Fusion* **5**, 261–269 (1980).

20. B. Coppi, "Theory of Plasmas near the Lawson Limit," Annual Sherwood Theory Conference on Plasma Physics and Controlled Fusion (Lake Tahoe, Nevada, April, 1984), Invited Paper A-3.

21. B. Coppi, S. Cowley, P. Detragiache, R. Kulsrud, and W. Tang, "Physics of Plasmas Close to the Lawson Limit," in *Plasma Physics and Controlled Nuclear Fusion Research 1984* (IAEA, Vienna, 1985), Vol. II, p. 93.

22. M. Greenwald *et al.*, "Pellet Fuelling Experiments in Alcator C," in *Plasma Physics and Controlled Nuclear Fusion Research 1984* (IAEA, Vienna, 1985), Vol. I, p. 45.

23. B. Coppi, H. P. Furth, M. N. Rosenbluth, and R. Z. Sagdeev, "Drift Instability due to Impurity Ions," *Physical Review Letters* **17**, 377 (1966).

24. B. Coppi, G. Laval, and R. Pellat, "Dynamics of the Geomagnetic Tail," *Physical Review Letters* **16**, 1207 (1966).

25. B. Coppi and M. N. Rosenbluth, "Model for the Earth's Magnetic Tail," in "The Stability of Plane Plasmas," *Proceedings of the ESRLN Study Group* (Frascati, Italy, December, 1967), ed. K. Schindler, Report ESRO SP-86 (European Space Research Organization, Paris, 1968), p. 1.

26. B. Coppi, G. Laval, R. Pellat, and M. N. Rosenbluth, "Collisionless Microinstabilities in Configurations with Periodic Magnetic Curvature," *Plasma Physics* **10**, 1–22 (1968).

27. B. Coppi, M. N. Rosenbluth, and S. Yoshikawa, "Localized (Ballooning) Modes in Multipole Configurations," *Physical Review Letters* **20**, 190 (1968).

28. B. Coppi, S. Ossakow, and M. N. Rosenbluth, "Low-Density Modes in Multipole Devices," *Plasma Physics* **10**, 571–580 (1968).

29. B. Coppi, M. N. Rosenbluth, and P. Rutherford, "Fluidlike Electron and Ion Modes in Inhomogeneous Plasmas," *Physical Review Letters* **21**, 1055 (1968).

30. P. Rutherford, M. Rosenbluth, W. Horton, E. Frieman, and B. Coppi, "Low-Frequency Stability of Axisymmetric Toruses," in *Plasma Physics and Controlled Nuclear Fusion Research 1968* (International Atomic Energy Agency, Vienna, 1969), Vol. I, p. 367.

31. B. Coppi, M. N. Rosenbluth, and R. N. Sudan, "Nonlinear Interactions of Positive and Negative Energy Modes in Rarefied Plasmas (I)," *Annals of Physics* (*New York*) **55**, 207 (1969).

32. M. N. Rosenbluth, B. Coppi, and R. N. Sudan, "Nonlinear Interactions of Positive and Negative Energy Modes in Rarefied Plasmas (II)," *Annals of Physics* (*New York*) **55**, 248 (1969).

33. B. Coppi, R. Galvao, R. Pellat, M. N. Rosenbluth, and P. H. Rutherford, "Resistive Internal Kink Modes," *Fizika Plazmy* **2**, 961–966 (1976) [*Soviet Journal of Plasma Physics* **2**, 533–535 (1976)].

34. G. Ara, B. Basu, B. Coppi, G. Laval, M. N. Rosenbluth, and B. V. Waddell, "Magnetic Reconnection and $m = 1$ Oscillations in Current Carrying Plasmas," *Annals of Physics (New York)* **112**, 443–476 (1978).
35. B. Coppi, "Profile Consistency: Global and Nonlinear Transport," *Physics Letters A* **128**, 193–197 (1988).
36. B. Coppi, "Compact Experiments for α-Particle Heating," *Comments on Plasma Physics and Controlled Fusion* **3**, 2 (1977).

Roscoe B. White
Plasma Physics Laboratory
Princeton University
Princeton, New Jersey 08544

Nonlinear Tokamak Dynamics

A mostly historical review is presented of the theoretical investigation of nonlinear tokamak dynamics, including ideal and resistive instabilities and also the tokamak phenomena of sawtooth oscillations, major disruptions, and fishbone modes.

INTRODUCTION

The investigation of the nonlinear consequences of the equations of magnetohydrodynamics (MHD) has from the beginning been motivated by the desire to explain the rapid, sometimes violent, tokamak behavior observed experimentally, with the hope of finding methods to control or avoid it. The use of resistive MHD theory in explaining the large-scale dynamics has been remarkably successful, perhaps more so than should have been expected. The methods used in analyzing this behavior have combined analytical and numerical techniques, guided by physical intuition, and the absence of any one of these three components would have delayed, if not completely thwarted, our arriving at the understanding of the principle large-scale tokamak phenomena that we presently possess.

Marshall Rosenbluth has been centrally involved in this development. He has contributed physical insight, crucial expansion schemes, numerical approximation methods, and analytical wizardry at a rate that often leaves collaborators numbed.

From Particles to Plasmas
Addison-Wesley Publishing Company, 1988

At the same time, his jovial demeanor and kindness have resulted in a list of collaborators that encompasses a large fraction of this generation of plasma physicists.

In this paper I will review, in a more or less historical sequence, those aspects of large-scale tokamak dynamics that are reasonably well understood. Interesting problems remain in this area, and unexplained features abound, but it has become reasonably clear that future work will consist in variations on the themes present in these descriptions and not in wholesale revision of them. Fortunately, the number of large-scale tokamak instabilities is not too large. The topics are conveniently divided into the ideal instabilities, including external and internal kinks, and the resistive instabilities. These modes, depending on the details of the current profile, are understood to be mainly responsible for the tokamak phenomena of the sawtooth oscillation and the major disruption. Finally, the more recently discovered phenomenon of low-mode-number ideal modes destabilized by trapped particle effects, i.e. the fishbone mode, will be discussed.

THE SURFACE KINK

The possibility of large helical distortions of the plasma surface was, from the start, a likely candidate for violent plasma disruption. If the aspect ratio R/r of a tokamak is regarded as large, the inverse aspect ratio $\epsilon = r/R$ provides a useful expansion parameter. To lowest order in ϵ, toroidal effects are negligible, and the tokamak is treated in the cylindrical approximation. A helical perturbation with helicity $\delta\varphi/\delta\theta = m/n$, i.e., the linear mode is of the form $\exp[i(m\theta - n\varphi)]$, produces a topological change in the magnetic field at the surface r where the magnetic field has exactly this helicity. In the large-aspect-ratio approximation the helicity of the magnetic field, $q(r)$, is given by $q(r) = rB\varphi/RB_\theta$; during typical tokamak operation, $q(r)$ is near unity on axis and is two or larger at the plasma edge. The magnetic field change consists in the formation of a magnetic island at the surface $q(r) = m/n$, with width proportional to the square root of the perturbation. In Fig. 1 is shown the distortion of the magnetic flux surfaces in the (r, θ) plane produced by such a perturbation, which extends helically in φ. The presence of the island gives a helical deformation to nearby flux surfaces and also to the plasma-vacuum surface.

Linear analysis, performed first by Shafranov[1] in 1970, showed that a simple equilibrium bounded by a vacuum region and a conducting wall is unstable to such an island if this surface lies in the vacuum region.

In ideal MHD, an island cannot form within the plasma because the plasma is stuck to the magnetic flux surfaces. In 1971, Rutherford, Furth, and Rosenbluth[2] examined the nonlinear behavior of such a system for small amplitudes and showed that in a constant current profile (shear-free) equilibrium, the nonlinear terms produce further destabilizing forces, and thus the final state might be expected to be highly distorted.

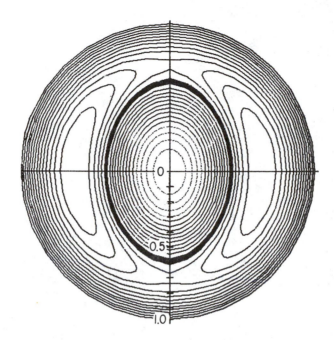

FIGURE 1   A helical perturbation.

Kadomtsev and Pogutse[3] showed in 1973 that in fact a shear-free plasma can possess a second equilibrium state with large helical vacuum bubbles present, with the energy of this bubble state lower than that of the initial circular equilibrium. The presence of a large bubble might cause a significant plasma volume to make contact with the tokamak wall, and such a process was suspected of being involved in tokamak disruption. Immediately it became of interest to discover whether these bubble states were accessible from an initial equilibrium, to determine how much shear was sufficient to prohibit the formation of such states, and to examine the temporal evolution associated with bubble formation.

This task involved numerical analysis, since it was hopeless to attempt to describe the evolution through highly distorted plasma shapes analytically. Fortunately the free energy reservoir for the helical deformation is large compared to toroidal effects, so the problem could be done in the cylindrical approximation. Consideration of a single helicity m/n then means that all quantities are functions only of r and $m\theta - n\varphi$, even in the full nonlinear evolution, which reduces the problem from three dimensions to two. The ideal MHD equations were simplified by an inverse aspect ratio expansion. Marshall Rosenbluth, with the assistance of Don Monticello, Hank Strauss, and myself, set out to construct a numerical scheme for

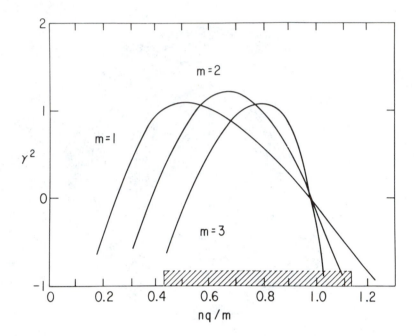

FIGURE 2   Kink mode growth rate. [From Ref. 5]

the examination of the full nonlinear evolution of an equilibrium linearly unstable to kink perturbations. First results were given at the 1974 IAEA meeting,[4] and a more detailed description[5] was published in 1976. At the same time a similar effort was under way at the Kurchatov Institute, led by Yuri Dnestrovskii.[6]

The time scale for the mode is the Alfvén time, determined by the field line tension and the plasma inertia. The linear growth rate is shown in Fig. 2, for the case of a shear-free plasma, with $q(r) = m/n$ within the plasma. The equilibrium is varied by a surface current which produces a discontinuity in q at the plasma surface, and the growth rate is shown as a function of the value of q just outside the plasma. The shaded region is the range of q for which the bubble state is energetically favorable. In the case of zero shear, the nonlinear evolution is reduced to following the motion of the plasma-vacuum interface.

Results showed that the plasma evolves rapidly to a bubble state in this case. The surface deformation for the case of the $m/n = 1/1$ mode is shown in Fig. 3. The absence of shear means that the plasma is free to change its shape quickly to achieve overall magnetic energy minimum. The evolution to the state labeled 3 occurs in approximately 10 Alfvén times. Cases of severe bubble formation could

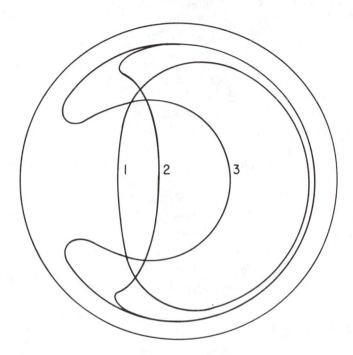

FIGURE 3   Bubble formation.

not be treated using a Fourier decomposition code, because of the very large number of harmonics involved.

This evolution to a bubble state is very effectively impeded by the presence of magnetic shear, and for a parabolic current profile no real bubble states exist, the plasma only assuming a relatively minor helical deformation. For most current profiles an important role of bubble states in the evolution of the plasma is ruled out, but the possibility exists for such deformations to occur in states of very low shear. We will return to this eventuality in the discussion of the sawtooth and the major tokamak disruption.

THE INTERNAL KINK

The $m/n = 1/1$ mode is unique because, to first approximation, the plasma displacement consists of a rigid motion of all plasma inside the surface given by $q = 1$, and no motion outside that surface. This and the very long wavelength of the mode mean that there is a minimum of field line bending that resists the displacement. The perturbed field vanishes at the $q = 1$ surface and a nonzero amplitude mode can occur without creating a magnetic island. The potential energy δW for plasma displacement is of order ϵ^4, of the same order as the toroidal effects. Bussac et al.[7] showed that the toroidal effects introduce a threshold in plasma pressure, making the mode stable for poloidal beta values of less than about 0.3, for typical current profiles.

The internal kink was a possible candidate for a description of the internal tokamak disruption, or sawtooth, having many of the correct qualitative features. Nonlinearly, however, Rosenbluth, Dagazian, and Rutherford[8] found that the mode saturates at too small an amplitude to account for the experimentally observed negative voltage spike. This analytic result, which was later confirmed with a nonlinear code by Park et al.,[9] effectively ruled out the internal kink from playing an important role in tokamak dynamics. It was relegated to a minor role until it was later discovered that it was capable of strongly interacting with high-energy trapped particles, which we will later discuss in connection with the fishbone mode.

RESISTIVE MODES

The linear theory of the tearing mode was developed by Furth, Killeen, and Rosenbluth.[10] This is an asymptotic boundary layer theory, with the solution away from the rational surface $q(r) = m/n$ given by the ideal MHD equations, and the solution in a narrow layer near the rational surface determined by the resistive diffusion of the plasma across the magnetic field, where the field lines tear, or reconnect, to form a magnetic island. The tearing layer decreases in width with increasing plasma temperature. The mode resembles a kink mode, as shown in Fig. 1, except that the rational surface lies within the plasma. An intuitively useful fact is that the reconnection is driven by the differential in the magnetic energy arising from the growth of the magnetic island, as first shown by Furth.[11] The magnetic energy differential is given by the quantity Δ', which is determined by the external ideal solution. It is largest for low values of m and n, reflecting the smaller amount of work required for field line bending. For m and n larger than certain values that depend on the profile, Δ' is negative and the mode is stable. Growth is impeded by the relatively slow rate of the reconnection, so the growth time is a hybrid of the Alfvén time and the resistive time.

Nonlinear theory for the tearing mode began in 1973 when Rutherford[12] showed that the mode ceases exponential growth and enters a domain of algebraic growth as soon as the island is larger than the tearing layer. This slowing-down of the mode is due to the formation of an inductive current flowing at the island O-point, parallel to the Ohmic current. This current, which opposes island growth, replaces the effect of inertia in the description of the evolution, and the plasma current becomes a function of the magnetic flux. The properties of the solutions of the ideal MHD equations that possess magnetic islands were studied extensively by Grad.[13]

In the early 1970's, Rosenbluth had collected three assistants at the Institute for Advanced Study to numerically explore the full nonlinear resistive evolution of tokamak profiles, using the inverse aspect ratio expansion first developed for the investigation of the kink mode. Bruce Waddell, Don Monticello, and I put together a numerical code for that purpose. The results of Rutherford were duplicated,[14] and it was soon shown[15] that the evolution of the magnetic island in the tearing mode to a saturated state could be predicted by a generalization of Δ' to the case of finite island width — i.e., the island state simply found a width determined by a magnetic energy minimum and then only slowly changed its width to follow the slower resistive evolution of the current profile. This allows for the possibility of stable configurations that possess magnetic islands and exhibit Mirnov oscillations.

By the late 1970's, results were being produced with a variety of numerical resistive codes that used the tokamak ordering expansion. Some of the results will be discussed in the subsequent sections on the sawtooth and the major disruptions. Waddell had gone to Oak Ridge, where he collaborated with Carreras and Hicks and began to work on the extension of the calculations to include many different helicities. In a real tokamak, helical symmetry could certainly assumed to be broken, and it was possible that nonlinear coupling of different helicities could be important at large mode amplitude. Monticello and White included diamagnetic terms to study nonlinear effects on mode rotation and the possible modification of the saturation results and to explore the possibility of feedback stabilization of the mode.[16]

In addition, new analytic results were extending the range of validity of the non-linear calculations and adding insight to the numerical results. Strauss[17] extended the formalism to higher order in the expansion parameter to be able to treat cases of high plasma beta. Drake and Lee[18] and Cowley, Kulsrud, and Hahm[19] examined different domains of collisionality and found modifications of the linear theory but showed that the results of Rutherford still held when the island size became larger than the tearing layer.

In a short time, several similar nonlinear resistive MHD codes were being run throughout the world. Numerical code development was continued by Dnestrovskii at Kurchatov. Biskamp developed a code at Garching. Sykes and Wesson began investigation of the sawtooth and the major disruption at JET. Kurita and Azumi developed a code at JAERI. Park and Monticello at Princeton incorporated the higher order analysis of Strauss to treat higher beta plasmas. J.K. Lee extended and modified the Princeton code at General Atomic. Drake *et al.* developed a code at Maryland and examined the sawtooth for flat and non-monotone q profiles. The advent of the CRAY supercomputer has also made it possible to develop codes that treat the full three-dimensional resistive MHD equations without employing an inverse aspect ratio expansion. Codes of this nature are in use at Oak Ridge and at JET.

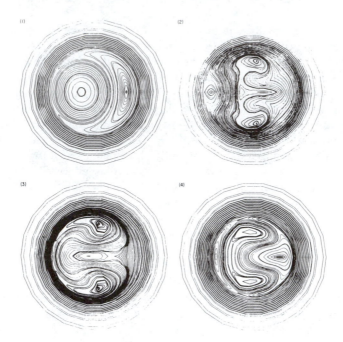

FIGURE 4   Evolution of the $m = 1$ mode. [From Ref. 22]

THE SAWTOOTH

Kadomtsev[20] developed a qualitative scenario for the experimentally observed sawtooth oscillation, involving the $m/n = 1/1$ mode. Again, only numerical analysis could discover whether the time scale for the process was in agreement with the experiments. Nonlinear evolution of the $m = 1$ mode showed[21-23] that the magnetic island grew far beyond the tearing layer width with no modification of the exponential growth rate given by linear analysis. The conditions necessary for the Rutherford derivation of algebraic growth are not satisfied.

There are a variety of different forms of soft X-ray signals observed during the periodic central temperature modulations referred to as sawtooth oscillations. The simple sawtooth, observed regularly on low-beta tokamaks with small $q = 1$ radii, consists of a $m/n = 1/1$ precursor oscillation, which grows in amplitude, followed by a rapid drop and flattening of the central temperature and a subsequent rise in temperature outside the $q = 1$ surface. Numerical simulations give a reasonable description of this phenomenon, with the complete reconnection of the $m = 1$ island giving both rapid thermal transport in the region inside $q = 1$ as well as the

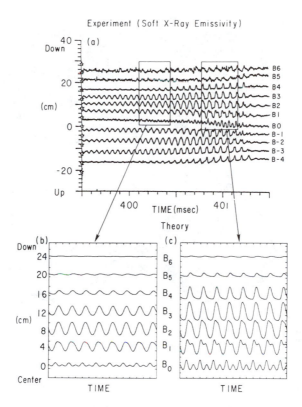

FIGURE 5   Soft x-ray signals obtained experimentally and theoretically.

rapid drop in the internal temperature as originally proposed by Kadomtsev. The evolution of the flux surfaces so obtained is shown in Fig. 4.

The simulations provide detailed information concerning the form of the diagnostic signals expected from soft X-ray emissivity and from magnetic fluctuations measured with Mirnov coils. An example of the comparison of the soft X-ray emissivity expected on an array of detectors is shown in Fig. 5. Shown are the experimental signals and those obtained by a nonlinear simulation using a numerical code (Wonchull Park, unpublished). Results of this nature permit a detailed quantitative check of the assumptions involved in the resistive MHD calculations. Similar agreement is found for the Mirnov signals.

Ohmic heating is then responsible for the slower rise in temperature after the reconnection, leading to a sawtooth form. Larger tokamaks and higher beta operation lead to behavior which, after the expected temperature scaling is taken into

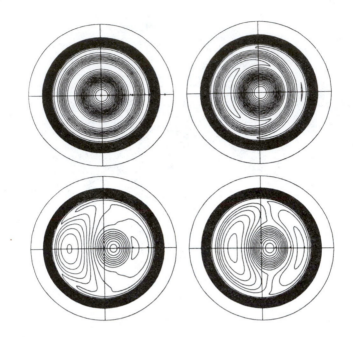

FIGURE 6   Evolution of a profile with two $q = 1$ surfaces.

account, was much more rapid and also more complicated. It is not as easily describable with the original Kadomtsev picture, but it appears that models involving very flat and even non-monotone q profiles, along with some ideal effects occurring with high beta and very low shear, can probably account for the variations, as has been shown by Wesson et al.,[24] Drake et al.,[25] and Park et al.[26] One major problem remaining is the indication in some experiments that the value of q on axis remains well below unity during the sawtooth process, indicating that complete reconnection does not occur. A model of a recurring sawtooth oscillation without a Kadomtsev-like reconnection has not been found.

The nonlinear evolution of a shear-free domain can be very rapid, as in the development of the surface kink. Similarly, the introduction of a hollow current profile with two $q = 1$ surfaces considerably changes the topology of reconnection during mode development. Shown in Fig. 6 is the evolution of an equilibrium with such a profile. Two $m = 1$ island states are seen to evolve, leading to diagnostic signals more complicated than in the simple Kadomtsev reconnection case. This area is still under active investigation and is quite important for the understanding of a fusion device, since the sawtooth has a large effect on heating and ignition scenarios. One of the puzzles is to explain the very rapid drop in central temperature observed

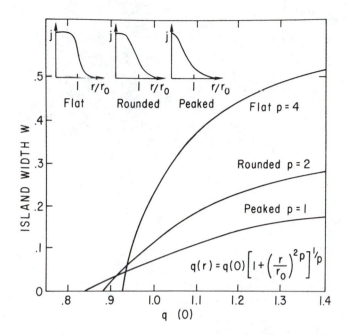

FIGURE 7 Saturated $m = 2$ island width. [From D.A. Monticello, R.B. White, and M.N. Rosenbluth, in *Plasma Physics and Controlled Nuclear Fusion Research 1978* (IAEA, Vienna, 1979), Vol. I, p. 605]

in large tokamaks, which can occur without any precursor $m = 1$ signal, but is often followed by a $m = 1$ successor oscillation. It is possible that the island state becomes unstable to an ideal perturbation, as suggested by Bussac and Pellat.[27] Another possibility is the development of extensive turbulence in the vicinity of the island separatrix, as suggested by Samain and Dubois.[28] Evidence for initial instability of the current sheet, perhaps related to the onset of such a turbulent state, has been found in computer simulations by Biskamp[29] and by Park.[30]

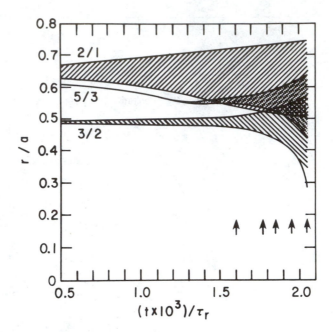

FIGURE 8 Multi-helicity evolution. [From Ref. 33]

THE MAJOR DISRUPTION

One of the important goals of the investigation of nonlinear MHD phenomena was the understanding of the major disruption. Experimental evidence indicated that the $m/n = 2/1$ mode played an important role in the onset of the disruption, because of the precursor oscillations observed with Mirnov coils and because of the tendency for disruption to occur when the value of q at the tokamak wall approached two. By investigating island saturation widths using a single helicity code, White, Monticello, and Rosenbluth[31] found conditions necessary for the occurrence of a large $m = 2$ island. Results are shown in Fig. 7.

Large islands occur only when the current profile is fairly flat and thus steep in the outer region, and when the value of q on axis rises above unity. The single helicity analysis is not valid for large islands, but is quite reliable for small islands. Hence this analysis isolates the necessary conditions for large island development, but it cannot explore the ensuing violent behavior.

At the same time, Waddell and collaborators at Oak Ridge were looking for significant nonlinear coupling of different helicities. Upon insertion of a flat profile with $q(0) > 1$, suggested by the single helicity analysis, they observed a nonlinear

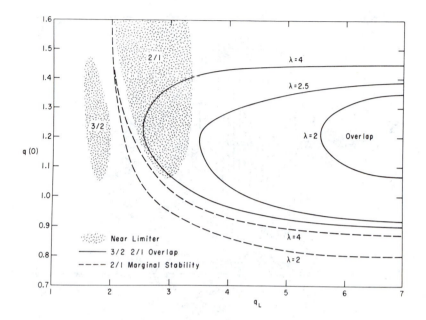

FIGURE 9 Domains of island overlap. [From Ref. 37]

destabilization of the $m/n = 3/2$ and $5/3$ modes when the large $m = 2$ island over-lapped the $q = 3/2$ and $5/3$ surfaces. The ensuing highly stochastic state provided a candidate for the turbulent final stages of the disruption. These results, reported by Callen et al.[32] and also by Carreras, Hicks, Holmes, and Waddell,[33] are shown in Fig. 8.

Toward the end of the simulations, the number of harmonics is very large. The final stage of the disruption is still a subject of study and speculation. The processes considered to be contributing to the rapid loss of plasma are: large scale stochasticity of the field, caused by the many modes excited; plasma contact with the limiter due to the large displacements of the flux surfaces, as simulated by Sykes and Wesson;[34] and, if the current profile is quite flat, resistive versions of the surface kink bubble formation, as analytically predicted by Zakharov[35] and simulated by Kurita et al.[36]

Current profile parameters which lead either to overlap of the 2/1 mode and the 3/2 mode or to contact with the limiter have been mapped out for a circular tokamak by Monticello and White.[37] In Fig. 9 are shown the domains of overlap and limiter contact as a function of q on axis and at the limiter for various q profiles, $q(r) = q(0)\left[1 + (r/a)^{2\lambda}\right]^{1/\lambda}$. In addition, the evolution of the saturated

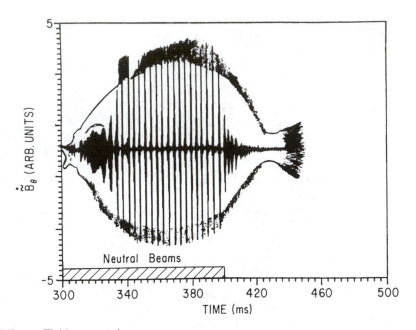

FIGURE 10   Fishbone mode.

island widths for the low (m, n) modes and their effect on transport during the evolution of the current profile has been examined by Wesson and Turner[38] and by Ivanov, Kakurin, and Chudnovskii.[39]

THE FISHBONE MODE

Although the interaction of trapped particles and MHD modes had been a subject of theoretical investigation for some time, an important parameter regime had been overlooked, and the first experiments involving neutral beam injection into a high-beta plasma therefore produced a surprise. A large amplitude $m = 1$ mode, with a real frequency matching the beam particle toroidal precession frequency, occurred in bursts that ejected the beam particles.

The mode was called the fishbone mode because of the structure of the magnetic signals, as seen in Fig. 10. The ejection mechanism was explained by White *et al.*[40] as a resonant interaction between the trapped particles and the field perturbation, with the trapped particles providing a free energy source. Next, Liu Chen, guided

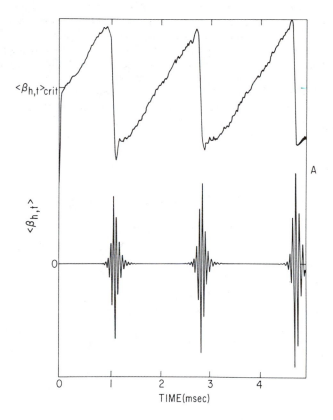

FIGURE 11   Simulation of the fishbone mode. [From Ref. 43]

by Rosenbluth, found a way to calculate the destabilization of the internal kink mode by the trapped particles, using a variational formulation of the equations describing the coupled system of an ideal MHD fluid and trapped high-energy particles. The change in energy of the system due to a plasma displacement is written as $\delta E = \delta W_{MHD} + \delta I + \delta W_{\kappa}$, where δW_{MHD} is the usual ideal internal kink potential, δI is the inertia term, and δW_{κ} is the kinetic term due to the trapped particles.

The ordering is such that to leading order, only the MHD terms contribute. Thus the first-order minimization leads to the eigenmode of the internal kink mode. This eigenmode can then be used to evaluate the trapped particle contribution, which is a resonant, non-Hermitian term, leading to a complex value for the frequency. There are two branches to the resulting dispersion relation. The MHD branch has small real frequency and is stabilized by the presence of the trapped particles. This effect had been previously found in the ballooning limit by Rosenbluth et al.[41] The trapped particle branch, with real frequency equal to the average trapped particle precession frequency, is unstable only if the plasma is near the threshold for the internal kink and if the trapped particle beta exceeds a threshold value, typically a fraction of a percent. Combining the destabilization of the mode with the resonant particle ejection process then allowed Chen, Rosenbluth, and White[42] to evaluate the dispersion relation for a model particle distribution and to simulate the full coupled system, obtaining good agreement with the experiments. These results are presented in Fig. 11, which shows the resulting magnetic signal and the trapped particle beta, as functions of time.[43]

REFERENCES

1. V.D. Shafranov, *Zh. Tekh. Fiz.* **40**, 241 (1970) [*Sov. Phys.-Tech. Phys.* **15**, 175 (1970)].
2. P.H. Rutherford, H.P. Furth, and M.N. Rosenbluth, in *Plasma Physics and Controlled Nuclear Fusion Research 1970* (IAEA, Vienna, 1971), Vol. I. p. 533.
3. B.B. Kadomtsev and O.P. Pogutse, *Zh. Eksp. Teor. Fiz.* **65**, 575 (1973) [*Sov. Phys.-JETP* **38**, 283 (1973)].
4. R. White, D. Monticello, M.N. Rosenbluth, and H. Strauss, in *Plasma Physics and Controlled Nuclear Fusion Research 1974* (IAEA, Vienna, 1975), Vol. I, p. 495.
5. M.N. Rosenbluth, D.A. Monticello, H.R. Strauss, and R.B. White, *Phys. Fluids* **19**, 1987 (1976).
6. Y.N. Dnestrovski, L.E. Zakharov, D.P. Kostomorov, A.C. Kukushkin, and L.F. Suzdaltseva, *Pis'ma Zh. Tekh. Fiz.* **1**, 45 (1975) [*Sov. Tech. Phys. Lett.* **1**, 18 (1975)].
7. M. Bussac, R. Pellat, D. Edery, and J. Soule, *Phys. Rev. Lett.* **35**, 1638 (1975).
8. M.N. Rosenbluth, R.Y. Dagazian, and P.H. Rutherford, *Phys. Fluids* **16**, 1894 (1973).
9. W. Park, D.A. Monticello, R.B. White, and S.C. Jardin, *Nucl. Fusion* **20**, 1181 (1980).
10. H.P. Furth, J. Killeen, and M.N. Rosenbluth, *Phys. Fluids* **6**, 459 (1963).
11. H.P. Furth, in *Propagation and Instabilities in Plasmas*, ed. W.T. Futterman (Stanford University Press, Stanford, CA, 1963), p. 87.
12. P.H. Rutherford, *Phys. Fluids* **16**, 1903 (1973).
13. H. Grad, in *Proc. Int. Congr. of Mathematicians*, Nice (Gauthier Villars, Paris, 1970), Vol. 3, p. 105.
14. R.B. White, D.A. Monticello, M.N. Rosenbluth, and B.V. Waddell, in *Plasma Physics and Controlled Nuclear Fusion Research 1976* (IAEA, Vienna, 1977), Vol. I, p. 569.
15. R.B. White, D.A. Monticello, M.N. Rosenbluth, and B.V. Waddell, *Phys. Fluids* **20**, 800 (1977).
16. D.A. Monticello and R.B. White, *Phys. Fluids* **23**, 366 (1980).
17. H.R. Strauss, *Nucl. Fusion* **23**, 649 (1983).

18. J.F. Drake and Y.C. Lee, *Phys. Rev. Lett.* **39**, 453 (1977).
19. S.C. Cowley, R.M. Kulsrud, and T.S. Hahm, *Phys. Fluids* **29**, 3230 (1986).
20. B B. Kadomtsev, *Fiz. Plazmy* **1**, 710 (1975) [*Sov. J. Plasma Phys.* **1**, 389 (1975]].
21. A. Sykes and J.A. Wesson, *Phys. Rev. Lett.* **37**, 140 (1976).
22. B.V. Waddell, M.N. Rosenbluth, D.A. Monticello, and R.B. White, *Nucl. Fusion* **16**, 528 (1976).
23. A.F. Danilov, Yu.N. Dnestrovskii, D.P. Kostomorov, and A.M. Popin, *Fiz. Plazmy* **2**, 167 [*Sov. J. Plasma Phys.* **2**, 93 (1976)].
24. J.A. Wesson *et al.*, in *Plasma Physics and Controlled Nuclear Fusion Research 1986* (IAEA, Vienna, 1987), Vol. II, p. 3.
25. D.A. Boyd *et al.*, in *Plasma Physics and Controlled Nuclear Fusion Research 1986* (IAEA, Vienna, 1987), Vol. I, p. 387.
26. W. Park, *Bulletin of the American Physical Society* **31**, 1527 (1986).
27. M.N. Bussac, D. Edery, R. Pellat, J.L. Soule, and M. Tagger, *Phys. Rev. Lett.* **109**, 331 (1985).
28. M. Dubois and A. Samain, *Nucl. Fusion* **20**, 1101 (1980).
29. D. Biskamp, *Phys. Fluids* **29**, 1520 (1986).
30. W. Park, D.A. Monticello, and R.B. White, *Phys. Fluids* **27**, 137 (1984).
31. R.B. White, D.A. Monticello, and M.N. Rosenbluth, *Phys. Rev. Lett.* **39**, 1618 (1977).
32. J.D. Callen, B.V. Waddell, B. Carreras, M. Azumi, P.J. Catto, H.R. Hicks, J.A. Holmes, D.K. Lee, S.J. Lynch, J. Smith, M. Soler, K.T. Tsang, and J.C. Whitson, in *Plasma Physics and Controlled Nuclear Fusion Research 1978* (IAEA, Vienna, 1979), Vol. I, p. 415.
33. B. Carreras, H.R. Hicks, J.A. Holmes, and B.V. Waddell, *Phys. Fluids* **23**, 1811 (1980).
34. A. Sykes and J.A. Wesson, *Phys. Rev. Lett.* **44**, 1215 (1980).
35. L.E. Zakharov, *Sov. Phys.-JETP Lett.* **31**, 714 (1981).
36. G. Kurita, M. Azumi, T. Takizuka, T. Tuda, T. Tsunematsu, Y. Tanaka, and T. Takeda, *Nuclear Fusion* **26**, 449 (1986).
37. R.B. White and D.A. Monticello, Princeton Plasma Physics Laboratory Report, No. PPPL-1674 (1980).
38. J.A. Wesson and M.F. Turner, *Nucl. Fusion* **22**, 1069 (1982).
39. N.V. Ivanov, A.M. Kakurin, and A.N. Chudnovskii, *Sov. J. Plasma Phys.* **10**, 38 (1984).
40. R.B. White, R.J. Goldston, K. McGuire, A.H. Boozer, D.A. Monticello, and W. Park, *Phys. Fluids* **26**, 2958 (1983).
41. M.N. Rosenbluth, S.T. Tsai, J.W. Van Dam, M.G. Engquist, *Phys. Rev. Lett.* **51**, 1967 (1983).
42. L. Chen, R.B. White, and M.N. Rosenbluth, *Phys. Rev. Lett.* **52**, 1122 (1984).
43. L. Chen *et al.*, in *Plasma Physics and Controlled Nuclear Fusion Research 1984* (IAEA, Vienna, 1985), Vol. II, p. 59.

L. Donald Pearlstein
Lawrence Livermore National Laboratory
University of California
Livermore, California 94550

Ion-Cyclotron Instability in Magnetic Mirrors

This paper reviews the stability to modes at the ion cyclotron frequency
and its harmonics. These modes have dominated the physics in minimum-
B, MHD-stabilized mirror machines. The free energy source is primarily
velocity space inversion and anisotropy, which are inherent to adiabatic
mirror confinement. The review considers both linear and nonlinear theories
and compares them with experiment.

INTRODUCTION

In this paper I shall review the role of ion-cyclotron frequency instability in magnetic
mirrors. Attention will be limited to those modes that are loss-cone driven and
anisotropy driven. The particular modes I will address are: the convective and
drift cyclotron loss-cone modes, and the anisotropy-driven Alfvén ion-cyclotron and
modified negative mass modes. I will first describe the energy sources that create
the drives for the instabilities. I will then enumerate the waves that can set up
in the plasma to tap these energy sources. Next, I will describe how the structure
of the configuration affects the analysis and properties of these instabilities. By

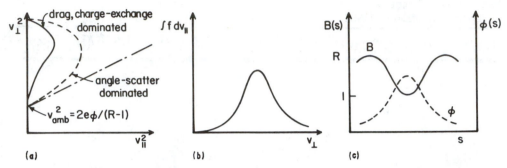

FIGURE 1 The mirror configuration: (a) sketch of phase space; (b) sketch of the distribution function averaged over $v_\parallel$ as a function of $v_\perp$; (c) sketch of the confining magnetic field B and the associated ambipolar potential ϕ.

structure I include the fact that: the mirror is of bounded extent; the cyclotron resonance is broadened by the nonuniform magnetic field; the ellipticity of the flux tube (fanning of the field lines) in minimum-B traps impacts the results; and the trapped ions bounce. I will then review the position of the nonlinear theories of these modes. Specifically, I will discuss quasilinear theory, explosive instabilities due to three-wave interaction and nonlinear Landau damping, and saturation due to nonlinear orbits.

PHASE SPACE

With adiabatic confinement, particles with "high" parallel velocities $v_\parallel$ are unconfined; in addition, because electrons are generally more collisional than ions, an ambipolar potential ϕ must be set up to hold in electrons. This means that "low" energy ions are no longer confined, since adiabatic confinement requires

$$v_\perp^2 (R-1) - v_\parallel^2 - 2e\phi \geq 0. \tag{1}$$

Such configurations exhibit anisotropy and inverted populations. As seen in Fig. 1, a normal loss-cone distribution filling the available phase space has a larger hole (the distribution is more inverted) than a drag- and charge exchange-dominated plasma, which has more anisotropy. This feature will prove to be important in the quasilinear analysis.

It was first pointed out by Harris[1] that the anisotropy ($T_\parallel < T_\perp$) led to inverse cyclotron damping and instability. Mikhailovskii[2] observed that the presence of drifts alone, in a Maxwellian plasma, could drive ion-cyclotron waves unstable. Rosenbluth and Post[3] showed that, with the adiabatic confinement of a mirror

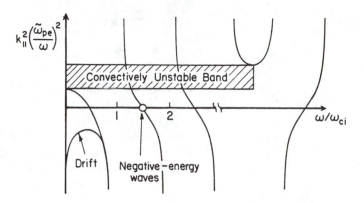

FIGURE 2 Sketch of the plasma response function, showing the resonances at the cyclotron frequency and its harmonics.

plasma and the loss-cone alone, there was instability due to this inverted population. Since this loss-cone feature is critical to mirror confinement, a great deal of activity was immediately generated to determine just how serious this prediction of instability was.

LOSS-CONE MODES

We now turn to the subject of loss-cone modes. In Fig. 2 are sketched the properties of the dielectric function, given by

$$k_\parallel^2 \frac{\tilde\omega_{pe}^2}{\omega^2} = k_\perp^2\, D(\omega, \omega_{ci}, \omega_{*i}, k_\perp, s).$$ (2)

In Eq. (2) the following definitions have been used:

$$\tilde\omega_{pe}^2 = \frac{k_\perp^2 c^2}{k_\perp^2 c^2 + \omega_{pe}^2}\,\omega_{pe}^2,$$ (3)

$$D = 1 + D_e + D_i$$ (4)

$$D_e = \frac{\omega_{pe}^2}{\omega_{ce}^2} - \frac{1}{k_\perp^2 \lambda_{Di}^2}\left[\frac{\omega_{*i}}{\omega} - \left(1 - \frac{\omega_{*i}}{\omega}\right)\frac{\beta}{2}\psi 1 - \frac{\omega}{\omega_{*i}}(1 - \psi)\right],$$ (5)

$$\psi = \int_0^\infty dx\, e^{-x} x\, \frac{\omega - \omega_{*e}}{\omega + \omega_{*e}\frac{\beta}{2}x}$$ (6)

$$D_i = \frac{\omega_{pi}^2}{k_\perp^2} \sum \int d^3v \left(n\omega_{ci} \frac{\partial f}{\partial v_\perp^2} - \frac{\partial f}{\partial x_\perp} \frac{k_\perp}{\omega_{ci}} \right)$$
$$\times i \int_{-\infty}^t dt' \exp\left(-i \int_{t'}^t dt'' \left(\omega - n\omega_{ci}(t'') \right) \right). \tag{7}$$

Essentially, Fig. 2 shows the coupling between the electron plasma oscillation (or shear Alfvén wave at moderate beta) to the ion cyclotron and drift wave in a loss-cone medium. When $\partial D/\partial\omega < 0$, the medium is negative energy and the crossings of the real $k_\parallel$-axis are stable negative energy waves. The shaded area in Fig. 2 corresponds to the Rosenbluth-Post convective loss-cone modes. The curve labelled "drift" shows the onset of drift cyclotron loss-cone instability (DCLC) when the curve falls below the axis. The dispersion relation with the various contributions is given in Eqs. (2)–(7). The two terms in the perpendicular electron response, Eq. (5), are the polarization drift and the electron drift wave with the full electromagnetic contributions[4] at $k_\perp \rho_e = 0$. The ion response, Eq. (7), shows the full resonance response and includes the drifts. In the analysis of these modes the ions are accurately represented by their electrostatic response and generally the Landau resonance can be ignored, $k_\parallel = 0$, since the frequency shift from resonance or the spatial variation in the cyclotron frequency dominate.[5] When the mode frequency approaches the midplane cyclotron frequency, the orbit integral can be important. The electrons are treated in the fluid limit (including drifts), and electromagnetic corrections are important. There is a response to the magnetic perturbations $\tilde{A}_\parallel$ and $\tilde{B}_\parallel$.[6] The expression in Eq. (2) has been obtained by eliminating these latter two perturbations in terms of the electrostatic piece, $\tilde{\phi}$.

Next, turn to the convective loss-cone mode.[3] This mode is characterized by large growth rates, $\gamma \gg \omega_{ci}$, and so the straight-line approximation can be used. The resultant dispersion relation is

$$k_\parallel^2 \frac{\omega_{pe}^2}{\omega^2} = k_\perp^2 \left(\frac{\omega_{pe}^2}{\omega_{ce}^2} + 1 \right) - \omega_{pi}^2 \int dv_\perp^2 dv_\parallel \frac{\partial f}{\partial v_\perp^2} \frac{\omega}{\sqrt{\omega^2 - k_\perp^2 v_\perp^2}} \tag{8}$$

$$\text{Im } k_\parallel \rho_i \left(\frac{m_i}{m_e} \right)^{1/2} \approx .07 \left(\frac{m_e}{m_i} + \omega_{ci}^2 \omega_{pi}^2 \right)^{-1/2}. \tag{9}$$

The first two terms in Eq. (8) contribute to the real part of the frequency and the last term, being a small perturbation, generates the imaginary contribution. It is the inverted population that gives negative dissipation to destabilize. This mode is convective, with predicted length restrictions comparable to mirror lengths; that is, the mode seemed to be borderline. There followed an immediate search for absolute instabilities (unstable "normal" modes). One of the first to be discovered, also by Post and Rosenbluth,[7] was the drift cyclotron loss-cone mode (DCLC). This mode is a flute ($k_\parallel = 0$) with growth rate the order of ω_{ci} and, hence, quite dangerous. The first analysis of the mode was electrostatic; when finite beta corrections are kept, a somewhat more optimistic stability criterion is obtained. This mode is generally more restrictive than the drift-cyclotron mode.

For the drift cyclotron loss-cone mode we generate the following dispersion relation:

$$\frac{m_e}{m_i} + \frac{\beta}{2k_\perp^2 a_i^2} - \frac{\omega_{ci}}{k_\perp R_p \omega}\left(1 + \frac{\beta}{2}\right) + \frac{\omega}{\omega_{ci}} \frac{F}{k_\perp^3 a_i^3} \cot \pi \frac{\omega}{\omega_{ci}} = 0. \tag{10}$$

Here the sum on Bessel functions has been replaced by the cotangent function in the limit $k_\perp a_i \gg 1$. Also, $F \approx 1$, and the terms proportional to β, the ratio of plasma pressure to magnetic pressure, are the electromagnetic terms mentioned earlier. The stability criterion can be obtained from a mini-max procedure with respect to $k_\perp$ and R_p/a_i, the radial scale length in units of the ion Larmor radius, leading to

$$\frac{a_i}{R_p} = 0.38 \left(\frac{m_e}{m_i} + \frac{\omega_{pi}^2}{\omega_{ci}^2}\right)^{2/3}, \quad k_\perp a_i = \left(\frac{m_e}{m_i} + \frac{\omega_{pi}^2}{\omega_{ci}^2}\right)^{-1/3} \tag{11}$$

whereas for finite β the stability criterion is[8]

$$\frac{c}{R_p \omega_{ci}} < \frac{2}{1 + \frac{\beta}{2}}\left(1 + \omega_{pe}^2 \omega_{ce}^2\right)^{1/2}, \quad \beta > 5\left(\frac{m_e}{m_i} + \frac{\omega_{pi}^2}{\omega_{ci}^2}\right)^{1/3}. \tag{12}$$

By comparison, the stability criterion for the drift-cyclotron mode[3] is

$$\frac{a_i}{R_p} < 2\left(\frac{m_e}{m_i} + \frac{\omega_{pi}^2}{\omega_{ci}^2}\right)^{1/2}, \quad k_\perp a_i \approx \left(\frac{m_e}{m_i} + \frac{\omega_{pi}^2}{\omega_{ci}^2}\right)^{-1/2}. \tag{13}$$

As $a_i/R_p \to 1$, the growth rates become larger than the cyclotron frequency and the mode goes over to its straight-line-orbit form (lower hybrid drift[9]).

The search for absolute modes then amplified. An early question was whether the electron plasma oscillation would convect to the mirror throat where it was Landau damped, or was there some reflection mechanism from thermal effects or nonlocal effects.[10] The conclusion from these investigations was that with collisional distributions the reflection was weak, essentially unaltering the original prediction. Next, there was considerable activity searching for absolutely (as opposed to convectively) unstable modes.[11,12] These were modes where the group velocity vanished in the complex $(\omega, k_\parallel)$ plane:, i.e.,

$$D(\omega, k_\parallel; k_\perp) = \frac{\partial D(\omega, k_\parallel; k_\perp)}{\partial k_\parallel} = 0. \tag{14}$$

Modes tended to be resonant near the cyclotron frequency and its harmonics (there were some exceptions where the real part of the frequency was near the half harmonic) with weak growth, $\gamma/\omega_{ci} \ll 1$. Since these modes had such weak growth,

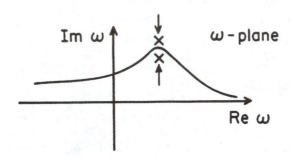

FIGURE 3   Picture of the complex frequency plane for complex $k_\parallel$, showing the "pole pinch."

they were quite sensitive to spatial inhomogeneities. It was then shown that, in a spatially varying medium, the aforementioned stable negative energy waves were reflected at turning points ($k_\parallel = 0$); thus the convective loss-cone mode was converted to an absolutely unstable mode, which is now called the axial loss-cone mode (ALC).[13]

I will now briefly review the theory of the absolute modes. To solve for the time asymptotic response of a system, examine the Fourier representation; to determine the time behavior, find the ω contour which lies above all unstable poles (Fig. 3). Now deform the $k_\parallel$ contour into the complex plane pushing the ω contour further up until it runs into a pole from on top, the pole "pinch," at which point

$$\epsilon(\omega, k_\parallel) = \frac{\partial \epsilon}{\partial k_\parallel} = 0. \tag{15}$$

To find such a state, generally we need the full plasma response for Maxwellian electrons and loss-cone ions. Using simple pole representations for the Z-functions, we obtain

$$\epsilon = \rho - \left(\frac{\kappa}{1 + i\eta^{1/3}\kappa} \right)^2 + \frac{\Omega + 2i\kappa}{(\Omega + i\kappa)^2} \tag{16}$$

with

$$\Omega = \frac{\omega - n\omega_{ci}}{n\omega_{ci}}\eta^{-1/3}, \quad \kappa = \frac{k_\parallel v_\parallel}{n\omega_{ci}}\eta^{-1/3}, \quad \rho = k_\perp^2 \lambda_{Di}^2 \frac{m_e}{m_i}\eta^{-2/3}. \tag{17}$$

The quantity η is defined as

$$\eta = 2v^2 \frac{m_e}{m_i} \int d^3v \frac{\partial f}{\partial v_\perp^2} J_n^2 \left(\frac{k_\perp v_\perp}{\omega_{ci}} \right). \tag{18}$$

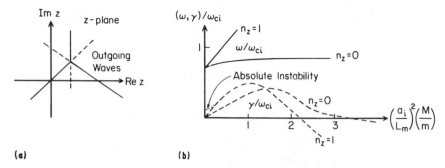

(a)   **(b)**

FIGURE 4   (a) Sketch of the Stokes and anti-Stokes lines around a turning point. Also depicted is the region in which boundary conditions require outgoing waves. (b) Sketch of the dimensionless frequency and growth rate as a function of the square of the dimensionless inverse magnetic scale length L_m (where $B = 1 + s^2/L_m^2 + \cdots$); depicted are the two lowest axial modes.

Stability requires

$$1.5\frac{\omega_{pi}^2}{\omega_{ci}^2}n^{-8/3}\left(\frac{m_i}{m_e}\right)^{1/3} = \frac{1.5}{\rho} < 1. \tag{19}$$

I now will sketch the calculation for the axial loss-cone modes. The basic technique is to expand about the infinite medium solution, viz., the stable negative energy waves, keeping first-order terms in $\delta\omega$ and s^2 and using the WKB phase integral to obtain the dispersion relation. The waves are seen to be evanescent near the origin and oscillatory farther out. As a consequence the turning points are on the 45° line. That is, we have normal modes in which energy is being radiated away; and because the standing waves are negative energy, we have instability. The basic features are that the negative energy waves are turning points in the WKB formalism and the boundary conditions correspond to "outgoing" waves (i.e., convection of energy out of the system). The structure of the wave equation is

$$\frac{d^2\psi}{ds^2} = -q_\parallel^2(\omega, s^2)\psi = -k_\perp^2 D(\omega, s^2)\psi. \tag{20}$$

To obtain the dispersion relation, expand about the negative energy solution

$$q_\parallel^2 = k_\perp^2\left(\frac{\partial D}{\partial\omega}\delta\omega + \frac{\partial D}{\partial s^2}s^2\right); \tag{21}$$

this leads to the phase integral $\oint q_\parallel ds = (2n + 1)\pi$, or

$$\delta\omega\frac{\partial D}{\partial\omega} = -(2n + 1)ik_\perp\left(\left|\frac{\partial D}{\partial s^2}\right|\right)^{-1/2}. \tag{22}$$

The phase level diagram and the eigenvalues are sketched in Fig. 4.

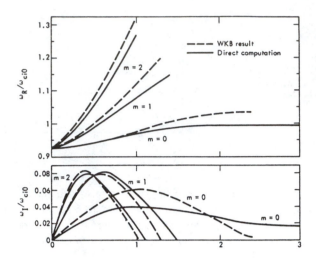

FIGURE 5 Eigenvalues versus normalized inverse scale length L_m [cf. Fig. 4(b)], computed from direct numerical integration and WKB approximation, where m refers to the axial mode number. [From Ref. 5]

To obtain quantitative results, we numerically solve the ordinary differential equation (ODE) along the field line, subject to outgoing wave boundary conditions.[5] The procedure used is called "shooting". We search for marginal stability (or sometimes the worst growth). Solving the ODE leads to $D(\omega, k_\perp; \text{parameters}) = 0$ for a specific ω and $k_\perp$, the worst case then being determined by forcing $d\gamma/dk_\perp = 0$, where γ is the growth rate. The differential equation has the form

$$\frac{d}{ds}\frac{\tilde{\omega}_{pe}^2}{\omega^2}\frac{d\phi}{ds} + F_{\text{fluid}}\phi = -(F_{\text{res}}^n - F_{\text{fluid}}^n)\phi = -\phi(0)\delta(s)\int ds V(s)$$

$$= \left[\omega_{ci}\tau_{\text{cor}} \approx \left(\frac{L_m}{v_\parallel}\omega_{ci}\right)^{2/3}\right] L_m\phi(0)\delta(s), \qquad (23)$$

where the right-hand side covers the orbit modification for modes resonant at the minimum cyclotron frequency or its harmonics. A modified WKB analysis based on Whittaker functions was also developed for this singular equation.[5] Generally, the lowest axial mode is flute-like at short enough scale lengths, whereas the higher modes are reasonably well described by the modified WKB analysis. Typical results are shown in Fig. 5.

Another instability considered was the Dory-Guest-Harris mode.[14] This flute mode is characterized by "narrow" unstable bands in $k_\perp$ and is stabilized by the field line fanning[15] over the axial extent of the mode:

$$k_\perp^2(s) = k_\perp^2(0)\exp[-2c(s)], \qquad (24)$$

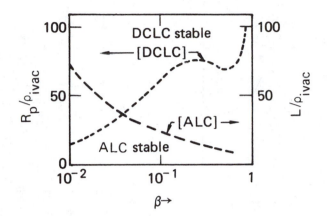

FIGURE 6 Stability boundaries for the axial loss-cone mode (ALC) and drift-cyclotron loss-cone (DCLC), with the dimensionless radius plotted as a function of the plasma beta.

where the exponential is the ellipticity of a quadrupole mirror.

The main results of these studies are shown in Fig. 6. The basic conclusion is that empty loss cones have a very uninteresting region of stability. The solution to this problem is to add to the unconfined region a small amount of warm plasma, either transiting through or, preferably, trapped in local potential wells. Whereas DCLC stability was found to require "large" radius, ALC stability was found to require "short" length:

$$\frac{L_m}{a_i}\left(\frac{m_e}{m_i}\right)^{1/2} \leq 1 \quad \text{for low } \beta \tag{25}$$

$$\frac{L_p}{a_i} \leq \frac{7.5}{\sqrt{\beta}} \quad \text{for high } \beta \tag{26}$$

Small amounts of warm plasma, $n_{\text{warm}}/n_{\text{hot}} \sim .05$ are stabilizing, but if this component is too cold, then it drives a two-component instability. Warm plasma stabilization was observed in the PR-6 and 2XIIB experiments, and was a key part of the designs for the TMX-U and MFTF-B devices.

Next we show results from the PR-6 experiment[16] and attempt to model the linear stability. The main feature of the experiment is that the presence of a potential well to trap warm ions, as depicted in Fig. 7(a), improves stability. Figure 8 shows the situation with an argon background: T_e is lowered, leading to a lower ambipolar potential and weaker loss-cone. Losses from the trapped plasma can then

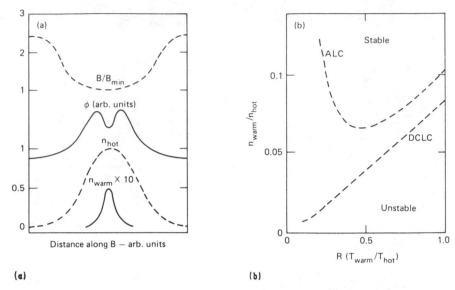

(a)   **(b)**

FIGURE 7   Theoretical modeling of the PR-6 Kurchatov experiment: (a) profiles of the magnetic field, floating potential, and calculated ion densities; and (b) warm plasma fraction required for stability as a function of the warm plasma temperature.

supply the required warm hydrogen plasma; this scenario would explain the gradual improvement. With a hydrogen background, there is an additional source of warm hydrogen plasma, hence the sudden onset of stability. Without the potential cup, there is only a gradual reduction in fluctuations, which is thought to be due to the reduced T_e and to a quasilinear relaxation of the distribution, to be described later. In the modelling, shown in Fig. 7(b), the loss cone was represented with a distribution function of the form

$$f(v_\perp^2) = \frac{R}{R-1} \left[\exp\left(-v_\perp^2\right) - \exp\left(-Rv_\perp^2\right) \right] . \tag{27}$$

where $1/R$ is a measure of the loss-cone. The modelling is qualitatively correct; however, the required warm plasma is higher than that estimated in the experiment. Figure 9 shows results from the 2XIIB experiment concerning the influence of an externally supplied plasma stream on stability.[17] The plasma can only be sustained with a stream; the rf fluctuation level ($\omega = \omega_{ci}$) is strongly reduced; and the energy hardens with the occurrence of high fluctuations. In Fig. 10 we show the results of modeling the linear stability with results from the differential equation. First, consider the graph in Fig. 10(a), which models 2XIIB. The strong influence of the stream on stability can be seen. With a warm plasma stream, the DCLC stability boundary corresponds to $\omega = \omega_{ci}$. This mode, the lowest axial mode in the system, is quite "flute-like" over the interaction region. The higher harmonics and the higher

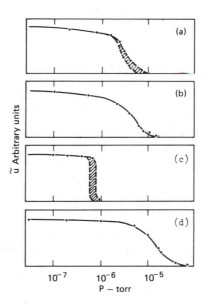

FIGURE 8   Influence of neutral gas on the DCLC threshold in the PR-6 experiment. The first two graphs show the results for argon fill. Oscillation amplitude as a function of argon pressure: (a) when there is a potential well generated by ECH to confine warm ions; (b) when there is no potential well. The two lower graphs show the stabilizing effect of hydrogen background gas. Oscillation amplitude as a function of hydrogen pressure: (c) when there is a potential well; (d) when there is no potential well. There is a marked improvement in stability when warm hydrogen ions can be trapped at the magnetic midplane. [From Ref. 16]

axial modes ($n_z = 1, 2, 3, \ldots$) are less unstable unless T_{warm} is too cold, at which point a two-component instability arises. These later curves can be computed with straight-line orbits since $\omega \gg \omega_{ci}$. In the configuration of Fig. 10(b), the ion pressure is peaked away from the midplane (i.e., "sloshing ions") and there is a barrier (i.e., a potential well at the midplane). The two curves are qualitatively the same as before. Now, however, the DCLC mode stabilizes at $\omega \approx 1.4\omega_{ci}$, the cyclotron frequency at the ion peak. In TMX-U this point corresponds to $\omega \approx 2\omega_{c_i}$, a frequency that is sometimes observed.

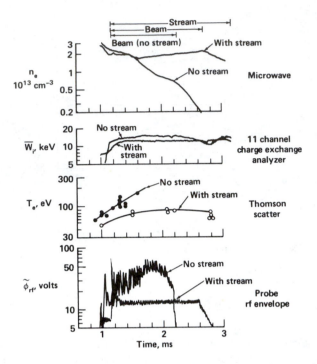

FIGURE 9   A comparison of plasma performance in the 2XII-B experiment with and without a plasma stream. Stream stabilization reduces rf, permits a density buildup, increases $n\tau$, but lowers T_e.

QUASILINEAR TREATMENT

We now turn to the quasilinear treatment of loss-cone modes. Galeev[18] first examined the quasilinear evolution of a loss-cone plasma. He assumed a plasma dominated by angle scattering. His analysis showed that the perpendicular diffusion, which is driven by the convective loss-cone modes, leads to confinement times of the order of a typical bounce time, considerably shorter than observed.

Later, the analysis was extended to account for electron drag, for charge exchange off of the neutral beam, and for the dominant mode being DCLC.[19] This analysis predicted lower loss rates in agreement with experimental observations. The analysis proceeded as follows. Stability requires a sufficient population of plasma in the loss-cone, on the order of $\Delta \approx .5\left(v_{\text{hole}}/v_{\text{hot}}\right)^2$. Furthermore, the dwell time of particles in the loss region, which had characteristic energies of the hole size, was considerably longer than a typical bounce time. To close the analysis, we need an equation for the electron temperature, since in this case (viz., drag and charge

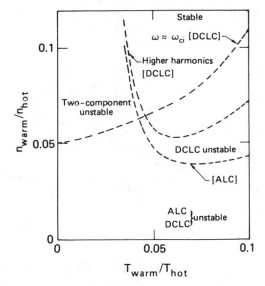

FIGURE 10(a) Theoretical stability boundaries for DCLC and ALC in the 2XII-B device. Shown is the fraction of warm plasma required for stabilization as a function of warm temperature ($R = 10, E_i = 13$ keV, $\beta \approx 0.5$).

exchange dominated) the hole size is determined by the ambipolar potential, which is determined by the electron temperature:

$$\frac{3}{2}\frac{d}{dt}nT_e = \frac{n_{\text{hot}}T_{\text{hot}}}{\tau_{\text{drag}}} - \eta\frac{nT_e}{\tau_L}. \tag{28}$$

In Eq. (28), η is the multiple of T_e carried away by each electron, where typically $\eta \sim 8$. In order to thermalize the plasma in the loss region, we require the diffusion coefficient to be

$$D = D(v_{\text{hole}})(v_{\text{hole}}v_\perp)^3, \quad D(v_{\text{hole}}) \geq \frac{v_{\text{hole}}^2}{\tau_{tr}} = \frac{v_{\text{hole}}^3}{L}. \tag{29}$$

If the warm plasma diffuses down from the trapped region, then

$$\frac{1}{\tau_L} = \frac{\Delta}{\tau_{tr}} \rightarrow D(v_{\text{hot}}) = \Delta\frac{v_{\text{hot}}^2}{\tau_L}. \tag{30}$$

On the other hand, if a cold stream is supplied, then the hot plasma must supply power lost by the stream:

$$P = \frac{n_{\text{warm}}v_{\text{hole}}^2}{\tau_L} \rightarrow D(v_{\text{hot}}) = \Delta\frac{v_{\text{hole}}^2}{\tau_L}. \tag{31}$$

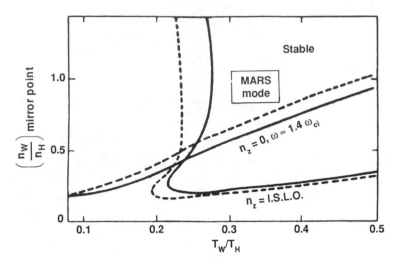

FIGURE 10(b) Theoretical stability boundaries for DCLC and ALC in the MFTF-B device. The vertical axis is the ratio of thermal-ion to sloshing-ion density at the mirror midplane. The horizontal axis is the ratio of thermal-ion to sloshing-ion energy. The difference between the solid and dashed curves represents variations due to a range of profiles. The square region depicts the nominal operating regime for the MARS mode.

Both of these low diffusion rates are adequate to thermalize plasma in the loss region, and we see that by supplying a stream the loss rate is reduced by v_{hole}^2/v_{hot}^2. Finally, from the electron energy balance equation, we obtain

$$T_e(keV) = 2.1 \times 10^{-5} n^{1/4} T_{hot}^{1/2}(keV), \qquad (32)$$

in good agreement with 2XIIB.

Next, an alternate derivation of the quasilinear equations that uses the magnetic moment and bounce invariant will be described. This procedure is useful since the lowest order distribution is only a function of these two invariants, μ and J, on each field line. Stochasticity, for the single wave that is followed, comes from the sum over bounces. At marginal stability, $\omega = \omega \equiv \omega_{ci}(0)$ and losses of hot particles are slow compared with bounce times. We can then bounce-average the background distribution function:

$$K = K_0 + \tilde{K} = \mu\Omega + J\omega_B(\mu) + \left(e\tilde{\phi}\exp\left(i\mathbf{k}_\perp \cdot \mathbf{x}_\perp - i\omega t\right) + c.c\right) \qquad (33)$$

$$-\frac{\partial f}{\partial t} = \frac{\partial}{\partial \mu}\dot{\mu}f + \frac{\partial}{\partial \theta}\dot{\phi}f + \frac{\partial}{\partial J}\dot{J}f + \frac{\partial}{\partial \xi}\dot{\xi}f$$

$$= \{f, K\} = \frac{\partial f}{\partial \theta}\frac{\partial K}{\partial \mu} - \frac{\partial f}{\partial \mu}\frac{\partial K}{\partial \theta} + \frac{\partial f}{\partial \xi}\frac{\partial K}{\partial J} - \frac{\partial f}{\partial J}\frac{\partial K}{\partial \xi}. \qquad (34)$$

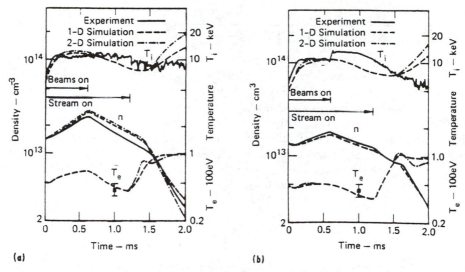

FIGURE 11 Comparison of the time history of plasma parameters in the 2XII-B experiment with that predicted by the quasilinear code. Dashed lines indicate theory: (a) 300 amp run; (b) 170 amp run.

The standard weak turbulence approximation leads to

$$\frac{\partial f_0}{\partial t} = \text{Im} \sum_{n,m} L_{n,m} \frac{\left|\tilde{\phi}\right|^2 J_n^2(k_\perp \rho) J_m^2\left(\frac{nJ}{4\mu}\right)}{D_{n,m} L_{n,m}} f_0 \tag{35}$$

with

$$D_{n,m} = \omega - n\hat{\Omega} - 2m\omega_B - \ell\hat{\omega}_D, \quad \hat{\Omega} = \Omega + \frac{J\omega_B}{2\mu} \tag{36}$$

$$L_{n,m} = \omega\left(\frac{\partial}{\partial\mathcal{E}} - \frac{1}{\Omega}\frac{\partial}{\partial\mu}\right) - \left(\frac{\omega}{\Omega} - n\right)\frac{\partial}{\partial\mu} + \ell\frac{\partial}{\partial\alpha_G} \tag{37}$$

$$D = \left|e\tilde{\phi}\right|^2 \tau_{\text{cor}} J_n^2(k_\perp \rho), \quad \tau_{\text{cor}} = \text{Im} \sum_m \frac{J_m^2\left(\frac{nJ}{4\mu}\right)}{D_{n,m}} \sim \frac{1}{\Omega}\left(\frac{L_m}{v_\parallel}\Omega\right)^{1/3} \tag{38}$$

We see that for a single frequency and wavenumber, stochasticity is due to overlapping bounce resonances.

Figure 11 shows a comparison of the quasilinear code results of Berk *et al.*[20] with the experimental results from the 2XIIB discharges. There are two code runs: a one-dimensional $(v_\perp)$ version and a two-dimensional $(v_\parallel, v_\perp)$ quasilinear version that has Coulomb collisions. As can be seen, the density and the ion temperature, T_i, agree quite well over most of the discharge. Late in time, there is more hardening of the ion energy in the experiment. The single measurement of T_e also agrees with the simulation.

OTHER NONLINEAR EFFECTS

Nonlinear effects, other than quasilinear, will now be discussed. Of the four nonlinear effects to be mentioned here, it is interesting to note that Marshall Rosenbluth was involved in developing the theory for three of them.

First consider explosive instabilities in media with positive and negative waves and dissipation.[21] The basic features are that for the three-wave interaction, the highest frequency must be of different sign of energy; and for nonlinear Landau damping, the highest frequency must be negative energy and the lowest frequency positive energy. Both of these conditions can be satisfied in mirrors with loss-cone distributions. The nonlinear growth rate is

$$\gamma_{NL} = \left(\frac{\mathcal{E}}{nT_i} \frac{\Omega^3}{\omega} \right)^{1/2}, \tag{39}$$

with $\mathcal{E}$ the mode energy. An important feature not treated in this work is the spatial variation of the resonance.

The next topic is nonlinear frequency shifts.[22] The nonlinear shift in the cyclotron frequency introduces a shift in the mode frequency that is stabilizing when it exceeds the linear growth rate. Theory predicts a saturation level for modes with $\omega \approx \omega_{*i} \approx n\omega_{ci}$ when

$$\frac{e\tilde{\phi}}{T_i} = \left(\frac{\pi}{8} \frac{4n^2 - 1}{n^2} \frac{m_e}{m_i} \right)^{1/2} \frac{a_i}{L}. \tag{40}$$

The third nonlinear effect is the prediction of enhanced scattering due to "stable" convective loss-cone modes:[23]

$$\frac{1}{n\tau} = \frac{1}{n\tau_{\text{drag}}} \left[1 + \frac{\omega_{ce}}{2\omega_{pe}} \left(1 + \frac{\omega_{pe}^2}{\omega_{ce}^2} \right)^{-3/2} \right]. \tag{41}$$

In Fig. 12, a comparison is shown between 2XII results and Eq. (41). The convective growth, which appears in an exponential, has been ignored in Eq. (41). It is believed that the relatively low energy (1 ~ 3 Kev) and the relatively high density of the plasma produced enough streaming through collisional processes to fill the loss-cone sufficiently to nearly give stabilization. With a minimum of quasilinear diffusion, the convective modes could easily have been marginally stable. Except at the higher energy and lower density where the plasma was less collisional (and consequently more unstable), agreement with theory is quite good.

Next, we turn to the question of superadiabaticity. The fundamental question is to determine the level of fluctuations at which one expects to find the μ-invariant destroyed. This calculation was due to Rosenbluth,[24] following an earlier paper by Aamodt.[25] The analysis proceeds as follows: If $\omega \approx n\Omega$ is assumed, then the magnetic moment (μ) experiences a kick each time the particle crosses the midplane. Provided the change in phase of the particle is large between successive crossings

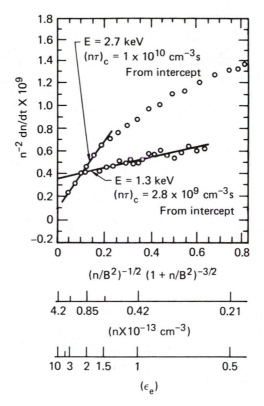

FIGURE 12 Comparison of 2XII data with the Baldwin-Callen scaling law.

$(\Omega \gg \omega_B)$ and provided the incremental change of phase due to the jump in μ is greater than unity, the motion is unstable (stochastic). Expand about the fixed points to obtain the standard map:

$$\tilde{\mu}_{m+1} = \tilde{\mu}_m \pm 3e\tilde{\phi}v_{\|}^{-2} \left(\frac{L^2 n}{\rho_i^2}\right)^{1/3} n\tilde{\psi}_m \qquad (42)$$

and

$$\tilde{\psi}_{m+1} = \tilde{\psi}_m - \frac{3\pi L}{2\rho_i}\mu_0^{-5/2}\tilde{\mu}_{m+1}, \quad \mu_0 = \left(\frac{Ln}{2\rho_i}\right)^{2/3}. \qquad (43)$$

A stability analysis then predicts the usual scaling for stochasticity,

$$\kappa = \frac{\partial\tilde{\phi}}{\partial\tilde{\mu}}\delta\tilde{\mu} \geq 4; \qquad (44)$$

however, a more careful calculation shows that onset of stochasticity occurs at $\kappa \approx 1$. Equation (44) can be rewritten to give

$$\frac{e\tilde{\phi}}{T_i} < \left(\frac{\rho_i}{L}\right)^{5/3} n^{-4/3} \tag{45}$$

as the condition for superadiabatic motion. Otherwise, when the motion is ergodic, we have diffusion:

$$\left(\frac{\Delta\mu}{\mu}\right)^2 = \left(\frac{e\tilde{\phi}}{T_i}\right)^2 \left(\frac{L^2 n}{\rho_i^2}\right)^{2/3} \frac{t v_{\parallel}}{L}. \tag{46}$$

The condition for this process to be more important than collisions is

$$\frac{e\tilde{\phi}}{T_i} > \left(\frac{\rho_i^2}{L^2 n}\right)^{1/3} \left(\frac{\nu L}{v_{\parallel}}\right)^{1/2}. \tag{47}$$

The following conclusions can be drawn from several simulations with regard to the various nonlinear processes. It should be emphasized that in the simulations, the geometry was frequently simplified so as to isolate the various processes. For a grossly unstable system, equilibrium settled to the quasilinear stable state; the process of relaxation was, in general, not simple. Quasilinear relaxation occurred; particle trapping occurred; and there was a nonlinear frequency shift.[26] If the parameters (wavenumbers, density, etc.) were chosen so as to restrict linear growth, this weakly unstable system exhibited nonlinear frequency shift saturation.[26] Explosive instabilities were seen in simulations in which linearly stable ion cyclotron modes were carefully initialized. The explosive instability was observed to grow in quantitative agreement with theory[27] that included nonlinear frequency shift effects. Ultimate saturation occurred when the ion orbits became strongly perturbed. This activity was never identified in simulations of unstable plasmas.

ANISOTROPY-DRIVEN MODES

We now turn attention to modes driven by anisotropy. An interesting mode observed in the DCXII experiment[28] and the PR-5 experiment[29] had a frequency close to Ω, the central cyclotron frequency, with low growth and long perpendicular wavelengths. Consequently, the mode could not be loss-cone driven. A new instability[30,31] driven by the extreme anisotropy present in both experiments was then discovered. Several ingredients enter the analysis of this mode. The mode requires coupling to bounce motion:

$$\left|\frac{\omega - \Omega}{\Omega}\right| \ll 1 \tag{48}$$

and

$$\frac{\Delta\hat{\Omega}}{\Omega} = \left|\frac{J\omega_B}{2\mu\Omega}\right| \ll 1, \quad \hat{\Omega} = \Omega + \frac{J\omega_B}{2\mu}. \tag{49}$$

Also, the instability requires strong anisotropy $(T_\parallel/T_\perp \ll 1)$. The instability is due to bunching. As the perpendicular energy increases, particles bounce closer to the minimum in the magnetic field and the average cyclotron frequency decreases. This instability was responsible for the anomalously short plasma lifetime observed in both the PR-5 and the DCX-II experiments. The anisotropy at threshold is given by[32]

$$\frac{T_\parallel}{T_\perp}\frac{\Omega}{\omega_B} < 0.77, \tag{50}$$

which condition is generally quite easy to satisfy.

ALFVEN ION-CYCLOTRON MODE

In the remainder I will concentrate on the Alfvén ion-cyclotron mode (AIC). As will be seen, this instability has influenced the performance of the TMX tandem mirror. With the change to "sloshing" ions in TMX-U, this mode was eliminated. I now indicate the history of this mode and show a sketch of its derivation.

This mode is also driven by anisotropy, $T_\perp/T_\parallel > 1$, and requires finite beta. It was discovered independently by Rosenbluth[33] and by Sagdeev[34] and was then analyzed with a mirror distribution.[35] Later this mode was examined in the strong anisotropy limit[36] $(T_\parallel \to 0)$. In this limit the growth rate is algebraic, rather than being driven by resonant particles. Reference 36 pointed out the importance of this mode to mirror confinement and gave an evaluation of plasma lifetimes from quasilinear theory. The dispersion relation for $k_\perp = 0$ has the form

$$\omega^2 - k_\parallel^2 c^2 - \omega_{pe}^2\left(\frac{\omega}{\omega-\omega_{ce}}\right) = \omega_{pi}^2\int d^3v\left[\frac{\omega - k_\parallel v_\parallel}{\omega - \omega_{ci} - k_\parallel v_\parallel} - \frac{1}{2}\frac{k_\parallel^2 v_\perp^2}{(\omega - \omega_{ci} - k_\parallel v_\parallel)^2}\right]f \tag{51}$$

which, in the limit $T_\parallel \to 0$, reduces to

$$0 = \kappa^2 + \left(\frac{\omega^2}{\omega-1} + \frac{1}{2}\frac{\kappa^2\beta_\perp}{(\omega-1)^2}\right), \quad \frac{\omega}{\Omega_{ci}} \to \omega \tag{52}$$

$$\lim \kappa = \frac{k_\parallel c}{\omega_{pi}} \to \infty, \quad \frac{\gamma}{\omega_{ci}} = \sqrt{\frac{\beta}{2}}. \tag{53}$$

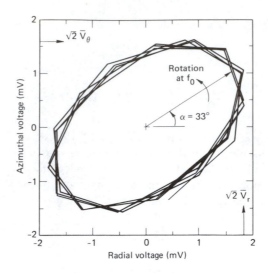

Azimuthal vs radial induced loop voltages. The
field vector rotates in the direction of ion gyration.
The tilt angle of the ellipse indicates a -65° phase
shift between B_r and B_θ (i.e., an elliptically
polarized wave).

FIGURE 13   Azimuthal versus radial induced loop voltages, indicating an elliptically polarized wave. [From Ref. 41]

End-wall ion energy analyzer

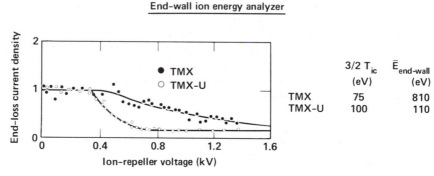

	$3/2\, T_{ic}$ (eV)	$\bar{E}_{end\text{-}wall}$ (eV)
TMX	75	810
TMX-U	100	110

FIGURE 14   Experimental results on AIC stability in the TMX and TMX-U devices. The
presence of the mode in the TMX end-plugs is seen by the increase in the endloss energy.
In TMX-U, this effect was absent. [From Ref. 42]

With the addition of finite $k_\perp$, there is a coupling to right- and left-circularly
polarized waves, and the dispersion relation is obtained from the solution of a two-
by-two determinant. Particle simulations support the conclusions of quasilinear
theory,[37] namely, that the anisotropy relaxes.

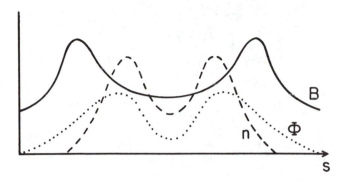

FIGURE 15   Trapping of warm ions in the midplane potential well.

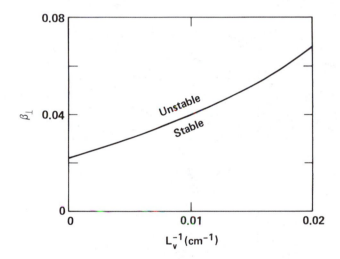

FIGURE 16   Stability boundary in $\beta_\perp$ versus L_v^{-1} (the vacuum magnetic scale length) for the TMX end cell, illustrating the stabilizing effect of axial nonuniformity. [From Ref. 40]

    To properly ascertain stability with respect to this mode, it is necessary to put in realistic distributions; the $T_\parallel \rightarrow 0$ limit is much too pessimistic. Analytic distributions that are product functions of pitch-angle and energy are chosen to match the Fokker-Planck results.[38,39] Spatial variation also plays an important role, just as it did in loss-cone modes. The procedure used to evaluate stability properties is the following.[40] Given $B(s)$ and the existence of constants like $\mathcal{E}$ and μ, one finds $f(v,\phi,s)$, the local distribution, that corresponds to $f(\mathcal{E},\mu)$. Knowing $f(v,\phi,s)$

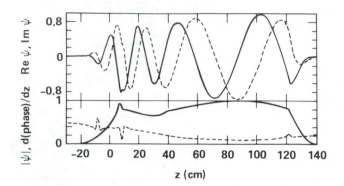

FIGURE 17   Axial eigenfunctions for a model of the MFTF-B axicell. (a) Real (—) and imaginary (- - -) parts of the eigenfunction. (b) Amplitude (—) and the derivative of the phase with respect to z(- - -, units: cm^{-1}) of the eigenfunction. [From Ref. 40]

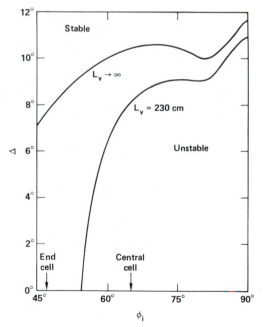

FIGURE 18   Stability boundaries as a function of pitch-angular width Δ and injection angle ϕ, for a uniform plasma ($L_v \rightarrow \infty$) and for a plasma with magnetic field that models the central cell of TMX-U ($L_v = 230$ cm, $B_{v_0} = 0.3$ T, $R_v \rightarrow \infty$).

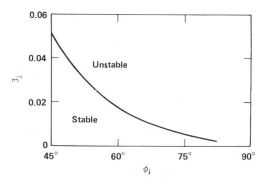

FIGURE 19   Stability boundary in $\beta_\perp$ versus injection angle ϕ_i, for the limit $\Delta = 0$ of the nonuniform plasma ($L_v = 230$ cm) of Fig. 18. The unstable region shrinks as Δ increases. [From Ref. 40]

one then calculates the local dispersion relation $D(\omega, k, s) = 0$ in the eikonal limit, which is relevant since $kL \gg 1$. Then extract $k(s, \omega)$ from $D = 0$ and apply WKB theory to obtain the global dispersion relation:

$$\oint ds\, k(s, \omega) = (2n + 1)\pi. \tag{54}$$

In Fig. 13 is shown some of the evidence for AIC instability in the TMX experiment.[41] Recall that for DCLC, the wave propagates in the ion diamagnetic drift direction, which is opposite to these observations. In addition, $k_\perp \rho_i \lesssim 1$ and the wave is nearly elliptically polarized in the sense of ion gyration. Both of these observations agree with AIC predictions and differ from DCLC observations. Figure 14 shows a comparison between TMX and TMX-U results.[42] The mode observed in TMX has the properties previously mentioned and was responsible for heating the ions. In TMX-U, this mode is largely absent. The reason for this improved stability is primarily the reduced anisotropy produced by "sloshing ion" injection. This improvement is due to several effects: the anisotropy is reduced; the pitch angle distribution (mapped from angled injection at the midplane) is spread out; $\beta_\perp$ is reduced; scale lengths at the anisotropy peak are reduced; and, as shown in Fig. 15, warm ions can be trapped in the potential well at the midplane of the mirror.

Finally, in Figs. 16–20, we show the status of the calculations on AIC stability and how they fit the theoretical picture described previously.[42] Figures 16 and 17 indicate improved stability with shorter scale lengths; Fig. 17 shows the eigenfunction. The improved stability with the spreading of the pitch angle distribution is shown in Fig. 18; and the improvement of stability with steeper pitch angle injection is shown in Fig. 19. In Fig. 20 we see the improvement of stability with

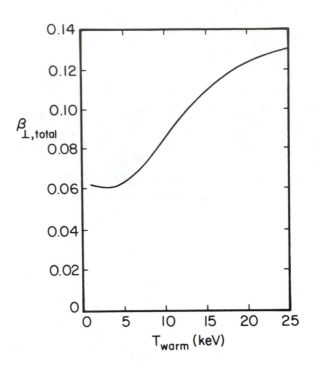

FIGURE 20 Stabilizing effect of isotropic warm plasma; stability is below the curve. [From Ref. 40]

the addition of warm plasma, provided $T_{\text{warm}} \geq T_{\|\text{hot}}$. These effects all add up to a prediction that the ions trapped in the plug region (i.e., the outer mirror) are stable, in agreement with the experimental observations.

ACKNOWLEDGMENTS

This work was performed under the auspices of the U.S. Department of Energy under contract number W-7405-ENG-48.

REFERENCES

1. E.G. Harris, *Journal of Nuclear Energy Pt. C: Plasma Physics* **2**, 138 (1961).
2. A.B. Mikhailovskii and A.V. Timofeev, *Zhurnal Eksperimental'noi i Teoreticheskoi Fiziki* **44**, 919 (1963) [*Sov. Phys.-JETP* **17**, 626 (1963).
3. M.N. Rosenbluth and R.F. Post, *Physics of Fluids* **8**, 547 (1965).
4. R.C. Davidson, N.T. Gladd, C.S. Wu, and J.D. Huba, *Physics of Fluids* **20**, 301 (1977).
5. H.L. Berk, L.D. Pearlstein, and J.G. Cordey, *Physics of Fluids* **15**, 891 (1972).
6. J.D. Callen and G.E. Guest, *Physics of Fluids* **14**, 1588 (1971).
7. R.F. Post and M.N. Rosenbluth, *Physics of Fluids* **9**, 730 (1967).
8. W.M. Tang, L.D. Pearlstein, and H.L. Berk, *Physics of Fluids* **15**, 1153 (1972).
9. R.C. Davidson and N.T. Gladd, *Physics of Fluids* **18**, 1327 (1975).
10. H.L. Berk and D.L. Book, *Physics of Fluids* **12**, 649 (1969); J.G. Cordey *et al.*, in *Plasma Physics and Controlled Nuclear Fusion Research 1968* (IAEA, Vienna, 1969), Vol. II, p. 267.
11. C.O. Beasley, Jr. and J.G. Cordey, *Plasma Physics* **10**, 411 (1968).
12. H.L. Berk, T.K. Fowler, L.D. Pearlstein, R.F. Post, J.D. Callen, C.W. Horton, and M.N. Rosenbluth, in *Plasma Physics and Controlled Nuclear Fusion Research 1968* (IAEA, Vienna, 1969), Vol. II, p. 151.
13. H.L. Berk, L.D. Pearlstein, J.D. Callen, C.W. Horton, and M.N. Rosenbluth, *Physical Review Letters* **22**, 876 (1969).
14. R.A. Dory, G.E. Guest, and E.G. Harris, *Physical Review Letters* **14**, 131 (1965).
15. D.E. Baldwin, *Physics of Fluids* **17**, 1346 (1975).
16. B.E. Kaneev, *Nuclear Fusion* **19**, 347 (1979).
17. F.H. Coensgen *et al.*, *Physical Review Letters* **35**, 1501 (1975).
18. A.A. Galeev, in *Plasma Physics and Controlled Nuclear Fusion Research 1965* (IAEA, Vienna, 1966), Vol. I., p. 393.
19. D.E. Baldwin, H.L. Berk, and L.D. Pearlstein, *Physical Review Letters* **43**, 1318 (1979).
20. H.L. Berk *et al.* in *Plasma Physics and Controlled Nuclear Fusion Research 1976* (IAEA, Vienna, 1977), Vol. III, p. 147.
21. B. Coppi, M.N. Rosenbluth, and R.N. Sudan, *Annals of Physics (New York)* **55**, 207 (1969).
22. R.E. Aamodt, Y.C. Lee, C.S. Liu, and M.N. Rosenbluth, *Physical Review Letters* **39**, 1660 (1977).
23. D.E. Baldwin and J.D. Callen, *Physical Review Letters* **28**, 1686 (1972).
24. M.N. Rosenbluth, *Physical Review Letters* **29**, 408 (1972).
25. R.E. Aamodt, *Physical Review Letters* **27**, 135 (1971).
26. B.I. Cohen, N. Maron, and G.R. Smith, *Physics of Fluids* **25**, 821 (1982); B.I. Cohen, G.R. Smith, N. Maron, and W.M. Nevins, *Physics of Fluids* **26**, 1851 (1983).
27. J.A. Byers, M.E. Rensink, and G.M. Walters, *Physics of Fluids* **14**, 826 (1971).

28. J.F. Clarke and G.G. Kelley, *Physical Review Letters* **21**, 1041 (1968).
29. T. Baiborodov, M.S. Ioffe, R.I. Sobelev, and I.I. Yushmanov, *Soviet Physics-JETP* **26**, 836 (1968).
30. B.B. Kadomtsev and O.P. Pogutse, in *Plasma Physics and Controlled Nuclear Fusion Research 1968* (IAEA, Vienna, 1969), Vol. II, p. 125.
31. J.F. Clarke and G.G. Kelly, in *Plasma Physics and Controlled Nuclear Fusion Research 1968* (IAEA, Vienna, 1969), Vol. II, p. 219.
32. J.D. Callen and W.C. Horton, *Physics of Fluids* **15**, 2306 (1972).
33. M.N. Rosenbluth, Risö Report. **18**, 189 (1960).
34. R.Z. Sagdeev and V.D. Shafranov, *Sov. Phys.-JETP* **12**, 130 (1961).
35. J.G. Cordey and R.J. Hastie, *Physics of Fluids* **15**, 2291 (1972).
36. R.C. Davidson and J.M. Ogden, *Physics of Fluids* **18**, 1045 (1975).
37. B.F. Otani, Ph. D thesis, University of California at Berkeley, 1986, submitted to *Physics of Fluids*.
38. D.C. Watson, *Physics of Fluids* **23**, 2485 (1980).
39. G.R. Smith, *Physics of Fluids* **27**, 1499 (1984).
40. G.R. Smith, W.M. Nevins, and W.M. Sharp, *Physics of Fluids* **27**, 2120 (1984).
41. T.A. Casper and G.R. Smith, *Physical Review Letters* **48**, 1015 (1982).
42. T.C. Simonen, in *Mirror-Based and Field-Reversed Approaches to Magnetic Fusion* (International School of Plasma Physics, Varenna, 1983), Vol. 1, p. 187.

Derek C. Robinson
Culham Laboratory
(UKAEA/EURATOM Fusion Association)
Abingdon, Oxon, OX14 3DB, United Kingdom

Toroidal Pinches

The theory of toroidal pinches—in particular, reverse field pinches and spheromaks—is reviewed. The key area of ideal and resistive stability of the pinch has been explored for nearly three decades, whereas the nonlinear consequences of these instabilities have only recently received attention. The spontaneous ability of pinches to relax and produce a reversed toroidal field was noted in early experiments, and numerous theoretical ideas were developed to explain this process. The theory of plasma relaxation is now well developed and provides a general basis for explaining many of the phenomena observed in pinches. The theory of anomalous transport is in its infancy, but the invariance properties of the underlying equations do provide important information on the transport scaling.

HISTORICAL INTRODUCTION

The theory of toroidal pinches really began some three to four decades ago. The theory of the pinch effect itself is much older. Evidence for the magnitude of the pinch effect was first noted early this century when currents of a few tens of kilo-amperes associated with lightning passed down a hollow lightning conductor. These early observations[1] demonstrated the unstable nature of the pinch effect in a current-carrying conductor. Research on toroidal pinches began in the late 1940's and was concentrated in a few countries, e.g., the UK, the USSR, and the USA. In 1946 G.P. Thomson filed a secret patent for a toroidal reactor using D-D (see Table 1 for parameters), which is interesting to compare with present reactor designs. Experiments were started at this time with a toroidal facility at Imperial College.[2] Thonemann also proposed the idea of thermonuclear reactions in a toroidal pinch, probably based upon the ideas of Tonks[3] and Bennett[4] in America.

In the early 1950's much of the theoretical work was classified, and we can look back on it with some interest when comparing the activities in the different countries. Pinch theory evolved along very similar lines in the USA, the USSR, and the UK. Thin-skin pinch models were topical at that time, and the theories of Rosenbluth,[5] Tayler,[6] and Shafranov[7] were very similar in this respect—at least they all got the same results! The work by Rosenbluth at Los Alamos was, however, a little different in that he also considered particle effects, using the adiabatic invariants of the motion. The result was an instability criterion involving anisotropic pressure which, if $p_\perp > p_\parallel$, was more unstable.[5]

TABLE 1 The Fusion Reactor of Thomson and Blackman

Quantity		Value
Fuel	D-D	
Dimensions	R/a	$= 1.30/0.30$ m
Discharge current	I_p	$= 500$ kA
Mean number density	n_e	$= 3.5 \times 10^{20}$ m^{-3}
*Electron temperature	T_e	$= 80$ keV
*Ion temperature	T_i	$= 500$ keV
Particle confinement time	τ_p	$= 65$ s
Output power	P_{TOT}	$= 9$ MW (Th)
Heating and current drive	RF	$= 2$ MW (at 3 GHz)
Neutron output		$= 1.9 \times 10^{19}$ s^{-1}

* The patent gives random speeds rather than temperatures.

A very significant step forward in the development of the theory of toroidal pinches occurred at the 1958 Geneva Conference on the Peaceful Uses of Atomic Energy. This was the first real opportunity for the countries actively involved in fusion research to jointly discuss their results in an open manner. There were numerous contributions on the theory of pinches. One reviewer[8] noted that "the theory papers showed that in the hydromagnetic approximation, complete stability could not be predicted for the magnetic field configurations so far observed in these pinch

discharges". However, the seeds of what was to come had already been sown. Rosenbluth, in his paper on the stability and heating in a pinch,[5] noted that if the sharp
surface layer of previous analyses was given finite thickness, this led to the important result that stability could only be achieved if the toroidal field was reversed
outside this layer. He also noted that stability would only be transient, due to diffusion and heating to high β, and that a new helical equilibria might be set up which
could in turn generate a reversed toroidal field as required for stability. Diffusion
would make the plasma unstable again, and the cycle would be repeated. This picture is not dissimilar to that which was observed much later in the fast reverse
field pinches built in the early 1970's. At the same time, the spontaneous reversal
of the toroidal magnetic field had been observed on the ZETA device,[9] and this
occasion can truly be said to be the birth of the reverse field pinch. Rosenbluth's
early study[5] of the pinch included a proposal for a reactor, whose parameters are
shown in Table 2.

It is a much smaller device than that originally proposed by Thomson and
Blackman in their secret patent, which is shown for comparison in Table 1.

TABLE 2 Possible Characteristics of a Diffusion-Limited, Self-Heated D-T Reactor
[From Ref. 5]

Quantity	Value
Major radius of torus	30 cm (arbitrary)
Minor radius of torus	6 cm
Initial plasma radius (r_o)	1.5 cm
Current	3×10^6 A
Pressure at wall	400 atm
Initial pinched density (ρ_o)	$1.3 \times 10^{17}/cm^3$
Burning temperature (T_p)	6 keV
Disassembly time	0.15 s
Total magnetic energy	4×10^6 joule
Losses (copper torus)	2.5×10^6 joule
Energy produced	3×10^7 joule
Temperature rise of copper surface due to radiation	500° C
Efficiency of burning	10%

At the present time when our large tokamaks approach conditions of substantial
α particle heating, it is interesting to compare these proposed parameters of some
three decades ago.

The early classification problems led to some difficulties in pinch research, and
Academician Lev Andreivich Artsimovich noted that "the question of whether a
given neutron belongs to the noble race of descendants of thermonuclear reactions

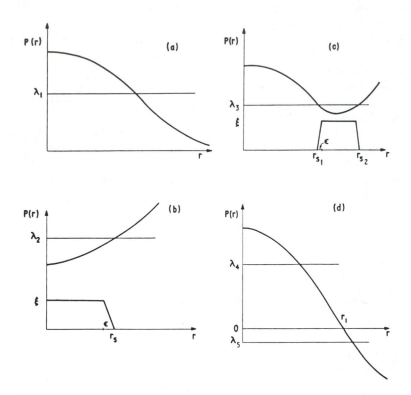

FIGURE 1   Four radial variations of the pitch of the magnetic field lines:
(a) that characterising a stabilised pinch with currents out to the wall;
(b) the variation similar to that of a tokamak;
(c) that associated with the stabilised pinch surrounded by a vacuum; and
(d) that associated with a reverse field pinch.

or whether it is a dubious offspring of a shady acceleration process is something
that may worry the press men but it should not, in the present stage of development
of our problem, ruffle the composure of the specialist." Artsimovich was, of course,
to play a very important role in the further development of pinch devices.

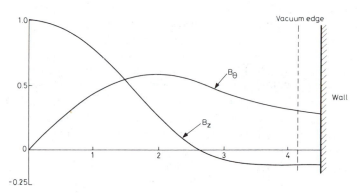

Pitch and pressure model equilibrium

FIGURE 2   Tearing-mode-stable reverse field configuration with a vacuum edge. [From Ref. 18]

STABILITY

The problems of stability encountered, both theoretically and experimentally in these early 'stabilised' pinch studies led researchers in the USA and the UK to study the hardcore geometry, for which MHD stability could be ensured. These hardcore experiments showed all the characteristics of instability previously seen in the stabilised pinch experiments and were not in agreement with simple hydromagnetic theory. This stimulated a re-examination of the theory, and a number of workers studied the effects of finite electrical conductivity. A number of authors quickly came up with what we now know as the tearing mode, in particular, Whiteman,[10] Rebut,[11] and Furth,[12] culminating in a fine paper on resistive instabilities by Furth, Killeen, and Rosenbluth.[13] This theoretical work dealt a hard blow to the pinch experiments as it seemed difficult to see how they could be stable, and many were discontinued. However, some work continued at a low level in the UK on the ZETA device, where it was noted that field reversal was accompanied by a quiescent period characterized by a striking reduction in the level of fluctuations. This prompted a careful examination of stability theory, and it was established that field reversal was a necessary condition for obtaining ideal MHD stability,[14] although at that time the issue of tearing mode stability remained. The radial variation of the pitch of the magnetic fields (see Fig. 1) was critical in controlling that stability—in particular, minima giving rise to instability. Work on pinches was also devoted to studies of the hydromagnetic turbulence present, which was later to turn out to be of considerable value in understanding the loss processes present in such devices.

At the Dubna conference in 1969, the results from the T-3 tokamak in the USSR on the electron temperature by Thomson scattering were presented, which were to have a profound effect on fusion research and its direction. At the same meeting, there were a number of presentations on the theory of toroidal pinches that led the theoretical reviewer, Marshall Rosenbluth, to note that the contributions by a particular author[15] were perhaps the most important of the conference! These theoretical results established the detailed conditions necessary for the ideal MHD stability of a reverse field pinch, indicating that central β values up to 30% were possible.[16] In addition, stimulated by toroidal stability calculations in the Soviet Union, results were presented for the on-axis stability criterion for a toroidal pinch, $R^2 q q''/2 < -4/9$, where q is the safety factor, which showed that the criterion was unchanged, provided $q(0) < 0.5$.

Work in the early 1970's shifted more towards growth rate calculations that allowed more detailed comparison with experiment, and the concept of sigma stability was introduced.[17] At the same time there was renewed interest in resistive instabilities. It was established that a wide range of tearing-mode-stable reversed field pinch configurations did exist,[18] with currents diminishing to zero near the conducting wall (Fig. 2). Thus the pinch could be stable without resorting to the high current density at the wall necessitated by the so-called Bessel function model (or minimum energy configuration). However, although the tearing and rippling mode (with high $\chi_\parallel$) could be stable, the resistive g mode, i.e., the mode driven by the pressure gradient, could not be stabilised. The tearing-mode-stable current distributions could support an ideal pressure limit of $\beta(0) \lesssim 17\%$ at the high pinch parameter, $\theta \lesssim 3$, with θ defined as $\theta = B_\theta(a)/\bar{B}_\phi$, although with a 4% vacuum region, this fell to 12%. The growth rate calculations of resistive modes were also fashionable at this time, particularly by Killeen and co-workers.[19] Such calculations did permit detailed identification of pitch minimum, kink, tearing, and g modes in fast pinch devices.

The resistive interchange, or g mode, arises in the pinch because the average curvature is unfavourable. It was thought that high temperature effects such as finite Larmor radius and perpendicular viscosity could easily stabilise the mode. Furthermore, even if the mode were unstable, the diffusion coefficient in the non-linear phase would be comparable to the classical value so that its effect would not be serious. Both conclusions, however, were found to be incorrect. In particular, the magnetic perturbations associated with the g mode lead to substantial transport because they produce magnetic islands which readily overlap and lead to the stochastic field lines. In addition it was found[20] that the electromagnetic mode with finite Larmor radius effects in cylindrical geometry was unstable at significant values of β (a few percent). Parallel viscosity, however, was found to be stabilising, and this led to critical pressure profiles which could be grossly stable when the average value of β was significantly below 10%.[21] Such modes were probably detected in the fast toroidal reverse field pinches: Fig. 3 shows the magnetic perturbation and its associated magnetic island structure.

The early work by Rosenbluth and others[22] on the diffusion due to stochastic fields produced by magnetic islands associated with resistive instabilities led to the

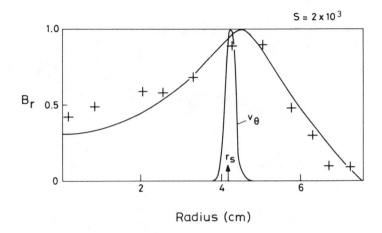

FIGURE 3(a)   Measured (+) and predicted (−) radial field perturbation for a resistive interchange, or g mode, in a pinch.

consideration in more detail of the losses associated with the magnetic perturbations
induced by current, or pressure-driven resistive modes, which initiated detailed
studies of their nonlinear behaviour in toroidal pinches. These were recognised as
probably determining the limiting value of β and the transport in a pinch and
also as being linked to the processes that control the reversed field generation, or
dynamo action, in the pinch.

Stabilised pinches, that is, toroidal pinches without field reversal, received early
attention; in particular, the force-free paramagnetic model was studied for its ideal
and resistive stability properties. For a pinch configuration parameter $\theta \lesssim 1.1$, it
was shown to be stable to both ideal and resistive tearing modes. It also supports
small plasma pressure, $\beta \sim 5\%$. Such a configuration has significant current density
at the conducting wall and cannot support a vacuum region between the edge of
the plasma and the conducting wall. Recently this configuration has received more
attention, particularly in TORIUT-6 where discharges with $q \sim 0.6$ were sustained,
apparently with relative stability.[23] This type of operation had previously been
noted in the setting-up phases of reverse field pinches, where "magic numbers" were
noted as the value of the safety factor passed through various rational values at the
boundary. These magic numbers were associated with bursts of strong MHD activity
that affected the impedance of the discharge and manifested itself in the current
and voltage waveforms. Ideal toroidal stability calculations for configurations with

$m = 1, k = - 0.5\,\text{cm}^{-1}$

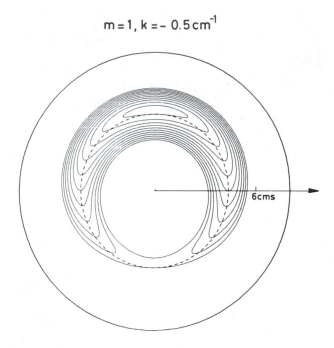

6cms

FIGURE 3(b)   The magnetic island produced by the resistive interchange mode in the pinch with $\delta B_r / B_\theta \sim 7\%$.

q in the region 0.3 to 0.7 do indicate regions of low growth when the region of pitch minimum falls close to an irrational q value.

A more recent theoretical question relates to thin shell operation, which is desirable in future reverse field pinches and spheromaks. This is particularly relevant to a reactor since the wall time constant is significantly shorter than the sustainment time of the pinch. Thin shell operation permits accurate control of plasma position using the vertical and radial magnetic fields and possible shift and tilt instability control using separate feedback coils. Configurations that are stable with a perfectly conducting wall become unstable with modes growing on the timescale for the helical field to penetrate through the resistive wall. If the configuration is only marginally stable, then growth on a hybrid timescale occurs, $\tau \sim \tau_W^{1/3}\tau_A^{2/3}$, where τ_W is the wall time constant and τ_A the Alfvén transit time.[24] Investigations of the effect of a resistive wall on both ideal and resistive modes[25] show that the wall destabilises both modes, but that the tearing mode growth can be stabilised by sufficient toroidal rotation. Ideal modes are destabilised when the rotation becomes close to Alfvénic as noted by Rosenbluth[26] long ago (who assumed a perfect wall) and are not affected by the wall. The reduction in mode rotation at finite amplitude due to wall dissipation suggests that there may be problems with nonlinear tearing modes in thin shell pinches, of which the SLINKY mode observed on the OHTE experiment may be an example.[27]

ANOMALOUS TRANSPORT

Some of the earliest work in this area was motivated by the failure of the experimental field configurations in pinches to satisfy the requirements of ideal and resistive MHD stability. This led to a general study[28] of the turbulence induced in the ZETA device by these instabilities. This work both theoretically and experimentally covered fully developed MHD turbulence, which was characterised by close proximity to two-dimensional turbulence. The measured triple correlation was close to that expected for such turbulence (see Fig 4). Reference 28 also considered the competition between convective transport losses and those associated with magnetic field fluctuations due to strong magnetic turbulence present in the pinch. Rusbridge[29] in his tangled discharge model showed that the fluctuations would lead to rapid heat loss along the field lines, which could explain the temperature profiles and also led to field configurations which produced reversal. The alternative explanation of the dynamo action in the pinch, involving a small number of helical instabilities through the $\langle \mathbf{v} \times \mathbf{b} \rangle$ correlation, was also considered sufficient to account for the observations.

The key problem in toroidal pinch research is the anomalous energy loss. The conventional approach has been to calculate the anomalous transport by identifying an instability and then performing a nonlinear calculation to obtain a saturated amplitude and deducing the resultant transport in this state. An alternative approach has also been adopted based upon the invariance properties of the equations describing the plasma.[30] Nearly all the theoretical approaches of anomalous loss in the pinch have been associated with resistive fluid turbulence. These losses can be either convective or associated with transport along the stochastic field lines, created by the resistive turbulence itself. The Rechester-Rosenbluth[31] expression

$$D_m = \left(\frac{\delta B_r}{B} \right)^2 L_c, \qquad (1)$$

where L_c is the correlation length, is a key element in these calculations. The correlation length is a complex issue [32] where in strong turbulence L_c can depend directly on D_m. In the collisional limit, some theories, including the invariance properties, give $\delta B_r / B \propto 1/S^{1/2}$ (where S is the Lundquist or magnetic Reynolds

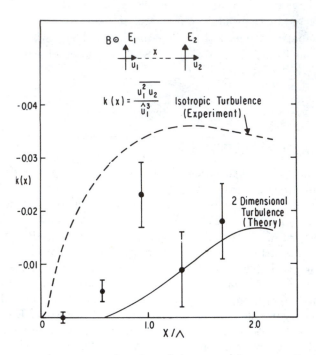

FIGURE 4   Triple correlation as a function of the separation normalised to the integral scale length (Λ). The theoretical curve for two-dimensional turbulence theory is shown and compared with the experimental results and with the experimental curve for isotropic turbulence from experiment. [From Ref. 28]

number, $S = \tau_\sigma/\tau_A$, with τ_σ the field diffusion time and τ_A the Alfvén transit time). For resistive g modes, the collisionless loss due to transport along stochastic magnetic fields is of the form

$$D \sim \eta \left(\frac{M}{m}\beta\right)^{1/2} f_o(\nabla p, s),$$

where f_o is a function of pressure gradient ∇p and shear s and η is the resistivity. The convective losses must have the form $D \sim \eta f_1(\nabla p, s)$; thus we have the result that conduction exceeds convection if $\beta > (m\beta_c/M)^{1/2}$, where β_c is the ideal ballooning limit in a tokamak or the Suydam limit in a pinch. If $\beta \sim 10\%$ and if $\beta/\beta_c > 0.05$, conduction losses will dominate. In the collisional limit, this expression becomes $\beta > (m\beta_c^3/M)^{1/4}$. As β_c is typically 10% in the pinch, then heat flow due to transport along stochastic magnetic field lines will be more important. The scaling of these losses in the pinch is such that $\beta \sim (m/M)^{1/6}\beta_c^{1/2} \sim 8\%$. An interesting point arising from these estimates is the relative performance of the tokamak and the reverse field pinch. If the tokamak and the pinch both have the same critical $\beta_c \sim 10\%$ and are subject to localised resistive pressure-gradient-driven turbulence, then the confinement in the reverse field pinch and the tokamak (with ohmic heating) should be the same!

The impact of the electromagnetic resistive g mode in the reversed field pinch has been studied including the effect of the resultant stochastic field lines on the confinement.[33] For β values $\sim 5\%$, the resistive g mode has island-forming parity and creates a substantial magnetic perturbation throughout the plasma. In this quasilinear treatment, each individual mode tends to saturate by flattening the equilibrium pressure gradient in the vicinity of the singular surface. Mixed-helicity interactions between the $m=1$ and $m=0$ modes play an important part in destroying magnetic surfaces, as indicated in Fig. 5. The fluctuation level is calculated to decrease with S in a manner similar to that given by the invariance arguments and also to increase with increasing plasma pressure. These modes give a temperature scaling $T \sim I^{0.7}$ where I is the current and it has been assumed that $I \propto N$, where N is the area density.[33,35] This is slightly more pessimistic than the temperature scaling $T \sim I$ implicit in the constant-β result for the low-β localised resistive g mode.

Diamond, Rosenbluth, and colleagues[34] studied the nonlinear interaction of tearing modes in the pinch and also estimated the thermal transport resulting from these modes. In this case, nonlinear generation and coupling to $m=2$ modes was advanced as the main mode saturation mechanism and the dynamo process was attributed to the $m=1$ tearing modes that sustained the configuration. With the use of renormalised turbulence theory (an approach similar to that used earlier in investigations of magnetic fluctuations on the ZETA device), it has been shown that coupling by interaction with neighbouring $m=1$ modes to stable $m=2$ modes balances the growth due to the relaxation of the current profile. This theory, like the electromagnetic g mode, yields $T \sim I^{0.7}$ by balancing the heat loss with ohmic heating and also assumes that $I \propto N$. The problem has been examined more recently for the nonlinear tearing mode in a reverse field pinch,[35] and a nonlinear saturation mechanism dependent on cascading to smaller scales was proposed and used to estimate the profile evolution. The interaction with the $m=0$ mode was also included. The behaviour appears to be similar to the inertial range in ordinary fluid turbulence, with a broadening of the magnetic spectrum from coupling of the different $m=1$ modes, leading to $m=0$, $m=1$, and $m=2$ activity. These new results indicate that $\delta B_r/B$ is virtually independent of S, with the linear instability growth rate being unrelated to the instability drive in the nonlinear regime. These results, like the electromagnetic g mode, indicate a temperature scaling $T \propto I^{0.7}$. Further examination[35] of resistive pressure-gradient-driven turbulence shows that more accurate calculations of the fluctuation level lead to a thermal diffusivity that is enhanced beyond the simple result given previously, leading to a slow degradation of the poloidal β as the plasma current increases. It has also been noted[35] that ion

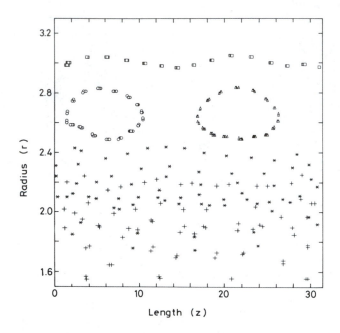

FIGURE 5   The destruction of magnetic surfaces in the radius/length plane for reverse field pinch excited by both $m = 0$ and $m = 1$ modes inside the reversal surface. The field reversal radius is at ~ 2.6. [From Ref. 33]

temperature-gradient-driven modes can arise in a reverse field pinch with a similar scaling.

It is quite possible that with future large devices and higher temperatures, the tearing mode behaviour associated with the dynamo action will become less important, that the pressure-gradient-driven turbulence will become weak due to collisionless effects, and that the $\beta_\theta \sim$ constant scaling will dominate.

RELAXATION AND RECONNECTION

A key element in toroidal pinch research is the relaxation of pinch configurations to a minimum energy state as propounded by Taylor.[36] Many of the theoretical predictions are in good agreement with the experimental data.[37] The self reversal of the toroidal reversed field in a pinch is explained as the natural tendency for the

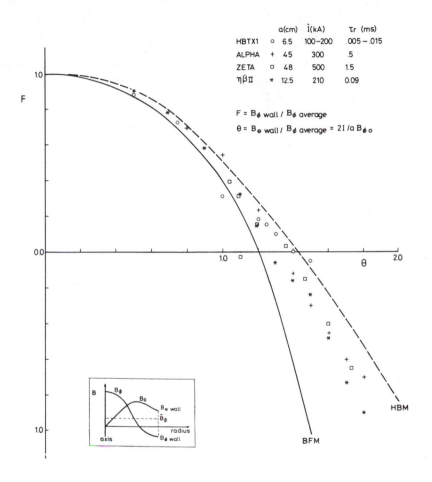

	a(cm)	Î(kA)	τr (ms)
HBTX1 o	6.5	100–200	.005 – .015
ALPHA +	45	300	.5
ZETA □	48	500	1.5
ηβ∏ *	12.5	210	0.09

$F = B_\phi$ wall $/ B_\phi$ average

$\theta = B_\theta$ wall $/ B_\phi$ average $= 2I/a B_{\phi o}$

Universal F – θ Curve, showing Data
from Four Machines.

FIGURE 6   The universal $F - \theta$ diagram showing data from four machines; BFM refers to the minimum energy configuration, and HBM to the minimum energy configuration including pressure up to a value limited by the Suydam criterion.

plasma to relax by a process involving dissipation and reconnection to a minimum energy state with a reversed toroidal magnetic field, if the pinch parameter is such that $\theta > 1.2$, where $\theta = B_\theta(a)/\bar{B}_\phi$; here $B_\theta(a)$ is the poloidal field at the wall, and

$\bar{B}_\phi$ is the average toroidal field. This final minimum energy state is characterised by a single invariant, the helicity $K_0 = \int_{V_0} \mathbf{A} \cdot \mathbf{B} d\tau$, where $\mathbf{A}$ is the vector potential, the volume V_0 here being bounded by a flux-conserving shell. (Note that in multiply connected domains, care must be exercised in evaluating K_0.) The relaxed state is then defined by a force-free configuration, $\mathbf{J} = \mu \mathbf{B}$, where μ is independent of position. In cylindrical geometry, this gives rise to the Bessel function model which has a reverse field for $\theta > 1.2$ and for which the minimum energy state becomes helical when $\theta > 1.56$. The ability of a plasma to relax to this state is portrayed by comparing the predictions with experiment using an $F - \theta$ diagram, in which the field reversal ratio $F = B_\phi(a)/\bar{B}_\phi$, $B_\phi(a)$ being the toroidal field component at the wall, is plotted as a function of θ, as shown in Fig. 6. The experimental points of all the toroidal reverse field pinches to date lie somewhat to the right of this minimum energy curve, in part due to the fact that the current density near the boundary tends to be small and also due to the fact that the configuration can support a significant plasma pressure. Tearing-mode-stable reverse field configurations have $\mu(r)$ decreasing with radius and give $F - \theta$ curves similar to the observations. Field diffusion will make the configuration move away from this minimum energy configuration, leading to instability, which tends to counteract the diffusion so that the configuration can be maintained; this explains the quasi-stationary nature of the reverse field pinch observed in the experiments. The nature of the processes involved in generating the field configuration and the associated plasma losses are not yet well understood and are clearly critical in determining the confinement properties of the system. Some theories have suggested that the scaling of the magnetic field fluctuations to provide this dynamo action is $\delta B/B \propto S^{-1/2}$. There is experimental evidence to support such a relationship.[38]

The helicity has a practical interpretation at constant toroidal flux: it is proportional to the volt-seconds stored in the discharge. The helicity has a rate of change given by $dK_1/dt = 2V_l\psi$, where ψ is the toroidal flux. A fascinating consequence of relaxation theory is the possibility of current drive by oscillating the driving voltages. If the toroidal and poloidal voltages are applied out of phase, then they generate helicity, which implies an effective unidirectional voltage which is then available to sustain the plasma current.[39] This is receiving careful attention experimentally and theoretically at the present time.[40]

Given the positive stability of the minimum energy state, then some window for finite stable β should be present. This has been calculated in a number of ways—in particular, by inflating the equilibrium in a flux-conserving manner to determine the ideal MHD β limit for such configurations. This is shown in Fig. 7, where values of central $\beta \sim 10\%$ are obtained.

The properties of the relaxed state are found to depend crucially on the topology, be it toroidal or spherical, and on the boundary conditions. This is particularly important when comparing the results from different experiments such as the reverse field pinch, OHTE,[41] multi-pinch,[42] and spheromaks. The process of reconnection almost certainly involves the presence of resistivity which, no matter how small, allows the topology of the field lines to be no longer preserved. As the resistivity becomes small, the region over which it acts becomes smaller, with

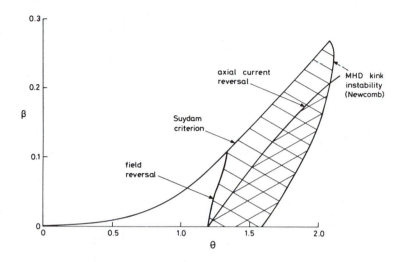

FIGURE 7 Stability diagram, showing the limiting β on axis as a function of θ for the minimum energy configuration.

larger gradients there. A very similar process is involved in resistive instabilities and magnetic reconnection at X-points.

The general theory of relaxed states in toroidal systems has also been discussed.[43] In the cylindrical relaxed state, when $\theta > 1.56$ the eigenfunction is not axisymmetric but consists of a symmetric part and a helical component with $m = 1$ and $Ka = 1.25$. This is true also in other cross-sections, although with more convoluted cross-sections such as the multi-pinch, the lowest eigenfunction may be axisymmetric. The multi-pinch has a cross section that is rather like that of the Doublet experiment. As μa is increased, the initial configuration is a symmetric one with B_ϕ the same sign everywhere, whereas when $\mu a \sim 2.2$ the configuration becomes anti-symmetric with B_ϕ of opposite sign in the two halves of the configuration. For this particular case, full reversal and current saturation are almost coincident. These particular features of the relaxed state have been clearly demonstrated in the multi-pinch experiment. Relaxation is also important in the spheromak,[44] shown in Fig. 8. There are toroidal surfaces as in the toroidal pinch, but the confining shell is now spherical. There is no central aperture for toroidal field coils; consequently the toroidal field is zero at the boundary, and thus it resembles a toroidal pinch at the point of field reversal. In the spheromak there is only a single invariant, K, rather than the two invariants in a toroidal device, K and ϕ, where ϕ is the toroidal flux. In this case the value of μ is determined by the shape

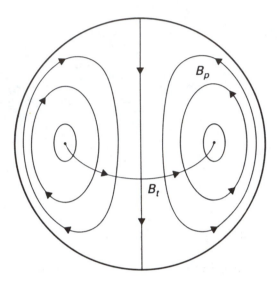

FIGURE 8   The classical spheromak configuration, showing poloidal and toroidal fields.

alone. For a sphere, $\mu a = 4.49$. The eigenvalue may not be axisymmetric, depending on the shape of the container. For example, for a cylinder of height h and radius a, with $h/a > 1.67$ there is a non-axisymmetric mode, viz., the "tilting" mode.[45] As in reverse field pinch discharges, the measured field configurations are in agreement with the relaxed state predictions and the $\mu(r)$ profiles are approximately constant. The flux core spheromak is and interesting variant of the conventional spheromak, which is obtained by introducing a core of externally produced flux along this line of symmetry. In this case the helicity is not well defined, and a relative helicity is introduced. This demonstrates that helicity can be injected or extracted from the flux core spheromak, which provides a means of sustaining the relaxed state against resistive decay. A good example of a flux core configuration is the CTX experiment,[46] where it has been possible to sustain the spheromak at high currents (~ 0.5 MA) for several milliseconds. The possibility that relaxed states can be maintained in this manner has led to investigations of new configurations where the source of the helicity is more distant from the confinement region.[47] It is very important to note that all relaxed states are stable against the perturbations that leave the helicity invariant which includes resistive tearing modes; thus, relaxed state theory provides a full nonlinear description of the resistive tearing mode.

Tokamaks also display examples of relaxation towards a minimum energy state, but they are intermittent and certainly non-unique in character. The inherent stability of the tokamak to both ideal and resistive modes seems, in general, to preclude a continuous relaxation towards a minimum energy state. Nevertheless, it does highlight the fact that reconnection may be quite different in different circumstances, even though the final state is a unique one.

The fast pinch devices were ideal facilities for testing stability theory and Taylor's relaxation theory. By fast programming, it is possible to force the plasma into unstable configurations and then observe the violent relaxation back towards the minimum energy state. Slow programming was found to accurately follow the $F - \theta$ trajectory of the minimum energy state. The relaxation could be studied in detail, and large amplitude helical kink stabilities were detected of a sufficient amplitude that the solenoid effect of the helix was large enough to produce the reversed field. This behaviour was in very good agreement with early three-dimensional simulations of self reversal.[48] These fast experiments also demonstrated that when the current was sustained, then the field reversal was also sustained, for times longer than the field diffusion time. It can thus be concluded that the tendency to relax to a minimum energy configuration can occur by a variety of phenomena: by large amplitude helical instabilities (as suggested long ago), by fine scale instabilities, and by turbulence.[49]

Self reversal has also been explained in terms of mean field equations that are applied to the MHD turbulence appropriate to the earth's dynamo problem, which can drive j_θ currents.[50] The turbulent motions leads to two effects:[51] the production of an electric field parallel to the magnetic field, called the α effect; and a turbulent resistivity called the β effect. Recent nonlinear tearing mode calculations[52] that calculate the $< \mathbf{v} \times \mathbf{b} >$ correlation do demonstrate the maintenance of reversal as well as giving an "eddy" resistivity,[53] which can be more important than classical resistivity. A variety of Ohm's laws for the mean magnetic field have been derived, leading to a coefficient of electric current viscosity[54,55] akin to the α effect.

There are a number of simulations using spectral codes in three-dimensions.[56,57] These calculations indicate a greater tendency to relax to the minimum energy configuration with compressible motions than than incompressible ones. However, incompressible calculations[57] with different boundary conditions (see Fig. 9) indicate an $F - \theta$ evolution in time that is very similar to that which occurs in the fast pinch experiments. For $S \sim 10^4$ and a nonuniform resistivity profile, the $F - \theta$ curve is very close to that observed experimentally. The calculations show that the dynamo effect arises from the mean average coupling between the radial field and axial velocity perturbations. The calculations also indicate that μ is nonuniform, unlike the minimum energy configuration, but varies with radius in a manner similar to that observed experimentally (Fig. 10).

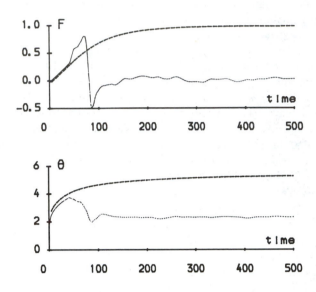

FIGURE 9 F and θ as a function of time from 3-D code calculations. The dashed line shows results of 1-D resistive diffusion. In this case, $S = 10^3$ and η is uniform.

SPHEROMAKS

Rosenbluth and Bussac[44] used the Taylor minimum energy principle to study the stability of the spheromak, which is the tight-aspect-ratio limit of the toroidal pinch. This is a force-free configuration with flux surfaces similar to the Hills vortex and is of particular note to astrophysicists. Such configurations could simplify the reactor engineering for a fusion device by removing the necessity for toroidal field coils; also, no toroidal blanket is required. Creating such configurations does pose rather special problems but the ingenuity of plasma physicists has led to elegant solutions. One method that has been highly successful is to use a coaxial plasma gun and inject into a flux-conserving chamber; combinations of θ- and Z-pinch discharges have also been successful, and a slow inductive process has been used in the S1 device.

The stability of the spheromak was studied by noting that the helicity $K = \int \mathbf{A} \cdot \mathbf{B} d\tau$ within the plasma volume is conserved for both ideal and resistive modes. It was found that an oblate spheromak was stable to MHD and resistive tearing modes if surrounded by a conducting wall that was within 15% of the edge of the plasma. Such configurations have very little shear, and the safety factor q decreases from 0.825 on axis to 0.72 at the edge. A prolate spheromak is found to be unstable to a tilting mode, whereas an oblate spheromak is unstable to a shift-mode. The

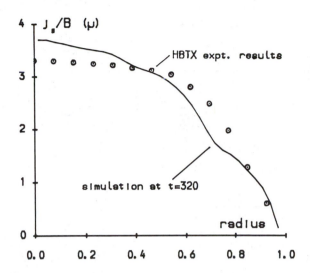

FIGURE 10 $J_{\parallel}/B$ as a function of radius from 3-D code calculations compared with a
measured profile from the HBTX experiment. In the calculations, $S = 10^3$.

tilting instabilities are observed experimentally, and reducing the height-to-radius
ratio stabilises the mode in agreement with theoretical predictions. As in the pinch,
the existence of a free boundary associated with a vacuum region surrounding the
plasma leads to new instabilities. The Mercier criterion can only be satisfied by
such configurations with very low values of $\beta \sim 0.2\%$. With the oblate spheromak,
more shear is possible and the β value rises to $\sim 1\%$. Resistive g modes will be
unstable because of the unfavourable curvature. However it should be noted that in
the spheromak, as in the pinch, the central values of β are smaller than the volume
average β. In particular, the engineering β taken out to the external field coils can
be greater than 10%. As with the pinch, a close-fitting conducting shell appears to
be an essential requirement for stability.

Spheromaks possess very similar properties to the reverse field pinch concerning
conservation of helicity during the strong relaxation phenomena that occur during
the setting-up phase. However the magnetic fluctuations ($\sim 10\%$) associated with
$n = 1$ modes appear to be larger than those present in the pinch. The volume average
β values obtained in the spheromak are comparable to those in pinches, but the
confinement time appears to be shorter,[58] possibly associated with a higher level
of magnetic fluctuations. This could be because the critical β for the onset of ideal
pressure-driven modes for this configuration is lower than in the pinch. Attempts
to improve the β value have been made by inserting a copper cylinder along the

axis of the spheromak; however, this has not led to an improvement in plasma parameters to date. A difficulty in the spheromak is that the flux conserver has a time constant such that significant flux is lost because of the finite resistivity of the wall. By analogy with the thin shell stability results in a pinch, this may also lead to difficulties with enhanced MHD activity. Attempts to detach the plasma from the flux conserver in the spheromak have been made by applying magnetic flux throughout the wall, although the external flux acts to destabilise the tilt mode, and the $n = 2$ mode is also destabilised.[59] A critical bias flux is indeed observed to produce instability, but the tilting mode could be stabilised using a central conductor.

CONCLUSIONS

Early studies of pinch dynamics and stability theory by Rosenbluth were very important in influencing the evolution of the toroidal pinch. He was the first worker to establish the importance of the reverse toroidal field for stability, and he then went on with Furth and Killeen to explain the unstable nature of the hard-core pinches in terms of resistive instabilities. This left the toroidal pinches in some disarray, and they were largely abandoned in the early 1960's in favour of mirror machines, θ pinches, and hard-core pinches. The 1960's, however, did see the consolidation of ideal stability theory for the diffuse pinch, with the importance of the reverse field becoming clear theoretically and experimentally. The Novosibirsk and Dubna conferences saw the emergence of the tokamak, but there was also progress with the ZETA reverse field pinch experiment with $\beta \sim 5\%$ and with the theory for RFP.

The early 1970's saw the start of a number of fast RFP experiments in response to both theory and experiment. These used fast programming rather than self-reversal, which was little understood at that time. Interest was further stimulated by the work of Taylor, who established that the RFP is a minimum energy state and thus is stable against MHD and tearing modes. These experiments were unable to achieve sustained high-temperature operation, and there was a return to the slow mode of operation as obtained on the ZETA device with considerably more success.

In the late 1970's Rosenbluth used Taylor's theory to introduce a new pinch configuration, the spheromak, which is being actively pursued as an alternative to the RFP and tokamak.

Throughout the pinch investigations, anomalous transport has been a problem. In particular, the influence of magnetic perturbations has been highlighted repeatedly by Rosenbluth (and others).

The further development of the theory of toroidal pinches, in conjunction with present and future larger experiments, is a vital element in the progress of the toroidal pinch —which I am sure will be forthcoming from the Institute of Fusion Studies at the University of Texas under the guidance of Marshall Rosenbluth.

REFERENCES

1. J.A. Pollock and S.H. Barraclough, *Proc. R.S. New South Wales*, Vol. XXXIX, p. 131 (1905) [courtesy of Dr. R.S. Pease].
2. S.W. Cousins and A.A. Ware, *Proceedings of the Physical Society (London)*, **B64**, 159 (1951).
3. L. Tonks, *Physical Review* **56**, 360 (1939).
4. W.H. Bennett, *Physical Review* **45**, 890 (1934).
5. M.N. Rosenbluth, in *Proc. 2nd Int. Conf. on Peaceful Uses of Atomic Energy*, Vol. 31, p. 85 (1958).
6. R.J. Tayler, *Proceedings of the Physical Society (London)* **B70**, 31 (1957).
7. V.D. Shafranov, *Atomnaye Energiya* 5, 709 (1956).
8. R.S. Pease, *Nature*, **182**, 1051 (1958).
9. E.P. Butt, R. Carruthers *et al.*, in *Proc. 2nd Int. Conf. on Peaceful Uses of Atomic Energy* **32**, 42 (1958).
10. K.J. Whiteman, Culham Laboratory Report CLM-P14 (1962).
11. P.H. Rebut, *Plasma Physics (Journal of Nuclear Energy*, Pt. C) **4**, 159 (1962).
12. H.P. Furth, *Bulletin of the American Physical Society* **6**, 193 (1961).
13. H.P. Furth, J. Killeen, and M.N. Rosenbluth, *Physics of Fluids* **6**, 459 (1963).
14. D.C. Robinson *et al.*, *Plasma Physics* **10**, 469 (1968)–2nd European Conference on Controlled Fusion and Plasma Physics (Stockholm, 1967).
15. D.C. Robinson *et al.*, (a) "Measurement of the Electron Temperature by Thompson Scattering in Tokamak T3," *Nature* **224**, 488 (1969); (b) "A Criterion for the Hydromagnetic 'Kink' Stability of a Cylindrical Pinch Discharge," Dubna Symposium (1969)," Abstract 52b, Report MATT-TRANS-91; (c) "Stability Close to the Magnetic Axis in a Toroid for Vanishing Pressure Gradient," Dubna Symposium (1969), Abstract 52c, Report MATT-TRANS-91.
16. D.C. Robinson, *Plasma Physics* **13**, 439 (1971).
17. J.P. Goedbloed, *Physics of Fluids* **17**, 908 (1974).
18. D.C. Robinson, *Nuclear Fusion* **18**, 939 (1978).
19. D.C. Robinson, J.A. Dibiase, A.S. Furzer, J. Killeen, and J.E. Nunn-Price, Culham Laboratory Report CLM-P710 (1983).
20. T.C. Hender *et al.*, *Computer Physics Comm.* **24**, 413–419 (1981).
21. R.J. Hosking *et al.*, in *Proc. of the 9th European Conference on Controlled Fusion and Plasma Physics* (Oxford), Paper BP6 (1979).
22. M.N. Rosenbluth, R.Z. Sagdeev, J.B. Taylor, G.M. Zaslavski, *Nuclear Fusion* **6**, 297 (1966).
23. N. Asakura *et al.*, in *Plasma Physics and Controlled Nuclear Fusion Research 1986*, (IAEA, Vienna, 1987), Vol. II, p. 433.
24. A.S. Furzer *et al.*, in *Proceedings of the 7th European Conference on Controlled Fusion and Plasma Physics* (Lausanne), Vol. 1, p. 114 (1975).
25. T.C. Hender, C. Gimblett, and D.C. Robinson, in *European Physical Soviet Conference* (Schliersee), Vol. 1, p. 61 (1986).
26. E. Gerjuoy and M.N. Rosenbluth, *Physics of Fluids* **4**, 112 (1961).

27. T. Tamano *et al.*, in *Plasma Physics and Controlled Nuclear Fusion Research 1986*, (IAEA, Vienna, 1987), Vol. II, p. 655.
28. D.C. Robinson and M.G. Rusbridge, *Physics of Fluids* **14**, 2499 (1971).
29. M.G. Rusbridge, *Plasma Physics* **19**, 499 (1977).
30. J.W. Connor and J.B. Taylor, *Physics of Fluids* **27**, 2676 (1984).
31. A.B. Rechester and M.N. Rosenbluth, *Physical Review Letters* **40**, 38 (1978).
32. J.A. Krommes *et al.*, *Journal of Plasma Physics* **30**, 11 (1983).
33. T.C. Hender and D.C. Robinson, in *Plasma Physics and Controlled Nuclear Fusion Research 1982* (IAEA, Vienna, 1983), Vol. III, p. 417.
34. Z.G. An, P.H. Diamond, M.N. Rosenbluth, in *Plasma Physics and Controlled Nuclear Fusion Research 1984* (IAEA Vienna, 1985), Vol. II, p. 231.
35. Z.G. An, P.H. Diamond *et al.*, in *Plasma Physics and Controlled Nuclear Fusion Research 1986* (IAEA, Vienna, 1987), Vol. II, p. 663.
36. J.B. Taylor, *Physical Review Letters* **33**, 1139 (1974).
37. J.B. Taylor, *Culham Laboratory Report* CLM-P765 (1985).
38. R.J. La Haye *et al.*, *Nuclear Fusion* **21**, 1235 (1981).
39. M. Bevir and J. Gray, in *RFP Theory Workshop*, Los Alamos (1980), Vol. 3, Paper A-3.
40. K.F. Shoenberg *et al.*, in *Plasma Physics and Controlled Nuclear Fusion Research 1986* (IAEA, Vienna, 1987), Vol. II., p. 423.
41. T. Tamano *et al.*, in *Proceedings of the Course on Mirror Based and Field Reversed Magnetic Fusion*, Vol. 2, p. 653 (1983).
42. R.J. La Haye *et al.*, *Bulletin of the American Physical Society* **29**, 1331 (1984).
43. T. Jensen and M. Chu, *Physics of Fluids* **27**, 2881 (1984).
44. M.N. Rosenbluth and M.N. Bussac, *Nuclear Fusion* **19**, 489 (1979).
45. J.H. Finn, W.M. Manheimer and E. Ott, *Physics of Fluids* **24**, 1336 (1981).
46. C.W. Barnes *et al.*, in *Plasma Physics and Controlled Nuclear Fusion Research 1986* (IAEA, Vienna, 1987), Vol. II, p. 519.
47. T.R. Jarboe *et al.*, *Comments on Plasma Physics and Controlled Fusion* **9**, 161 (1985).
48. A. Sykes and J. Wesson, *Physical Review Letters* **37**, 140 (1976).
49. D. Montgomery, L. Turner, and G. Vahala, *Physics of Fluids* **21**, 757 (1978).
50. C.G. Gimblett *et al.*, in *VII European Conference on Controlled Fusion and Plasma Physics*, (Lausanne), p. 103 (1975).
51. H.K. Moffat, *Magnetic Field Generation in Electrically Conducting Fluids* (Cambridge University Press, Cambridge, 1978).
52. D.D. Schnack *et al.*, *Physics of Fluids* **28**, 321 (1985).
53. A. Bhattacharjee *et al.*, in *Plasma Physics and Controlled Nuclear Fusion Research 1986* (IAEA, Vienna, 1987), Vol. II, p. 711.
54. A.H. Boozer, *Journal of Plasma Physics* **35**, 133 (1986).
55. H.R. Strauss, *Physics of Fluids* **28**, 2786 (1985).
56. A.Y. Aydemir *et al.*, *Physics of Fluids* **28**, 898 (1985).
57. P. Kirby, *Culham Laboratory Report* CLM-P802 (1987).
58. B.L. Wright *et al.*, in *Plasma Physics and Controlled Nuclear Fusion 1986* (IAEA, Vienna, 1987), Vol. II, p. 519.

59. N. Satomi *et al.*, in *Plasma Physics and Controlled Nuclear Fusion 1986* (IAEA, Vienna, 1987), Vol. II, p. 529.

Malvin H. Kalos
Courant Institute of Mathematical Sciences
New York University
New York, New York 10012

Monte Carlo Solution of Transport Equations on Highly Parallel Computers

The essence and significance of the Monte Carlo algorithm of Metropolis, Rosenbluth, Rosenbluth, Teller, and Teller is described. New computers with highly parallel architecture are ideally suited for Monte Carlo transport calculations.

INTRODUCTION

It is a pleasure to have been invited to join in honoring Marshall Rosenbluth. It is particularly pleasurable since I have the opportunity to discuss his remarkable contribution to Monte Carlo methods. I think of it as a universal algorithm for Monte Carlo sampling. Unquestionably it is a deep and enduring contribution to computational science, from the point of view of both practice and theory. It is and will continue to be very widely used. I would argue, in fact, that it is the single most elegant and powerful algorithm in common use today, not just in Monte Carlo. That is a strong statement, especially to people who do much computing. Anyone who improves a computation so as to run a million times faster may justifiably be proud of himself. If the power of an algorithm is measured by the ratio of computer time that would be needed without it to the time in which the same problem can be solved with it, then this Monte Carlo algorithm saves hundreds of orders of magnitude in computing time. Put another way, with its use on contemporary computers, a very large range of important problems is now accessible that would otherwise be impossible.[1,2]

The history of Monte Carlo is quite interesting.[1-4] The use of random sampling methods dates back to needle throwing experiments of Comte de Buffon (1777) and a treatment of the Boltzmann equation by Lord Kelvin (1901). In the 1930's, Fermi studied neutron diffusion and transport by carrying out unpublished numerical sampling experiments. However, the systematic development of the mathematical methods that were given the name Monte Carlo is usually associated with scientists who worked on nuclear weapons at Los Alamos in the 1940's. In particular, von Neumann, Ulam, and Metropolis[5] recognized and popularized the idea that sampling random points with the aid of a high-speed digital computer is a powerful tool for doing multi-dimensional integration in a complicated space. For the class of problems in which the weighting in phase space is intractible, an advanced algorithm for sampling representative points was invented by Metropolis, Rosenbluth, Rosenbluth, Teller, and Teller.[6] The M(RT)^2 algorithm proved to be so useful that in statistical mechanics, the term "Monte Carlo" refers almost exclusively to this technique and its generalizations.[7]

THE M(RT)^2 ALGORITHM

The M(RT)^2 algorithm is concerned with the Monte Carlo estimation of multi-dimensional integrals or sums. The general form can be written as the average of some function $f(x)$:

$$\langle f(x) \rangle = \frac{\int_\Omega f(x)p(x)\,dx}{\int_\Omega p(x)\,dx} \tag{1}$$

with respect to $p(x)$, an unnormalized probability measure or distribution function. Here, x should be thought of as a point in a space Ω of very large dimensions; typically, modern-day Monte Carlo problems involve spaces whose dimensions range from scores to millions. Doing integrals in such spaces is otherwise quite daunting.

In the original application of the M(RT)^2 algorithm to the classical statistical mechanics many-body problem for hard disks in two dimensions[6] and hard spheres in three dimensions,[8] the distribution function $p(x)$ was the canonical Boltzmann distribution,

$$p(x) = e^{-\beta U(x)} \tag{2}$$

where U is the energy and $\beta = 1/kT$, with T the temperature and k the Boltzmann constant. However, one of the elegant things about this algorithm is that, although it is often considered to be associated with the Boltzmann distribution, it has nothing directly to do with any specific distribution. Instead, it is practical for any distribution that is extremely complicated to sample.

The fundamental theorem for the Monte Carlo estimation of integrals is as follows. Suppose that an integral of the form given in Eq. (1) is required to be evaluated. First, it is necessary to carry out an algorithm which "samples" the distribution, that is, which generates a sequence of random variables $x_1, x_2, \ldots$, whose distribution function is $p(x)$. Thus, the probability that any x lies in any domain in the full space is simply the integral of $p(x)$ over that domain. Once the distribution can be sampled in this way, then a "naive" Monte Carlo estimate of the integral is simply given by the arithmetic average of the function x over the random variables that have been generated:

$$\langle f(x) \rangle \cong \frac{1}{N} \sum_{n=1}^{N} f(x_n) \tag{3}$$

Thus in principle the integration is reduced to the straightforward process of sampling.

In practice, straightforward sampling procedures break down for complicated distribution functions, like the Boltzmann distribution, in which every variable is connected to every other variable. This extremely difficult problem is trivially resolved by the $M(RT)^2$ algorithm.

The essence of the $M(RT)^2$ algorithm is as follows. Start from some point, call it x_0, in the phase space. Then, invent an artificial dynamics to govern the evolution of the system, such that successive configurations $x_1, x_2, \ldots$ are generated whose associated probability distribution functions asymptotically approach the desired $p(x)$. The artificial dynamics has nothing to do with the actual dynamics by which atoms or molecules interact with each other in nature. In fact, the procedure uses the essence of the behavior of statistical systems which approach an equilibrium whose properties are independent of the kinetics of the system. The algorithm is simply a description for a random walk: At each step n where $x = x_n$, one makes a completely random change in the position and thereby generates a possible next value for x, namely, x'_{n+1}. Then, the function $p(x)$ is computed at the new position. If $p(x)$ increases, i.e., $p(x'_{n+1})/p(x_n) > 1$, then one sets $x_{n+1} = x'_{n+1}$ so that the system is actually put at this new point. However, if $p(x)$ decreases, then either one puts the system at the new point with a probability equal to the ratio of the probability function evaluated at the two points, or else one returns the system to the previous position. Put another way, if $p(x'_{n+1})/p(x_n) > 1$, then x_{n+1} will always equal x'_{n+1}. If $p(x'_{n+1})/p(x_n) < 1$, then with probability $p(x'_{n+1})/p(x_n) > \xi$, where ξ is a random number whose value lies between 0 and 1, one puts $x_{n+1} = x'_{n+1}$ and otherwise $x_{n+1} = x_n$. The algorithm thus involves an element of rejection; however, if a new value x'_{n+1} is not accepted, one simply reuses the previous value. This procedure is iterated repeatedly. Under very general conditions, the asymptotic distribution of this random walk is the prescribed function $p(x)$. The only requirement is that the random walk result in an ergotic system, which is almost always the case.

In particular, for the case of a Boltzmann distribution, we have

$$\frac{p(x'_{n+1})}{p(x_n)} = \exp\left[-\beta U(x'_{n+1}) + \beta U(x_n)\right] = \exp[-\beta \Delta U] \qquad (4)$$

where ΔU is the difference in energy between the two states. If the random walk brings the system to a state of lower energy ($\Delta U < 0$), the move is allowed and the system is put in its new position. If the energy increases, the move is allowed with probability $\exp(-\beta \Delta U)$ or else rejected. Whether or not the proposed move is accepted, the next position is used in computing averages with respect to $\exp[-\beta U]$.

Note that this procedure guarantees that the probability distribution $p(x)$ will be sampled, but only asymptotically. Therefore, some first steps, say L, of the random walk should be discarded until convergence to the asymptotic limit is considered to begin:

$$\langle f(x) \rangle \cong \frac{1}{N} \sum_{n=L}^{N+L-1} f(x_n). \qquad (5)$$

Determining an appropriate value for L involves substantial trial and error. The principle disadvantages of the method are this necessity to omit a part of the calculation that cannot be accurately estimated in advance, plus the associated problem that the positions x_n are not statistically independent. The latter has two negative consequences: statistical errors are hard to compute in the face of the unknown sequential correlation, and the correlations can become very large in the neighborhood of a "critical point" in a statistical system.

The $M(RT)^2$ algorithm is trivial to program and carry out, but with it a great deal of physics can be powerfully simulated. It is extensively used in its original context of equilibrium statistical mechanics. What is even more remarkable is that it has come to be used in a much wider variety of applications. One such large class of applications has to do with sampling Feynman paths in complex time—i.e., with paths with Wiener measure. In fact, this algorithm, or variants of it, is the standard algorithm used in many computations of quantum chromodynamics.[8] It is also used in calculating properties of quantum many-body systems and in many other applications in physics, chemistry, biology, and so forth. Somewhat recently, a remarkable new application has been made of it: viz., stochastic optimization. This is a random walk procedure which optimizes an arbitrary function. The spirit of this procedure is to think of the function to be optimized as the analogue of the potential energy function in statistical mechanics. As the temperature is gradually lowered while the random walk is being executed, the minimum of the function will be found. By casting some interesting problems in the form of minimizations, one then obtains a very powerful computational algorithm. It has also been applied to computer vision: a special-purpose device called the Boltzmann machine was built to analyze computer vision, precisely with the use of this algorithm. In short, the $M(RT)^2$ algorithm is important, widely used, and—not in the least—extremely elegant.

MONTE CARLO TRANSPORT

The main part of this lecture concerns transport processes and Monte Carlo. It is simply a stochastic numerical technique for solving the linear Boltzmann equation. It is also useful to think of Monte Carlo here as a direct simulation of natural stochastic processes. It can be applied to the transport of radiation and of neutral and charged particles. We assume that the nature of the medium, as well as the probability distribution functions that describe the interaction of particles or radiation passing through that medium, are understood. One then prepares a stochastic description of those interactions, which translates into a stochastic program using random or pseudo-random numbers. The first step is to specify and sample the source, in terms of distribution functions for the initial values of the parameters that describe the radiation or particles. Next, the passage of the radiation or particles between significant interactions is viewed as a probability distribution function for the position of a significant next interaction, conditional on its current direction and energy; this is called "tracking the path." Finally, one describes and samples the nature of the possible interactions with the medium.

This Monte Carlo transport method can be directly applied to radiation in a plasma, to the transport of protons or neutrons out of a plasma, and to the transport of neutrons and other radiation in any reactor outside of a plasma. The physics of these interactions is assumed to be understood—which in many cases it is. These are translated into probability distributions, which are then sampled by various numerical techniques. The structure of such calculations is a random walk in (up to) seven dimensions: three positions, three velocities, and one time dimension. Earlier, the range of application was indicated as the linearized Boltzmann equation. In fact, it is possible to relax that constraint almost arbitrarily. To treat non-linear effects, the random walks may be allowed to interact, the easiest assumption being that they interact with each other weakly through integrated quantities, such as average properties of a gas. For example, the radiation may deplete some component of the gas, or it may enhance some isotope in a reactor and thereby change the probability distribution for other random walkers that follow. All of this can be built into the calculation.

It is also helpful to think of Monte Carlo transport as what is called an "event driven" simulation, viz., a simulation in which the state of a system is specified at those times at which significant parameters change. The collisions are the events. The nature of the next significant event is calculated from the previous state, and the system is followed from significant new state to significant new state, with intermediate stages being constant or trivially calculable. The kind of computer programs that accomplish this can be intricate in the sense that they contain many data-dependent and random number-dependent branches. Many computer scientists and even some numerical analysts do not recognize these as "true" numerical calculations; they think that such calculations must involve linear algebra and often they do not concern themselves with the kind of programming languages and, above all, the kind of computer architectures that are friendly to Monte Carlo.

Before going on to discuss computer architecture, let me comment a little more about the internal structure of the programming. This has several steps, as follows: (1) As was mentioned earlier, one first selects the source variables, from random distributions. (2) The radiation is next followed to a collision, either in straight lines for neutral radiation or in a diffusive Moliere path, with energy loss along the path, for charged particles. (3) Then, depending on which of these two cases applies, one uses an appropriate technique to select some next event, which could be an escape of this particle or of the radiation from the domain of interest; or it could be the termination of the random walk through a natural absorption process; or it could be a collision, in which case the history continues; or, in certain very important cases, it could be branching, which creates another history. The last are natural processes by which the radiation or particles multiply; the most efficient calculations can also simulate such processes, in which, for example, one particle turns into two particles. The calculation involving steps (2) and (3) is iterated either until the history ends or for some specified time interval Δt, and thereafter the entire calculation, involving steps (1)–(3), is iterated over many histories. Along the way, if the histories interact through integrated quantities, these quantities and the dependent parameters must be recomputed and their effect absorbed back into the calculation. Finally, or along the way, one must calculate the type of averages that are desired. Sometimes the quantities of interest include fluctuations, as well as simply the mean values. It is essential also to compute statistical errors to provide measures of the reliability of the results.

PARALLEL COMPUTER ARCHITECTURE

As a computational technique, Monte Carlo transport methods are fairly mature. Computer architecture, on the other hand, is evolving rapidly. Because of very large scale integration (VLSI), sophisticated building blocks for computers are becoming quite inexpensive. The current price of memory for use in building a computer is about $200 a megabyte. Twenty years ago it was about a million dollars a megabyte; several years from now it will be down to $50 a megabyte. Large-scale logic chips exist. At present, microprocessors run three million instructions per second (3 mips) and about half a million floating point operations (0.5 megaflops), all on one or two very small 32-bit chips. In a few years, this capability will increase to twenty million instructions per second, with 10 megaflops, and might level off at between fifty and a hundred million instructions per second per chip. The computer-aided design (CAD) technology for designing custom chips to organize off-the-shelf components into sophisticated computers is maturing rapidly. Thus there is an opportunity to build a computer according to his specialized needs.

The question then arises: What kind of architecture would be best for transport Monte Carlo? To answer this question, one needs to consider carefully his aim for using transport Monte Carlo in the first place, along with the general features of the algorithm.

Let me mention some high level ways in which a new computer could be organized.

1. SIMD versus MIMD: One way is to combine together many computers or many processors, all of which do the same operation at the same time. This lockstep operation is called SIMD (Single stream of Instructions applied to Multiple streams of Data). The alternative is an arrangement in which each of the computers act more or less autonomously—i.e., each one follows its own "instruction stream." This is called MIMD (Multiple streams of Instructions applied to Multiple streams of Data).
2. Distributed memory versus shared memory: The memory can be organized in such a way that some memory is closely coupled with each processor, and the aggregate of that memory constitutes the memory of all of the machine. An example of this is the Hypercube machine. Alternatively, one can try to organize the memory so that it is shared: all processors have roughly equal access to all of the memory of the machine.
3. Few powerful processors versus many small processors: One could attempt to couple together some of the most powerful processors that can be built. For example, present-day CRAY supercomputers have two or four processors, and future CRAY machines will have bigger ensembles. Another company, called ETA, is currently building an even larger ensemble of very powerful processors. However, these machines derive their basic computing power from vector operations, which means that many additions and multiplications must all be done essentially at the same time. The alternative is to try to obtain the computing power from a large number of small processors.

The Ultracomputer project, with which I am associated, is pursuing an architecture in which the choices of MIMD, shared memory, and many small processors have been made.[9] We are also cooperating with IBM, which is building a powerful prototype (called the RP3) with a similar, but enhanced, architecture. In any such project, one must identify the target for which the machine is to be used. It would be quite legitimate to design a machine that is easy to build, or that achieves high peak performance, or that is optimized for some algorithm. We, however, are attempting to build a broadly useful computer that fulfils the aims of being a machine that is highly effective for a wide variety of applications, that is easily programmable in traditional style and to which existing programs will be easily portable, and that is able to support a modern multiprocessing operating system.

The important point to note is that this machine has exactly the properties—viz., speed, memory, and flexibility and portability—that are needed for doing transport Monte Carlo on a large and powerful scale. Most importantly, the shared memory scheme provides flexibility and reasonable efficiency, because global data can be easily accessed. Many Monte Carlo calculations are characterized by a rather intricate description of the geometry of the medium, an intricate description of the physics of the interactions, and also a complex description of the body of answers. One wishes to know particle fluxes in space, direction, time, and energy—as well as the errors associated with all of them. Approximately 100,000 words per nuclide are needed to describe accurately the interactions between neutrons and matter, plus perhaps 100,000 words to characterize a moderately complicated geometry, and nearly 10,000,000 words for fluxes and errors. This can easily add up to millions of words of information, produced at random by all of the processors. It is much easier to organize it as a program for a shared memory machine. Also, when walks interact, the necessary information can be shared very rapidly. Particle branching creates additional work, which ideally is shared by other processors so that, on the average, all of them do about equal amounts of work. This is called "load balancing." In our architecture, load balancing is quite straightforward.

There do remain technical problems associated with this kind of machine. One is "network latency": i.e., if a shared memory is used, a penalty is paid in the time required for information to flow through the communications network between processors and the memory. Nevertheless, we think that an effective machine with shared-memory MIMD architecture can be built. Let me again remark that this idea of MIMD, in which each processor can follow its own instruction string, is essential for transport Monte Carlo, because the flow of each calculation is totally uncorrelated with the flow of any other calculation. If one is constrained to use a vector machine or an SIMD parallel machine that operates in lock step, then the difficulty of programming will be substantially multiplied and the flexibility of Monte Carlo calculations—one of their major assets—is substantially reduced.

Transport Monte Carlo does have disadvantages. It seems slow, at least as compared with deterministic solutions in few dimensions; it has statistical error; and sometimes it is difficult to compare nearby problems (i.e., to carry out some sort of perturbation theory). However, its advantages are that it is an extremely flexible algorithm that can incorporate a completely accurate treatment of the cross sections and all other physics, of the geometry of the media, of the geometry and physics of detectors, and of fluctuation phenomena. It can also provide an objective treatment of error. Transport Monte Carlo is employed for maximum fidelity to nature, to experiments, and to engineering problems. These advantages should be preserved and enhanced in future computers and computational methods. In my opinion, the easiest way to retain these advantages is with an MIMD machine.

Figure 1 shows a schematic diagram of the general architecture of the NYU Ultracomputer. The figure shows a collection of processors (as many as 4,096) and a large number of memory units, seen as shared memory, all organized together. A so-called "Omega" communication network, much like a telephone switching system, connects every processor to every memory unit. The components called caches constitute local memory that is invisible to the user. Caches are memory organized so as preferentially to keep frequently used information. By means of such caches, the latency delay for information flowing through this kind of network is greatly minimized; the important information is kept in these caches close to the processors. In transport Monte Carlo, the most-often-used information describing the physical problem will probabilistically be favored for retention in the caches. One component

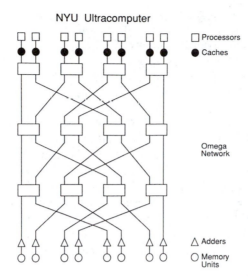

FIGURE 1 Schematic diagram of the NYU Ultracomputer architecture.

in this architecture is unprecedented: We propose to put adders between the network and the memory units, so that additions can be carried out at the memory end. The function of these adder elements is to support a new instruction called "Fetch-and-Add," which we have introduced to computers. Fetch-and-Add is a universal synchronization primitive, one that can be used to organize the cooperation of all the processors. What a Fetch-and-Add does is to ask that a datum be brought from memory, but at the same time it specifies that some increment be added to that datum, with the result being put directly back into memory. This produces an extremely flexible way of organizing dynamic control over unpredictable processes. The Fetch-and-Add provides the basis for a distributed operating system and for automatic, efficient load balancing.

In order to support memory access for a very large number of processors efficiently, the network must have a bandwidth that grows linearly with the number of processors. For this, the switches cannot be passive. Rather, some memory and a certain elementary processing power must be provided at each switch. In a sense, the switches become communication processors. These "smart switches" can store and forward conflicting messages in order to maintain the bandwidth and can combine messages (e.g., Fetch-and-Add) to a single location in order to handle "hot spots." With this enhancement, linear bandwidth in the network can be ensured.

Like many other proposals for parallel architectures, this machine has the feature that its hardware is modular. Although machines with parallel architectures will come in a variety of sizes, they can look and feel and smell exactly the same to a user. This will be a very important feature: When everything from workstation to the highest performance supercomputer can use the same software, computing productivity will be very much enhanced. Nowadays a microprocessor has some hundreds of thousands of transistors on it, and the microprocessors in a few years will have a couple of million transistors on them. Nevertheless, these components will be bought off the shelf, for a cost in the neighborhood of fifty dollars each. As technology changes, these computers can continue to look the same, even though they will be becoming even faster and cheaper. Thus as technology changes, programmers and programs will not have to adapt at new generations.

Finally, I would like to state my belief that shared memory machines represent the best hope for high performance, highly parallel computers that are relatively easy to program, machines for which existing sequential programs have some chance of being adapted, machines for which techniques for novel data structures and load-balancing techniques will not have to be researched for every new class of program. My impression is that most people who do programming are not interested in solving computational science problems—interesting as they may be—every time they want to do a new simulation of, say, a plasma.

ACKNOWLEDGEMENTS

It is a pleasure to thank James W. Van Dam for his invaluable assistance in the preparation of this paper. The work was supported by the Applied Mathematical Sciences subprogram of the Office of Energy Research, U. S. Department of Energy, under contract number DE-AC02 76ER03077.

REFERENCES

1. M.H. Kalos and P. A. Whitlock, *Monte Carlo Methods. Vol. I: Basics* (John Wiley, New York, 1986).
2. J.M. Hammersley and D. C. Handscomb, *Monte Carlo Methods* (Methuen and Co., London, 1964).
3. N. Metropolis, "Monte Carlo: In the Beginning and Some Great Expectations," in *Monte Carlo Methods and Applications in Neutronics, Photonics, and Statistical Physics* (Lecture Notes in Physics, No. 240) (Springer-Verlag, Berlin, 1985), edited by R. Alcouffe, R. Dautry, A. Forster, and G. Ledanois, pp. 62-70; N. Metropolis, "The Beginning of the Monte Carlo Method," in *Los Alamos Science* (Los Alamos National Laboratory, special issue, No. 15, 1987).
4. W.W. Wood, "Early History of Computer Simulations in Statistical Mechanics," in *Molecular Dynamics Simulation of Statistical Mechanical Systems* (Proc. International School "Enrico Fermi") (North-Holland, 1986), pp. 3-14.

5. N. Metropolis and S. Ulam, "The Monte Carlo Method," *Journal of the American Statistical Association* **247**, 335-341 (1949).
6. N. Metropolis, A.W. Rosenbluth, M.N. Rosenbluth, A.H. Teller, and E. Teller, "Equation of State Calculations by Fast Computing Machines," *Journal of Chemical Physics* **21**, 1087–1092 (1953).
7. W.W. Wood and J.J. Erpenbeck, "Molecular Dynamics and Monte Carlo Calculations in Statistical Mechanics," *Ann. Rev. Chem. Phys.* **27**, 319–348 (1976).
8. M.H. Kalos, ed., *Monte Carlo Methods in Quantum Problems* (D. Reidel Publishing Company, Dordrecht, 1984).
9. Allan Gottlieb, "An Overview of the NYU Ultracomputer Project," in *Experimental Parallel Computing Architectures*, edited by Jack J. Dongarra, (North Holland, Amsterdam, 1987), pp. 22–95.

Frederick L. Hinton
GA Technologies Inc.
San Diego, California 92138

Classical Dissipation and Transport in Plasmas

This paper reviews the subject of classical and neoclassical transport, including Fokker-Planck theory, with an emphasis on Marshall Rosenbluth's contributions. He has worked with many people on transport and I was fortunate to have been one of them. The paper is organized into four main parts, dealing with plasma kinetic theory, classical transport, neoclassical transport, and the present state of the subject.

FOKKER-PLANCK EQUATION

The calculation of plasma transport properties requires, first of all, an equation for describing the velocity distribution functions of the particles in the plasma. This equation differs from the Boltzmann equation for a neutral gas because of the long-range nature of the Coulomb force, since most Coulomb scattering events produce very small deflections. One way to obtain an appropriate kinetic equation is to expand the Boltzmann collision term in powers of the velocity deflection. By doing this, Rosenbluth, MacDonald, and Judd[1] derived an equation which is of the Fokker-Planck form. This equation describes the velocity space version of Brownian motion, for charged particles in a plasma.

The Rosenbluth, MacDonald, and Judd equation is

$$\frac{\partial f_a}{\partial t} + \mathbf{v} \cdot \nabla f_a + \frac{e_a}{m_a}\left(\mathbf{E} + \frac{\mathbf{v}}{c} \times \mathbf{B}\right) \cdot \frac{\partial f_a}{\partial \mathbf{v}} = \sum_b C_{ab}\left(f_a, f_b\right), \tag{1}$$

where the effect on species 'a' of collisions with all species 'b' is given by the Fokker-Planck term

$$C_{ab} = \frac{\partial}{\partial \mathbf{v}} \cdot \left[-\mathbf{A}_{ab} f_a + \frac{1}{2}\frac{\partial}{\partial \mathbf{v}} \cdot (\mathbf{D}_{ab} f_a)\right]. \tag{2}$$

The dynamic friction vector $\mathbf{A}_{ab}$ and the velocity diffusion tensor $\mathbf{D}_{ab}$ can be written in the form

$$\mathbf{A}_{ab} = \frac{4\pi e_a^2 e_b^2 \ln \Lambda}{m_a^2}\left(1 + \frac{m_a}{m_b}\right)\frac{\partial h_b}{\partial \mathbf{v}} \tag{3}$$

$$\mathbf{D}_{ab} = \frac{4\pi e_a^2 e_b^2 \ln \Lambda}{m_a^2}\frac{\partial^2 g_b}{\partial \mathbf{v} \partial \mathbf{v}}, \tag{4}$$

where the functions g_b and h_b are called Rosenbluth potentials and are given by

$$g_b(\mathbf{v}) = \int d^3 v' f_b(\mathbf{v}')|\mathbf{v} - \mathbf{v}'|, \tag{5}$$

$$h_b(\mathbf{v}) = \int d^3 v' \frac{f_b(\mathbf{v}')}{|\mathbf{v} - \mathbf{v}'|}. \tag{6}$$

The Coulomb logarithm, which results from cutting off a divergent integral over impact parameters, is defined as $\ln \Lambda = \ln(\lambda_D/b_0)$ where $\lambda_D = (T/4\pi n_e e^2)^{1/2}$ is the Debye shielding distance and $b_0 = e^2/3T$ is the "distance of closest approach." Because of the long-range force, many charged particles interact simultaneously, but the effective interaction range should be the Debye shielding distance.

Much work was done in the 1960's to provide a more basic derivation of the Fokker-Planck equation, and other more complicated kinetic equations were derived. Only the Rosenbluth, MacDonald, and Judd form of the Fokker-Planck equation has been used extensively in transport theory, however, because of the intractability of the theory with more complicated equations.

CLASSICAL TRANSPORT THEORY

In a report published by General Atomic in 1957, Rosenbluth considered collisional diffusion and ohmic heating in a stabilized pinch.[2] Moment equations were obtained by taking moments of the Fokker-Planck equation, and similarity solutions of these were obtained numerically, with the use of the IBM 650 computer at GA. In this report, and in a paper by Rosenbluth and Kaufman,[3] the transport coefficients for a collision-dominated plasma were calculated by using a method similar to that used by Chapman and Cowling for neutral gases, with the mean-free-path assumed to be short. In addition, the gyrofrequency was assumed to be larger than the collision frequency, which made the calculation of cross-field transport coefficients much more tractable than the general case. They showed that thermal transport was dominated by the ions and was much faster than magnetic field relaxation.

Many of the features of neoclassical transport theory are present also in classical transport theory, as formulated by Rosenbluth and Kaufman, and I will begin with a review of this theory. After transforming to coordinates in a moving reference frame, $\mathbf{v}' = \mathbf{v} - \mathbf{u}$, the Fokker-Planck equation becomes

$$\frac{\partial f}{\partial t} - \frac{\partial \mathbf{u}}{\partial t} \cdot \frac{\partial f}{\partial \mathbf{v}'} + (\mathbf{v}' + \mathbf{u}) \cdot \left[\nabla' f - (\nabla \mathbf{u}) \cdot \frac{\partial f}{\partial \mathbf{v}'} \right]$$

$$+ \frac{e}{m} \left(\mathbf{E}' + \mathbf{v}' \times \frac{\mathbf{B}}{c} \right) \cdot \frac{\partial f}{\partial \mathbf{v}'} = C , \tag{7}$$

where $\mathbf{E}' \equiv \mathbf{E} + \mathbf{u} \times \mathbf{B}/c$ is the electric field in the moving frame and C is the collision term. (Of course, this equation actually represents two or more coupled equations, for electrons and ions, but this detail need not concern us yet.) The distribution function is expanded in a power series in the basic small parameter, which is $\epsilon \equiv \rho/L$, the ratio of gyroradius to a typical length scale. The electric field in the moving frame is taken to be small: $cE'/B \sim \epsilon v_{\text{th}}$, where v_{th} is the thermal speed. For convenience, the collision frequency is assumed to be of the same order as the gyrofrequency: $\nu \sim \Omega$.

Then the zeroth-order equation is

$$\Omega \mathbf{v}' \times \hat{b} \cdot \frac{\partial f_0}{\partial \mathbf{v}'} = C(f_0, f_0), \tag{8}$$

The only solutions of this equation have the Maxwellian form, with the temperature T common to all particle species. The mean velocity is also common to all species and has been chosen to be zero in the moving frame. The first-order equation is

$$- \Omega \mathbf{v}' \times \hat{b} \cdot \frac{\partial f_1}{\partial \mathbf{v}'} + C^\ell f_1$$

$$= \frac{\partial f_0}{\partial t} - \frac{\partial \mathbf{u}}{\partial t} \cdot \frac{\partial f_0}{\partial \mathbf{v}'} + (\mathbf{v}' + \mathbf{u}) \cdot \left[\nabla' f_0 - (\nabla \mathbf{u}) \cdot \frac{\partial f_0}{\partial \mathbf{v}'} \right] + \frac{e}{m} \mathbf{E}' \cdot \frac{\partial f_0}{\partial \mathbf{v}'} , \tag{9}$$

where C^ℓ is the linearized Fokker-Planck operator. By substituting the Maxwellian expression for f_0 into this equation, the right-hand side can be expressed in terms of n, T, and $\mathbf{u}$. The moment equations obtained from this equation, corresponding to conservation of particles and energy (the latter summed over particle species, using energy conservation in collisions), are found to be

$$\frac{\partial n}{\partial t} + \nabla \cdot (n\mathbf{u}) = 0, \tag{10}$$

$$\frac{\partial T}{\partial t} + \mathbf{u} \cdot \nabla T = -\frac{2}{3} T\nabla \cdot \mathbf{u} , \tag{11}$$

When these equations are used in the first-order equation to eliminate $\partial n/\partial t$ and $\partial T/\partial t$, the equation becomes

$$-\Omega \mathbf{v}' \times \hat{b} \cdot \frac{\partial f_1}{\partial \mathbf{v}'} + C^\ell f_1$$

$$= \mathbf{v}' \cdot \left[\mathbf{A}_1 + \left(\frac{v'^2}{v_{\text{th}}^2} - \frac{5}{2}\right) \mathbf{A}_2\right] f_0 + \left(\mathbf{v}'\mathbf{v}' - \frac{1}{3}v'^2 \mathbf{I}\right) : \mathbf{A}_3 f_0 . \tag{12}$$

The terms on the right-hand side that drive the transport are

$$\mathbf{A}_1 = \frac{\nabla n}{n} + \frac{\nabla T}{T} + \frac{m}{T}\left(\frac{\partial \mathbf{u}}{\partial t} + \mathbf{u} \cdot \nabla \mathbf{u} - \frac{e}{m}\mathbf{E}'\right) , \tag{13}$$

$$\mathbf{A}_2 = \frac{\nabla T}{T}, \qquad \mathbf{A}_3 = \frac{m}{T}\nabla \mathbf{u} . \tag{14}$$

Before considering the solution of this equation, we note the following. The time derivative of $\mathbf{u}$ cannot be simply eliminated by using the momentum equation obtained from Eq. (12), because it contains moments of f_1. However, by adding the momentum equations for all species, we obtain the equation which determines $\partial \mathbf{u}/\partial t$:

$$\frac{1}{c}\mathbf{j} \times \mathbf{B} = \nabla P + \rho \left(\frac{\partial \mathbf{u}}{\partial t} + \mathbf{u} \cdot \nabla \mathbf{u}\right) , \tag{15}$$

where P is the total pressure and ρ is the mass density, and where $\mathbf{j}$ can be eliminated by using Maxwell's equation,

$$\frac{4\pi}{c}\mathbf{j} = \nabla \times \mathbf{B} - \frac{1}{c}\frac{\partial \mathbf{E}}{\partial t} . \tag{16}$$

Normally, the second term, involving $\mathbf{E}$, is neglected. The magnetic field is determined by Faraday's law, to this order:

$$\frac{\partial \mathbf{B}}{\partial t} = \nabla \times (\mathbf{u} \times \mathbf{B}) . \tag{17}$$

The above equations, which determine the time and space dependence of n, T and $\mathbf{u}$, are, of course, the ideal MHD equations, which contain no dissipative effects. Dissipation is contained in the more exact moment equations, which contain moments of f_1. Dissipative corrections to the ideal MHD equations are important in boundary layer situations, such as near rational magnetic surfaces in connection with kink-tearing instabilities in tokamaks. Here, I discuss only the slow evolution of MHD equilibrium configurations due to dissipation, which is the subject of transport.

In order to solve the first-order equation for f_1, cylindrical coordinates in velocity space are used, $\mathbf{u}' = v_\parallel \hat{b} + v_\perp (\hat{e}_1 \cos \zeta + \hat{e}_2 \sin \zeta)$, and f_1 is written in terms of its gyro-averaged part and its gyrophase-dependent part, as $f_1 = \bar{f}_1 + \tilde{f}_1$. By averaging Eq. (12) over the gyroangle ζ, we obtain the equation for $\bar{f}_1$:

$$C^\ell \bar{f}_1 = v_\parallel \hat{b} \cdot \left[\mathbf{A}_1 + \left(\frac{v'^2}{v_{\text{th}}^2} - \frac{5}{2} \right) \mathbf{A}_2 \right] f_0 + \left(v_\parallel^2 - \frac{v_\perp^2}{2} \right) \left(\hat{b}\hat{b} - \frac{1}{3}\mathbf{I} \right) : \mathbf{A}_3 f_0 . \quad (18)$$

The parallel particle fluxes require the solution of this equation. For example, the current density and the electron heat flux were obtained by Spitzer and Harm[4] by solving the electron version of this equation with $\mathbf{A}_3 = 0$.

By subtracting Eq. (18) from Eq. (12), we obtain the equation for $\tilde{f}_1$. The perpendicular fluxes can be calculated analytically when the collision frequency is much smaller than the gyrofrequency, which is usually the case of interest for plasmas used in fusion research. By expanding in powers of the small parameter ν/Ω, viz., $\tilde{f}_1 = \tilde{f}_1^{(0)} + \tilde{f}_1^{(1)} + \cdots$, we have

$$\Omega \frac{\partial \tilde{f}_1^{(0)}}{\partial \zeta} = \mathbf{v}_\perp \cdot \left[\mathbf{A}_1 + \left(\frac{v'^2}{v_{\text{th}}^2} - \frac{5}{2} \right) \mathbf{A}_2 \right] f_0$$

$$+ \left[v_\parallel \hat{b} \mathbf{v}_\perp + \mathbf{v}_\perp \hat{b} v_\parallel + \mathbf{v}_\perp \mathbf{v}_\perp - \frac{v_\perp^2}{2}(\mathbf{I} - \hat{b}\hat{b}) \right] : \mathbf{A}_3 f_0 , \quad (19)$$

and

$$\Omega \frac{\partial \tilde{f}_1^{(1)}}{\partial \zeta} = -C^\ell \tilde{f}_1^{(0)} . \quad (20)$$

The first equation is readily integrated to give

$$\tilde{f}_1^{(0)} = \frac{\mathbf{v}_\perp \times \hat{b}}{\Omega} \cdot \left[\mathbf{A}_1 + \left(\frac{v'^2}{v_{\text{th}}^2} - \frac{5}{2} \right) \mathbf{A}_2 \right] f_0$$

$$+ \frac{1}{\Omega} \left[\left(v_\parallel \hat{b} + \frac{\mathbf{v}_\perp}{4} \right) (\mathbf{v}_\perp \times \hat{b}) + (\mathbf{v}_\perp \times \hat{b}) \left(v_\parallel \hat{b} + \frac{\mathbf{v}_\perp}{4} \right) \right] : \mathbf{A}_3 f_0 , \quad (21)$$

from which the collision-independent perpendicular particle flux can be obtained:

$$\boldsymbol{\Gamma}_\perp^{(0)} \equiv \int d^3 v \, \mathbf{v}_\perp \tilde{f}_1^{(0)} = \frac{c}{eB} \hat{b} \times \left[\nabla(nT) + mn\mathbf{u} \cdot \nabla \mathbf{u} - ne\mathbf{E}' \right] . \quad (22)$$

By making use of $\mathbf{v}_\perp = (\partial/\partial\zeta)(\mathbf{v}_\perp \times \hat{b})$, integrating by parts, and using the equation for $\tilde{f}_1^{(1)}$, we obtain the collisional perpendicular particle flux, i.e., the diffusion flux:

$$\Gamma_\perp^{(1)} \equiv \int d^3v\mathbf{v}_\perp \tilde{f}_1^{(1)} = -\frac{\hat{b}}{\Omega} \times \int d^3v\mathbf{v}'C^\ell \tilde{f}_1^{(0)} \,. \tag{23}$$

The integrals are readily evaluated, but only some general properties are of interest here. The diffusion flux satisfies the ambipolarity condition, as a result of momentum conservation in collisions. Note that this is independent of the flow velocity $\mathbf{u}$ or the electric field in the flowing frame, $\mathbf{E}'$. The diffusion flux is due only to unlike-species collisions, since like-species collisions conserve momentum.

Similar expressions can be obtained for the heat flux and the momentum flux, as moments involving the collision term and the distribution function to zeroth order in the collision frequency:

$$\mathbf{q}_\perp \equiv \int d^3v \left(\frac{m}{2}v'^2 - \frac{5}{2}T\right)\mathbf{v}_\perp f_1$$

$$= \frac{5cnT}{2eB}\hat{b} \times \nabla T - \frac{T}{\Omega}\hat{b} \times \int d^3v \left(\frac{v'^2}{v_{\text{th}}^2} - \frac{5}{2}\right)\mathbf{v}_\perp C^\ell \tilde{f}_1^{(0)} \,, \tag{24}$$

$$\mathbf{P} \equiv m\int d^3v\mathbf{v}'\mathbf{v}'f_1 = m\int d^3v\left[v_\parallel^2 \hat{b}\hat{b} + \frac{v_\perp^2}{2}\left(\mathbf{I} - \hat{b}\hat{b}\right)\right]\bar{f}_1$$

$$+ m\int d^3v\left[v_\parallel \hat{b}\mathbf{v}_\perp + \mathbf{v}_\perp \hat{b}v_\parallel + \mathbf{v}_\perp\mathbf{v}_\perp - \frac{v_\perp^2}{2}\left(\mathbf{I} - \hat{b}\hat{b}\right)\right]\tilde{f}_1^{(0)}$$

$$+ \frac{m}{\Omega}\int d^3v\left[\left(v_\parallel\hat{b} + \frac{\mathbf{v}_\perp}{4}\right)(\mathbf{v} \times \hat{b}) + (\mathbf{v} \times \hat{b})\left(v_\parallel\hat{b} + \frac{\mathbf{v}_\perp}{4}\right)\right]C^\ell \tilde{f}_1^{(0)} \,. \tag{25}$$

The collisional fluxes depend upon collisions between both like and unlike particle species. For a pure plasma, with only one ion species, the ion-electron collisions make a negligible contribution to the ion fluxes. The integrals for the perpendicular fluxes can be evaluated, and the standard results found in the well-known article by Braginskii[5] are obtained.

NEOCLASSICAL TRANSPORT THEORY

When the mean free path of the particles is not short, the magnetic field geometry has effects on collisional transport which are not included in classical transport theory. The theory which includes these effects is called "neoclassical."

Galeev and Sagdeev[6] discovered important modifications of classical diffusion and heat conduction in toroidal confinement geometries at long mean-free-path. These are due to magnetically trapped particles and their "banana" orbits. Other new effects related to bananas were also discovered: the trapped particle pinch effect, independently by Ware[7] and Galeev,[8] and the bootstrap current, independently by

Bickerton, Connor, and Taylor[9] and Galeev.[8] The complete Onsager matrix of transport coefficients in the banana regime was calculated by Rosenbluth, Hazeltine, and Hinton,[10] who made use of a variational principle to include accurately the effects of like-particle collisions. The calculation was extended to include the higher collision frequency "plateau" regime by Hinton and Rosenbluth,[11] who used variational and numerical methods. The results showed that diffusion and thermal conductivity coefficients increase monotonically with collision frequency, even in the plateau regime. Diffusion of angular momentum determines the radial electric field in tokamaks. To lowest order in the gyroradius, conservation of angular momentum ensures that the diffusion is ambipolar, independent of the radial electric field. By going to fourth order in the gyroradius, Rosenbluth, Rutherford, Taylor, Frieman, and Kovrizhnykh[12] obtained an equation that determined the evolution the the radial electric field and the plasma rotation. By deriving the drift kinetic equation for a rotating reference frame, Hinton and Wong[13] generalized neoclassical theory to arbitrary rotation speeds and treated the angular velocity gradient and the temperature gradient driving terms together. They also included the coupling between angular momentum transport and energy transport. I will describe this more recent theory because it includes the previous work as the small rotation limit and because it more closely resembles the classical transport theory described in an earlier section.

THE DRIFT KINETIC EQUATION

The transport theory appropriate to long mean-free-path conditions begins with the Fokker-Planck equation, again transformed to a moving reference frame. Again expanding in powers of ϵ, but this time taking the collision frequency to be small ($\nu/\Omega \sim \epsilon$), we find the zeroth-order equation to be

$$\Omega \mathbf{v}' \times \hat{b} \cdot \frac{\partial f_0}{\partial \mathbf{v}'} = 0 , \tag{26}$$

which only says that f_0 is independent of gyrophase. The first-order equation is

$$-\Omega \mathbf{v}' \times \hat{b} \cdot \frac{\partial f_1}{\partial \mathbf{v}'} = \frac{\partial f_0}{\partial t} - \frac{\partial \mathbf{u}}{\partial t} \cdot \frac{\partial f_0}{\partial \mathbf{v}'} + \Lambda f_0 - C(f_0, f_0) , \tag{27}$$

where the operator Λ is defined by

$$\Lambda f_0 = (\mathbf{v}' + \mathbf{u}) \cdot \left[\nabla' f_0 - (\nabla \mathbf{u}) \cdot \frac{\partial f_0}{\partial \mathbf{v}'} \right] + \frac{e}{m} \mathbf{E}' \cdot \frac{\partial f_0}{\partial \mathbf{v}'} . \tag{28}$$

The equation that determines f_0 is obtained by averaging over gyrophase, after the equation is transformed to cylindrical velocity coordinates. The result is

$$\frac{\partial f_0}{\partial t} - \frac{\partial \mathbf{u}}{\partial t} \cdot \hat{b} \frac{\partial f_0}{\partial v_\parallel} + (v_\parallel \hat{b} + \mathbf{u}) \cdot \nabla f_0$$

$$+ \left[\frac{e}{m} \mathbf{E}' - (v_\| \hat{b} + \mathbf{u}) \cdot (\nabla \mathbf{u}) \right] \cdot \hat{b} \frac{\partial f_0}{\partial v_\|}$$

$$+ \frac{v_\perp^2}{2} (\nabla \cdot \hat{b}) \left(\frac{\partial f_0}{\partial v_\|} - \frac{v_\|}{v_\perp} \frac{\partial f_0}{\partial v_\perp} \right)$$

$$- \frac{v_\perp}{2} \left[\nabla \cdot \mathbf{u} - \hat{b} \cdot (\nabla \mathbf{u}) \cdot \hat{b} \right] \frac{\partial f_0}{\partial v_\perp} = C(f_0, f_0) \ . \tag{29}$$

This may be recognized as the kinetic equation used in guiding center theory, with the addition of the collision term on the right-hand side. Although many types of phenomena may be described by this equation, I shall discuss here only the case where the magnetic field is assumed to change slowly compared with the ion transit frequency, v_{th}/L. Then it follows that $\mathbf{E}' = -\nabla\Phi$, to this order. (Note that we are not assuming low β.) To be consistent with the long time scales on which transport occurs, I consider a toroidal magnetic confinement device, in particular an axisymmetric one such as a tokamak, where the magnetic field is given by $\mathbf{B} = I\nabla\phi + \nabla\phi \times \nabla\psi$. It can then be shown that the flow velocity $\mathbf{u}$ must be tangent to the equilibrium magnetic surfaces, $\psi = \text{const}$.

It is then possible to show that f_0 approaches a steady state, in a time of the order of an ion collision time, which is a Maxwellian with temperature constant on a magnetic surface, $T = T(\psi)$, and that the poloidal flow velocity component damps on this time scale, leaving a purely toroidal flow, $\mathbf{u} = \omega(\psi)R\hat{e}_\phi$, where ω is the angular velocity. The density variation on a magnetic surface is given by a Boltzmann factor, $n = N(\phi)\exp[-(e\tilde{\Phi}/T) + (m\omega^2 R^2/2T)]$, where $\tilde{\Phi}$ is the poloidally varying part of the electrostatic potential.

The first-order equation can be solved for $\tilde{f}_1$. Since the collision terms vanish on Maxwellians, the result does not contain collisional effects. The result can be used to calculate the perpendicular current density, which is found to satisfy the MHD equilibrium equation

$$\frac{1}{c} \mathbf{j} \times \mathbf{B} = \nabla P - \rho \omega^2 \mathbf{R} \ , \tag{30}$$

where P is the total pressure and ρ is the mass density.

The effects of collisional dissipation and transport are contained in $\bar{f}_1$, which satisfies the drift kinetic equation, obtained by going to the next order in the ϵ expansion. The next order equation is

$$\Omega \frac{\partial \tilde{f}_2}{\partial \zeta} = \Lambda f_1 + \frac{e}{m} \mathbf{E}^A \cdot \frac{\partial f_0}{\partial \mathbf{v}'} - C^\ell f_1 \ , \tag{31}$$

where $\mathbf{E}^A$ is the inductive part of $\mathbf{E}'$. By averaging over gyrophase, we obtain the drift kinetic equation,

$$\bar{\Lambda}\bar{f}_1 - C^\ell \bar{f}_1 = -\overline{\Lambda \tilde{f}_1} + \frac{e}{T} E_\|^A v_\| f_0 \ . \tag{32}$$

By evaluating the term involving $\tilde{f}_1$ with the forms of f_0 and $\mathbf{u}$ determined previously and using the assumed axisymmetry, we find this equation to be

$$v_\| \hat{b} \cdot \nabla \bar{f}_1 - C^\ell \bar{f}_1 = -v_\| \hat{b} \cdot \nabla(\alpha_1 A_1 + \alpha_2 A_2 + \alpha_3 A_3) f_0 + \frac{e}{T} E_\|^A v_\| f_0 \ . \qquad (33)$$

Here $\bar{f}_1$ is taken to be a function of the independent velocity variables ε and μ, the energy and magnetic moment in the moving frame, and σ, the sign of $v_\|$, where

$$\varepsilon = \frac{1}{2}(v_\|^2 + v_\perp^2) + \frac{e}{m}\tilde{\Phi} - \frac{\omega^2 R^2}{2} \ , \quad \mu = \frac{v_\perp^2}{2B} \ . \qquad (34)$$

The electrostatic potential in the rotating frame has been written as the sum of magnetic-surface averaged and poloidally varying parts: $\Phi = \langle \Phi \rangle + \tilde{\Phi}$.

The driving terms are

$$A_1 = \frac{1}{N}\frac{\partial N}{\partial \phi} + \frac{e}{T}\frac{\partial}{\partial \phi}\langle \Phi \rangle + \frac{1}{T}\frac{\partial T}{\partial \phi} \ , \qquad (35)$$

$$A_2 = \frac{1}{T}\frac{\partial T}{\partial \psi} \ , \quad A_3 = \frac{1}{\omega}\frac{\partial \omega}{\partial \psi} \ . \qquad (36)$$

The functions of velocity that multiply the driving terms are

$$\alpha_1 = \frac{mc}{e}\left(\frac{I v_\|}{B} + \omega R^2\right) \ , \qquad (37)$$

$$\alpha_2 = \left(\frac{m\varepsilon}{T} - \frac{5}{2}\right)\alpha_1 \ , \qquad (38)$$

$$\alpha_3 = \frac{mc\omega}{e v_{\text{th}}^2}\left[\left(\frac{I v_\|}{B} + \omega R^2\right)^2 + \mu\frac{|\nabla \psi|^2}{B}\right] \ , \qquad (39)$$

in which $I = RB_\phi$ and $v_\| = [2(\varepsilon - \mu B - e\tilde{\Phi}/m + \omega^2 R^2/2)]^{1/2}$.

We note two important properties of Eq. (33). The equation is linear, so $\bar{f}_1$ depends linearly on the driving terms, A_1, A_2, A_3, and $E_\|^A$. However, it depends on them in a nonlocal way, since $\hat{b} \cdot \nabla \bar{f}_1$ appears on the left-hand side of the equation. Thus, for example, the parallel current density at a given spatial point depends on the parallel inductive electric field at all points on the same magnetic surface, in general.

The second-order fluxes of particles, energy, and angular momentum, which determine the slow time dependence of the density, temperature, and angular velocity, may be obtained from moments involving the collision terms and the first-order distribution function. The magnetic-surface-averaged particle flux is

$$\Gamma \equiv \left\langle \int d^3v \mathbf{v} \cdot \nabla\psi f_2 \right\rangle = -\frac{c}{e}\left\langle \int d^3v m R v_\phi C^\ell(f_1 - f_s) \right\rangle \ , \qquad (40)$$

where f_s is the Spitzer function, defined here as a solution of the equation

$$C^\ell f_s = \frac{e}{m}\mathbf{E}^A \cdot \frac{\partial f_0}{\partial \mathbf{v}} \ . \tag{41}$$

By using $f_1 = \bar{f}_1 + \tilde{f}_1$ and similarly for f_s, we find

$$\Gamma = -\left\langle \int d^3v\, \alpha_1 C^\ell (\bar{f}_1 - \bar{f}_s) \right\rangle$$

$$- \frac{mc}{e}\left\langle \int d^3v\, R\hat{e}_\phi \cdot \mathbf{v}_\perp C^\ell \tilde{f}_1 \right\rangle + c\left\langle n\frac{\mathbf{E}^A \times \hat{b}}{B} \cdot \nabla \psi \right\rangle , \tag{42}$$

where the brackets $\langle\ \rangle$ denote the average and α_1 is defined above. The first term is the neoclassical flux, the second is the classical collisional flux, and the third is the usual $E \times B$ pinch effect.

This expression shows that the particle flux is ambipolar, independent of the angular velocity ω or the electric field, as a consequence of collisional momentum conservation and charge neutrality. Only unlike-species collisions contribute to the collisional particle flux; it is much smaller than the ion heat flux, which is determined by the ion-ion collision rate.

The angular momentum flux is

$$\Pi \equiv \left\langle \int d^3v\, mRv_\phi \mathbf{v} \cdot \nabla \psi f_2 \right\rangle = -\frac{mc}{e}\left\langle \int d^3v\, \frac{m}{2}R^2 v_\phi^2 C^\ell (f_1 - f_s) \right\rangle . \tag{43}$$

Again using $f_1 = \bar{f}_1 + \tilde{f}_1$ and similarly for f_s, we find that the term depending on $\bar{f}_1$, the neoclassical term, is

$$\bar{\Pi} = -\frac{T}{\omega}\left\langle \int d^3v\, \alpha_3 C^\ell (\bar{f}_1 - \bar{f}_s) \right\rangle . \tag{44}$$

The energy flux is

$$Q \equiv \left\langle \int d^3v \left(\frac{m}{2}v^2 + e\tilde{\phi} \right) \mathbf{v} \cdot \nabla \psi f_2 \right\rangle , \tag{45}$$

where the electrostatic energy is included as well as the kinetic energy. The heat flux q can be defined by

$$q \equiv Q - \omega\Pi - \frac{5}{2}T\Gamma$$

$$= \left\langle \int d^3v (\frac{m}{2}v^2 + e\tilde{\phi} - \frac{m}{2}\omega^2 R^2 - \frac{5}{2}T)\mathbf{v} \cdot \nabla \psi f_2 \right\rangle . \tag{46}$$

The neoclassical part of the heat flux is given by

$$\bar{q} = -T\left\langle \int d^3v\, \alpha_2 C^\ell (\bar{f}_1 - \bar{f}_s) \right\rangle . \tag{47}$$

Thus, all of the neoclassical fluxes are expressed in terms of collisional moments of the solution of the drift kinetic equation, $\bar{f}_1$.

ION TRANSPORT

In the ion drift kinetic equation, the ion-electron collision term and $E_\parallel^A$ can be assumed to be smaller by $(m_e/m_i)^{1/2}$ than the other terms and may be neglected. Then it can be shown that the term α_1 can be removed from the equation by a transformation and that α_2 and α_3 can be replaced with

$$\beta_2 = \frac{Iv_\parallel}{\Omega}\left(\frac{m\varepsilon}{T} - \frac{5}{2}\right) , \tag{48}$$

$$\beta_3 = \frac{\omega}{\Omega v_{\text{th}}^2}\left(\frac{I^2 v_\parallel^2}{B} + \mu|\nabla\Psi|^2\right) . \tag{49}$$

The neoclassical angular momentum flux and heat flux are given by

$$\bar{\Pi} = -\frac{T}{\omega}\left\langle\int d^3v\ \beta_3 C_{ii}^\ell \bar{f}_1\right\rangle , \tag{50}$$

$$\bar{q} = -T\left\langle\int d^3v \beta_2 C_{ii}^\ell \bar{f}_1\right\rangle . \tag{51}$$

Because of the linearity of the drift kinetic equation, the distribution function can be written as

$$\bar{f}_1 = (g_2 A_2 + g_3 A_3)f_0 , \tag{52}$$

and so the heat flux and momentum flux are linear in the driving terms $A_2 \equiv (1/T)\partial T/\partial\psi$ and $A_3 \equiv (1/\omega)\partial\omega/\partial\psi$:

$$\bar{q} = -T(L_{22}A_2 + L_{23}A_3) , \tag{53}$$

$$\bar{\Pi} = -\left(\frac{T}{\omega}\right)(L_{32}A_2 + L_{33}A_3) . \tag{54}$$

It can be shown that

$$L_{32} = -L_{23} . \tag{55}$$

That is, the effect of a gradient of toroidal angular velocity on heat flow is opposite to the effect of a temperature gradient on momentum flux. The signs depend on the magnetic field geometry; with up-down symmetric geometries, the cross terms are zero: $L_{32} = -L_{23} = 0$.

The value of the ion thermal conductivity in the banana regime is still of current interest. The banana regime is defined by the condition that the effective trapped ion collision frequency be smaller than the trapped ion bounce frequency. Then the distribution function can be expanded in powers of the ratio of collision frequency to bounce frequency: $\bar{f}_1 = f^{(0)} + f^{(1)} + \cdots$. The zeroth-order distribution function is determined by the equations

$$v_{\|}\hat{b} \cdot \nabla(f^{(0)} + \beta_2 A_2 f_0) = 0 , \tag{56}$$

$$\oint \frac{d\ell}{v_{\|}} C_{ii}^{\ell} f^{(0)} = 0 , \tag{57}$$

in which

$$\beta_2 \equiv \frac{I v_{\|}}{\Omega_i}\left(\frac{v^2}{v_i^2} - \frac{5}{2}\right) ,$$

(with the centrifugal and electrostatic potentials now neglected). Moreover, because of axisymmetry, we have

$$\oint \frac{dl}{v_{\|}} \cdots = \oint \frac{dl_p B}{B_p v_{\|}} \cdots .$$

The neoclassical heat flux can be written as

$$\bar{q} = -\frac{T}{A_2}\left\langle \int d^3 v \hat{f}^{(0)} C_{ii}^{\ell} f^{(0)} \right\rangle , \tag{58}$$

in which we use the notation $\hat{f} \equiv f/f_0$. This expression is variational: when used with Eq. (56) as a constraint, it yields Eq. (57) as the Euler equation. Its usefulness here stems from its quadratic dependence on $f^{(0)}$.

Analytic solution is possible in the large aspect ratio limit, $\delta \equiv r/R \ll 1$. This is because pitch-angle scattering plays a predominant role in the transport process in that limit. The linearized collision operator can be written (now dropping the subscript ii) as

$$C^{\ell} = C^{(0)} + C^{(1)} , \tag{59}$$

where $C^{(0)} = \nu_{ii}(v)\mathcal{L}$, with $\mathcal{L}$ the pitch-angle operator given by

$$\mathcal{L}f = \frac{1}{2}\frac{\partial \xi}{\partial \xi}(1 - \xi^2)\frac{\partial f}{\partial \xi} , \tag{60}$$

where $\xi \equiv v_{\|}/v$. Here $\nu_{ii}(v)$ is the ion-ion pitch-angle scattering collision frequency. The remainder of the collision operator, which includes the energy scattering and integral operators, is denoted by $C^{(1)}$. The ion thermal conductivity can be obtained correctly to order $\delta^{1/2}$ (i.e., neglecting only terms of order δ) by neglecting $C^{(1)}$. Since this result is of some importance, it will now be demonstrated in detail.

We write

$$f^{(0)} = \left[g^{(0)} + g^{(1)}\right] A_2 , \tag{61}$$

where

$$g^{(0)} = -\frac{I v_\parallel}{\Omega_i} \left(\frac{v^2}{v_i^2} - y \right) f_0 + G^{(0)} , \tag{62}$$

$$g^{(1)} = \frac{I v_\parallel}{\Omega_i} \left(\frac{5}{2} - y \right) f_0 + G^{(1)} , \tag{63}$$

with $\hat{b} \cdot \nabla G^{(0)} = 0$ and $\hat{b} \cdot \nabla G^{(1)} = 0$. Note that the term proportional to $\frac{5}{2} - y$ makes no contribution in the following. The function $g^{(0)}$ is defined to be a solution of

$$\oint \frac{dl}{v_\parallel} C^{(0)} g^{(0)} = 0 , \tag{64}$$

and $G^{(1)}$ satisfies

$$\oint \frac{dl}{v_\parallel} \left(C^\ell G^{(1)} + C^{(1)} g^{(0)} \right) = 0 . \tag{65}$$

The parameter y is as yet undetermined, but will be determined by the condition that the dominant contribution to $\bar{q}$ is obtained by neglecting $C^{(1)}$ and $g^{(1)}$.

Then the ion thermal conductivity is given by

$$L_{22} = L_{22}^{(0)} + L_{22}^{(1)} , \tag{66}$$

where

$$L_{22}^{(0)} = -\left\langle \int d^3 v \hat{g}^{(0)} C^{(0)} g^{(0)} \right\rangle , \tag{67}$$

while $L_{22}^{(1)}$ involves $C^{(1)}$ and $G^{(1)}$ can be written as

$$L_{22}^{(1)} = -\left\langle \int d^3 v \hat{g}^{(0)} C^{(1)} g^{(0)} \right\rangle + \left\langle \int d^3 v \hat{G}^{(1)} C^\ell G^{(1)} \right\rangle . \tag{68}$$

Here the notation $\hat{g}^{(0)} \equiv g^{(0)}/f_0$ is used.

The integrals may be written out explicitly as follows, using the explicit form of the Fokker-Planck collision operator:

$$\int d^3 v \hat{g} C^\ell g = -\frac{1}{2} \int d^3 v f_0(v) \nu_{ii}(v)(1 - \xi^2) \left(\frac{\partial \hat{g}}{\partial \xi} \right)^2$$

$$- \frac{1}{2} \int d^3 v f_0(v) D_\parallel(v) \left(\frac{\partial \hat{g}}{\partial v} \right)^2$$

$$+ \frac{\Gamma_i}{2} \int d^3 v f_0(v) \frac{\partial \hat{g}}{\partial \mathbf{v}} \cdot \int d^3 v' f_0(v') \left(\frac{\mathbf{I}}{u} - \frac{\mathbf{uu}}{u^3} \right) \cdot \frac{\partial \hat{g}}{\partial \mathbf{v'}} , \tag{69}$$

in which $\nu_{ii}(v) = D_\perp(v)/v^2$ is the pitch-angle scattering collision frequency, $D_\perp(v)$ and $D_\parallel(v)$ are the components of the velocity space diffusion tensor, and $\mathbf{u} \equiv \mathbf{v} - \mathbf{v}'$ and $\Gamma_i \equiv (4\pi e^4 \ln \Lambda)/m_i^2$. The velocity gradient is to be written in terms of the pitch-angle variable ξ and the magnitude of velocity, v, as

$$\frac{\partial \hat{g}}{\partial \mathbf{v}} = (\hat{b} - \xi \hat{v}) \frac{1}{v} \frac{\partial \hat{g}}{\partial \xi} + \frac{\hat{v} \partial \hat{g}}{\partial v} \ , \tag{70}$$

where $\hat{v} = \mathbf{v}/v$.

We first calculate $L_{22}^{(0)}$. The solution of Eq. (64) is

$$g^{(0)} = -V(v) \left[\frac{\xi}{B} - \frac{1}{2} \Theta_u \int_\lambda^{\lambda_c} \frac{d\lambda}{\langle \xi \rangle} \right] \ , \tag{71}$$

where $\lambda = (1 - \xi^2)/B$,

$$V(v) \equiv \frac{IB}{\Omega_i} v \left(\frac{v^2}{v_i^2} - y \right) f_0 \ , \tag{72}$$

$\lambda_c = 1/B_{\max}$, and Θ_u is a step function which is 1 in the untrapped region $\lambda < \lambda_c$ and 0 in the trapped region $\lambda > \lambda_c$. The pitch-angle derivative is

$$\frac{\partial g^{(0)}}{\partial \xi} = -\frac{2\xi}{B} \frac{\partial g^{(0)}}{\partial \lambda} = -\frac{V}{B} \left[1 - \frac{\Theta_u \xi}{\langle \xi \rangle} \right] \ . \tag{73}$$

Then, by substituting this expression into Eq. (67), we find

$$L_{22}^{(0)} = - \left(\frac{IB}{\Omega_i} \right)^2 \pi J \int v^4 dv \nu_{ii} \left(\frac{v^2}{v_i^2} - y \right)^2 f_0 \ , \tag{74}$$

where

$$J = \left\langle \int_0^{1/B} \frac{\lambda d\lambda}{\xi} \right\rangle - \int_0^{\lambda_c} \frac{\lambda d\lambda}{\langle \xi \rangle} = 1.38(2\delta)^{1/2} \ , \tag{75}$$

for circular cross section magnetic surfaces, using $B \simeq B_0/(1 + \delta \cos \theta)$.

A comparison of that term in Eq. (68) for $L_{22}^{(1)}$ which contains $C^{(1)}$ with the expression for $L_{22}^{(0)}$, which contains $C^{(0)}$, can be made by substituting Eq. (71) for $g^{(0)}$ into that term. It is not necessary to evaluate the integrals to determine their dependence on δ; this can be done as follows. The solution of Eq. (64) for $g^{(0)}$ is easily shown to have the following properties:

$$g^{(0)} = O(\delta^{1/2}) \ , \tag{76}$$

$$v \frac{\partial g^{(0)}}{\partial v} = O(\delta^{1/2}) \ , \tag{77}$$

and

$$\frac{\partial g^{(0)}}{\partial \xi} = \begin{cases} O(1) , & \text{for } |\xi| \lesssim \delta^{1/2} , \\ O(\delta) , & \text{otherwise} . \end{cases} \tag{78}$$

The reason is basically that Eq. (57) has no effect on $g^{(0)}$ in the trapped particle region, which allows its pitch-angle derivative to be relatively large there, but it allows $G^{(0)}$ to tend to cancel the first term in Eq. (62) in the untrapped region, which makes the magnitude of $g^{(0)}$ relatively small. Also, the v-derivative is not large because v appears only as a parameter.

Thus, when $\partial g^{(0)}/\partial \xi$ appears squared and integrated, as it does in $L_{22}^{(0)}$, we get a term which is $O(1)$, integrated over the trapped particle region whose width is $\Delta \xi \sim \delta^{1/2}$, so the result is $O(\delta^{1/2})$. However, when $\partial g^{(0)}/\partial v$ appears squared and integrated, we get a term which is $O(\delta)$ over the whole interval $-1 \leq \xi \leq 1$, and the result is $O(\delta)$. Also, in the double integral terms, the pitch-angle derivative terms are not squared before integration, so each integral contributes a term of order $\delta^{1/2}$, the same contribution made by the v-derivatives. But the double integral is of the same order as the square of one of the single integrals and hence is of order δ. Therefore, only the pitch angle derivatives contribute, to order $\delta^{1/2}$.

We now note that, without specifying y, the term containing $G^{(1)}$ in Eq. (68) for $L_{22}^{(1)}$ can not be shown to be smaller than $L_{22}^{(0)}$ given by Eq. (67), because changing the value of y in Eq. (62) is equivalent to adding a term to $G^{(1)}$ that is proportional to $G^{(0)}$; with the pitch-angle derivatives in Eq. (68), this gives a term comparable to $L_{22}^{(0)}$, in general. To find the correct value of y, we proceed as follows.

We define the function h by

$$G^{(1)} = g_* + h , \tag{79}$$

where g_* is a solution of

$$C^\ell g_* = -C^{(1)} g^{(0)} . \tag{80}$$

Provided that this equation has a solution, h satisfies the equation

$$\oint \frac{dl}{v_\|} C^\ell h = 0 . \tag{81}$$

Since $\hat{b} \cdot \nabla G^{(1)} = 0$, h satisfies the same equations as does $f^{(0)}$, except that the driving term $\beta_2 A_2 f_0$ is replaced by g_*. Now, an approximate solution of Eq. (80) can be found, of the form $g_* \approx v_\| F_*(v)$ where $F_* \sim \delta^{1/2}$, and so h is smaller than $g^{(0)}$ by one order in $\delta^{1/2}$, i.e., $h = O(\delta)$. In $L_{22}^{(1)}$, $G^{(1)}$ may thus be replaced by g_*. Since derivatives of g_* are of the same order as g_* itself, and they appear quadratically in $L_{22}^{(1)}$, it follows that $L_{22}^{(1)} = O(\delta)$.

The parameter y is determined by the solubility condition on Eq. (80), which is

$$\int d^3 v \, v_\| C^{(1)} g^{(0)} = 0 . \tag{82}$$

When the momentum conservation property of C^ℓ is used, this can be written as

$$\int d^3v\, v_\parallel C^{(0)} g^{(0)} = 0 \; , \tag{83}$$

which gives $y = 1.33$. Thus, the correct value of y is determined by forcing $C^{(0)}$ to satisfy momentum conservation.

When expressed in terms of the usual units, using r as the minor radius coordinate, we obtain the result for the ion heat flux:

$$q_i = -n_i \chi_i \frac{\partial T_i}{\partial r} \; , \tag{84}$$

where

$$\chi_i = 0.66 \delta^{1/2} \frac{\rho_{i\theta}^2}{\tau_i} \; . \tag{85}$$

This result was first obtained by Rosenbluth, Hazeltine, and Hinton.[10]

The other diagonal transport coefficient, L_{33}, is the viscosity, which was calculated by Rosenbluth, Rutherford, Taylor, Frieman, and Kovrizhnykh[12] for the case of small rotation speeds, $\omega R \ll v_i$. Their calculation was extended to include finite rotation speeds by Hinton and Wong,[13] as follows. A variational expression for the viscosity is

$$L_{33} = -\left\langle \int d^3v (-\beta_3 + \hat{G}_3) C^\ell (-\beta_3 f_0 + G_3) \right\rangle \; , \tag{86}$$

where G_3 satisfies $\hat{b} \cdot \nabla G_3 = 0$. This expression can be minimized by choosing a trial function G_3 that will approximately cancel the β_3 terms everywhere in velocity space, with differences that are of the order of δ. We consider only the limit $B_p \ll B_\phi$, so that

$$\beta_3 \simeq \frac{\omega}{v_i^2} \frac{mc}{e} I^2 \frac{(v^2 - v_\perp^2)}{B^2} \; . \tag{87}$$

Then, since the term proportional to v^2 does not contribute in the above integral, it is clear that the trial function $G_3 = -2\mu B_0 A$ has the desired properties, where A is a constant to be determined and B_0 is the field on the magnetic axis. The variational expression was minimized to obtain A; substituting it in to obtain L_{33} and taking the large-aspect-ratio limit of the result yields an expression which is independent of rotation speed, and equal to the result obtained by Rosenbluth, Rutherford, Taylor, Frieman, and Kovrizhnykh.[12] Thus, the viscosity is not significantly larger for arbitrarily large rotation than in the small rotation case. The role of trapped ions is negligible in either case, because angular momentum transport is caused primarily by untrapped ions.

In the usual units, the momentum flux is given by

$$\Pi_{\phi r} = -m_i n_i \mu_i \frac{\partial u_\phi}{\partial r} \; , \tag{88}$$

where

$$\mu_i = 0.1 \, \delta^2 \frac{\rho_{i\theta}^2}{\tau_i} \; . \tag{89}$$

ELECTRON TRANSPORT

For electrons, we assume $\omega R \ll v_e$ and neglect α_3. We use the fact that m_e/m_i is very small and that the ion distribution function enters the electron equation only through $u_{i\parallel}$, approximately. After making some transformations, the electron equation takes the form

$$v_\parallel \hat{b} \cdot \nabla f - C_e f = -v_\parallel \hat{b} \cdot \nabla \left(\alpha_1 A_1' + \alpha_2 A_2 + \alpha_3 A_3 \right) f_0 \; , \tag{90}$$

where

$$C_e = C_{ee}^\ell + \nu_{ei} \mathcal{L} \; , \tag{91}$$

with $\mathcal{L}$ the pitch angle scattering operator. All the driving terms now appear in the same way; their definitions are the same as given previously, except that

$$A_1' = A_1 - \frac{\Omega m}{IT_e} u_{i\parallel} \; , \tag{92}$$

$$\alpha_3 = \hat{g}_s B \qquad \text{and} \qquad A_3 = \frac{\left\langle E_\parallel^A B \right\rangle}{\langle B^2 \rangle} \; , \tag{93}$$

in which $g_s = \bar{f}_s / E_\parallel^A$, where $\bar{f}_s$ is the gyroaveraged Spitzer function.

The current density can be written in terms of the transformed electron distribution function:

$$j_\parallel = -e \int d^3 v v_\parallel \bar{f}_{1e} + e \int d^3 v v_\parallel \bar{f}_{1i}$$

$$= j_s - e \int d^3 v v_\parallel f \; , \tag{94}$$

where

$$j_s = \sigma_s \frac{B \left\langle E_\parallel^A B \right\rangle}{\langle B^2 \rangle} \; , \tag{95}$$

with σ_s the Spitzer conductivity. By means of the equation for $\bar{f}_s$, the expression for $j_\parallel$ can be transformed into a form very similar to the expressions for particle and heat fluxes, in terms of the collision operator:

$$j_\parallel - j_s = T \int d^3v \hat{f} C_e g_s = T \int d^3v \hat{g}_s C_e f = \frac{T}{B} \int d^3v \alpha_3 C_e f . \qquad (96)$$

Now using the linearity of the equation, we can write the solution in the form

$$f = g_1 A_1 + g_2 A_2 + g_3 A_3 , \qquad (97)$$

so that the fluxes are

$$\Gamma = -(L_{11}A_1 + L_{12}A_2 + L_{13}A_3) , \qquad (98)$$

$$\frac{q_e}{T} = -(L_{21}A_1 + L_{22}A_2 + L_{23}A_3) , \qquad (99)$$

$$\frac{\langle (j_\parallel - j_s)B \rangle}{T} = -(L_{31}A_1 + L_{32}A_2 + L_{33}A_3) , \qquad (100)$$

where the transport coefficients are given by

$$L_{mn} = \left\langle \int d^3v \alpha_m C_e g_n \right\rangle . \qquad (101)$$

The physical meanings of some of the terms in these equations are as follows. The term proportional to L_{13} is the trapped particle pinch effect, in which a particle flux is proportional to the parallel inductive electric field. As explained by Ware,[7] this is due to the inward bounce-averaged radial drift of bananas when their turning points are shifted slightly by the electric field. The term proportional to L_{23} is the corresponding heat flux. The terms proportional to L_{31} and L_{32} give the bootstrap current, a parallel current driven by gradients of density and temperature normal to magnetic surfaces. The term "bootstrap" originates with the possibility of a tokamak with no external current drive (except at the magnetic axis), as pointed out by Bickerton, Connor and Taylor.[9] Physically, it is due to the parallel motion of particles in their banana orbits, in the presence of radial gradients, along with the friction with passing particles. The term proportional to L_{33} is the reduction of the current due to trapped electrons not carrying any Ohmic current.

The Onsager relations are the statement of the symmetry of the transport coefficient matrix. They can be proved for general collisionality, but the proof is particularly simple in the banana regime, when the collision frequency is much smaller than the bounce frequency. Then the functions g_m satisfy the equations

$$v_\parallel \hat{b} \cdot \nabla (g_m + \alpha_m f_0) = 0 , \qquad (102)$$

$$\oint \frac{dl}{v_\parallel} C_e g_m = 0 . \qquad (103)$$

It follows that the transport coefficients can be written as

$$L_{mn} = -\left\langle \int d^3 v \, \hat{g}_m \, C_e g_n \right\rangle = L_{nm} , \qquad (104)$$

in which the symmetry is a simple consequence of the self-adjointness of the collision operator. Thus, for example, the trapped particle pinch coefficients equal the bootstrap current coefficients: $L_{13} = L_{31}$, $L_{23} = L_{32}$. This expression for L_{mn} is variational, because when Eq. (102) is used as a constraint on the trial functions, the Euler equation is Eq. (103).

In the limit of large aspect ratio, the banana regime transport coefficients may be calculated analytically, by the same method described earlier for the ion thermal conductivity. It can be shown that the transport coefficients are given, to order $\delta^{1/2}$, by

$$L_{mn}^{(0)} = -\left\langle \int d^3 v \, \hat{g}_m^{(0)} \, C_e^{(0)} g_n^{(0)} \right\rangle , \qquad (105)$$

with

$$C_e^{(0)} = [\nu_{ee}(v) + \nu_{ei}(v)] \mathcal{L} , \qquad (106)$$

where $\mathcal{L}$ is the pitch-angle scattering operator and $g_m^{(0)}$ is the solution of Eq. (103) with C_e replaced by $C_e^{(0)}$. The neglected terms are of order δ. Note that since the operator C_e includes electron-ion collisions, which do not conserve momentum, there is no parameter analogous to y, which was used in the ion problem.

The transport coefficients depend quantitatively on the proportion of electron-electron to electron-ion collisions. This may be surprising in the case of the particle diffusion coefficient since the particle flux is due only to electron-ion collisions. It can be understood as follows.

The above result can be written as

$$L_{mn}^{(0)} = \left\langle \int d^3 v \, \alpha_m \, C_e^{(0)} g_n^{(0)} \right\rangle$$

$$= \left\langle \int d^3 v \, \alpha_m \, C_e \left(g_n^{(0)} + f_* \right) \right\rangle , \qquad (107)$$

where f_* is a solution of

$$C_e f_* = -C_{ee}^{(1)} g_n^{(0)} , \qquad (108)$$

with $C_{ee}^{(1)} = C_e - C_e^{(0)}$ the non-pitch angle scattering part of the electron-electron collision operator. In the second form of Eq. (107), with the exact C_e, the part of the distribution function driven by the non-pitch angle scattering part of the collision operator appears, which depends upon electron-electron collisions quantitatively. In the case $m = 1$ (corresponding to the particle flux), since $\alpha_1 \propto v_\parallel$, we have

$$L_{1n}^{(0)} = \left\langle \int d^3 v \alpha_1 \nu_{ei} \mathcal{L} \left(g_n^{(0)} + f_* \right) \right\rangle , \qquad (109)$$

where electron-electron collisions do not contribute directly, because of momentum conservation, but they do contribute indirectly through the function f_*. For example, in the usual units, we have

$$L_{13} = -1.46 \left(1 + \frac{0.67}{Z} \right) \frac{\delta^{1/2} n_e c}{B_\theta} , \qquad (110)$$

in which the term proportional to $1/Z$ is the electron-electron contribution.

The banana regime transport coefficients, with the correct numerical coefficients, were calculated first by Rosenbluth, Hazeltine and Hinton.[10]

PRESENT STATUS OF NEOCLASSICAL THEORY
ION THERMAL CONDUCTIVITY

Neoclassical theory was summarized in 1976 by Hinton and Hazeltine[14] in an article in *Reviews of Modern Physics*. Modifications of the pure plasma results due to impurities were summarized by Hirshman and Sigmar,[15] in a 1981 *Nuclear Fusion* review paper. A number of other effects have been found which are of quantitative importance. Some of these effects are summarized as follows.

Work done by Bolton and Ware[16] showed that finite aspect ratio effects increased the ion thermal conductivity. The result was incorporated into a simple fit formula (of the type used by Hinton and Hazeltine[14]) by Chang and Hinton,[17] which gives an enhancement of roughly a factor of two. A later modification of this formula by Chang and Hinton[18] to include impurities gives a further enhancement which is larger than a simple Z_{eff} multiplier.

Comparison of existing theory with experiment shows good agreement under some conditions and only order-of-magnitude agreement under others. In Alcator C, with pellet injection, the agreement was good, while with gas fueling, the ion thermal diffusivity was deduced to be considerably larger than neoclassical, typically by a factor of four.[19] A possible mechanism has been proposed: viz., turbulence due to the "eta-i" mode, which would be expected to be suppressed with pellet injection, leaving only neoclassical ion thermal transport. In Doublet III, measurements of the radial ion temperature profile led to radial profiles of the ion thermal diffusivity which were significantly different from the profiles expected from neoclassical theory; the magnitude of the diffusivity was considerably larger in the region outside the half minor radius.[20]

In addition to the effects mentioned above, there are other neoclassical effects which may be of quantitative importance. The effects of rotation and poloidal electric fields were first considered by Kovrizhnykh[21] and by Hazeltine and Ware.[22]

More recently, C.S. Chang[23] has shown that the banana regime ion thermal conductivity increases with increasing magnitude of the effective trapping potential (the combined centrifugal and electrostatic potentials), although the increase is generally less than a factor of two.

Energetic ions can have very important effects, especially in plasmas with high power neutral beam heating, but the necessary quantitative theoretical work has not yet been done. Because of finite ion orbit excursions, the distribution function is expected to be considerably distorted from a Maxwellian when the average ion energy is a strong function of radius, even for ohmically heated plasmas, as pointed out by Ware.[24] This effect could be considerably enhanced with neutral beam heating or ion cyclotron heating, which directly affect the ion tail. Since it is distortions from Maxwellian that drive the ion neoclassical transport, the magnitude of the transport is expected to be significantly affected when these additional distortions are included. The effects of ripple or stationary magnetic islands on ion thermal transport in tokamaks have not been thoroughly explored. Approximate analytic calculations do indicate that they increase dramatically with ion temperature, however. These effects are expected to be considerably larger for the tail of the ion distribution function than for the bulk.

I recently showed that poloidal asymmetries in the boundary conditions at the edge of the plasma, of the kind that would be expected with divertor operation, can have a significant effect on the ion transport rate near the plasma edge.[25] In particular, the direction of the ion grad-B drift relative to the position of the X-point, with single null operation was shown to be quite important. Wagner[26] found that this correlates well with the threshold power needed for the H-mode of improved confinement in ASDEX, which indicates that the ions are playing an important (although not understood) role in the dominant electron thermal transport. If the direction is chosen as it was in the Doublet III device, (in which the H-mode was not obtained), the predicted effect is to increase the ion thermal transport and raise the H-mode power threshold. The toroidal field direction was reversed in the new Doublet-III-D device, and the H-mode was obtained with power comparable to that needed for the H-mode in ASDEX.[27] When the direction was temporarily changed back to the previous direction, the H-mode could not be obtained even with more power.

BOOTSTRAP CURRENT

In the past, attempts to observe the bootstrap current in a tokamak have failed. However, the theory discussed here has been worked out under the assumption of an axisymmetric and quiescent background, which is usually unrealistic. Recently, the bootstrap current was identified in TFTR,[28] in the "super-shot" plasmas when sawteeth were not present, under conditions closer to those assumed theoretically, and confinement times were better than normal.

SUMMARY

The results of the neoclassical theory of transport are still being used to give the lower limit on the transport rates in tokamaks, which would apply if instabilities and turbulence could be suppressed. So far, only the ion thermal conductivity and the current density have been found experimentally to agree with this theory, and only under special conditions. The electron thermal conductivity has been found experimentally to be much larger than the neoclassical prediction.

ACKNOWLEDGMENT

This is a report of work sponsored by the U.S. Department of Energy under Contract No. DE-AC03-84ER53158.

REFERENCES

1. M.N.Rosenbluth, W. MacDonald, and D. Judd, *Phys. Rev.* **107**, 1 (1957).
2. M.N. Rosenbluth, "Diffusion and Ohmic Heating in Plasma Magnetic Field Systems," GA Technologies Report GA-99 (April 15, 1957).
3. M.N. Rosenbluth and A.N. Kaufman, *Phys. Rev.* **109**, 1 (1958).
4. L. Spitzer Jr. and R. Harm, *Phys. Rev.* **89**, 977 (1953).
5. S.I. Braginskii , in *Reviews of Plasma Physics*, edited by M.A. Leontovich, (Consultants Bureau, New York, 1965), Vol. 1, p. 205.
6. A.A. Galeev and R.Z. Sagdeev, *Sov. Phys.-JETP* **26**, 233 (1968).
7. A.A. Ware, Phys. Rev. Lett. **25**, 916 (1970).
8. A.A. Galeev, *Sov. Phys.-JETP* **32**, 752 (1971).
9. R.J. Bickerton, J.W. Connor, and J.B. Taylor, *Nature Phys. Sci.* **229**, 110 (1971).
10. M.N. Rosenbluth, R.D. Hazeltine, and F.L. Hinton, *Phys. Fluids* **15**, 116 (1972).
11. F.L. Hinton and M.N. Rosenbluth, *Phys. Fluids* **16**, 836 (1973).
12. M.N. Rosenbluth, P.H. Rutherford, J.B. Taylor, E.A. Frieman, and L.M. Kovrizhnikh, in *Plasma Physics and Controlled Nuclear Fusion Research 1970* (IAEA, Vienna, 1971), Vol. I, p. 495.
13. F.L. Hinton and S.K. Wong, *Phys. Fluids* **28**, 3082 (1985).
14. F.L. Hinton and R.D. Hazeltine, *Rev. Mod. Phys.* **48**, 239 (1976).
15. S.P. Hirshman and D.J. Sigmar, *Nucl. Fusion* **21**, 1079 (1981).
16. C. Bolton and A.A. Ware, *Phys. Fluids* **26**, 459 (1983).
17. C.S. Chang and F.L. Hinton, *Phys. Fluids* **25**, 1493 (1982).
18. C.S. Chang and F.L. Hinton, *Phys. Fluids* **29**, 3314 (1986).
19. M. Greenwald *et al.*, in *Plasma Physics and Controlled Nuclear Fusion Research 1986* (IAEA, Vienna, 1987), Vol. I, p. 139.
20. R.J. Groebner, *Nucl. Fusion* **26**, 543 (1986).
21. L.M. Kovrizhnykh, *Sov. Phys.-JETP* **35**, 709 (1972).
22. R.D. Hazeltine and A.A. Ware, *Phys. Fluids* **19**, 1163 (1976).
23. C.S. Chang, *Phys. Fluids* **26**, 2140 (1983).

24. A.A. Ware, *Phys. Fluids* **27**, 1215 (1984).
25. F.L. Hinton, *Nucl. Fusion* **25**, 1457 (1985).
26. F. Wagner, *Nucl. Fusion* **25**, 1490 (1985).
27. J.L. Luxon *et al.*, in *Plasma Physics and Controlled Nuclear Fusion Research 1986* (IAEA, Vienna, 1987), Vol. I, p. 159.
28. R.J. Hawryluk *et al.*, in *Plasma Physics and Controlled Nuclear Fusion Research 1986* (IAEA, Vienna, 1987), Vol. I, p. 51.

John M. Dawson
Department of Physics
University of California, Los Angeles
Los Angeles, California 90024

Rosenbluth-Liu Amplitude Limit for the Beat Wave Accelerator

The controversy about the correct prediction for the frequency shift in a beat wave accelerator is reviewed and then resolved in favor of the Rosenbluth-Liu result. Simulations also show agreement with this theory.

INTRODUCTION

In the early 1970's with the advent of strong programs on laser fusion, there was a great deal of theoretical work on parametric instabilities. Like so many other areas of plasma physics, Marshall Rosenbluth was right there working on the subject and writing some papers that have become classics. One of these was his paper[1] with C.S. Liu on the "Excitation of Plasma Waves by Two Laser Beams." I rather suspect that Marshall regards this as a minor paper compared to many of the other

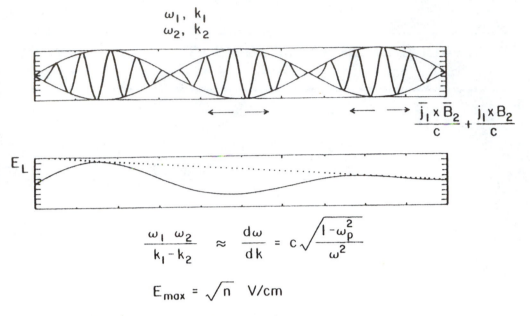

$$\frac{\omega_1 \, \omega_2}{k_1 - k_2} \approx \frac{d\omega}{dk} = c\sqrt{1 - \frac{\omega_p^2}{\omega^2}}$$

$$E_{max} = \sqrt{n} \quad V/cm$$

FIGURE 1 The beat wave excitation mechanism.

things he has done, but like so many other things it has come to be a classic; it has become a fundamental paper in beat wave accelerator research.

A few years ago some controversy developed about the result; there were a number of different papers on the subject all giving different answers. It turned out that the Rosenbluth-Liu answer was correct. However, the resolution of the controversy involves some subtle points of physics, which I would like to describe.

REVIEW

First I might briefly summarize what Rosenbluth and Liu did. The situation is illustrated in Fig. 1. They looked at two colinear laser beams propagating through an infinite uniform plasma. The two waves had frequencies ω_1 and ω_2 and wave numbers k_1 and k_2. A plasma, like any medium, is nonlinear, and the beating produces a force on the electrons in the direction of propagation, which is at frequency $\omega_1 - \omega_2$ and wave number $k_1 - k_2$. The force comes from $[\mathbf{j}_1 \times \mathbf{B}_2 + \mathbf{j}_2 \times \mathbf{B}_1]/c$, where the subscripts refer to the appropriate wave; also, we take $\omega_1 > \omega_2$. This force is sometimes called the ponderomotive force. If its frequency matches the plasma frequency, ω_p, then it resonantly drives plasma oscillations and their amplitude builds up to a large value. What Rosenbluth and Liu did was to calculate how the plasma

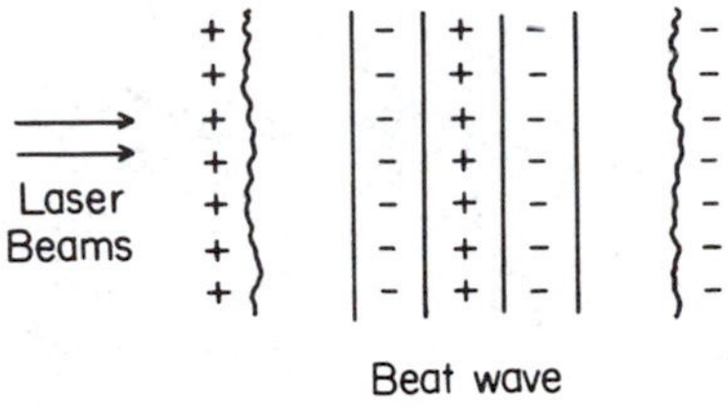

Beat wave

FIGURE 2 Plasma polarizes as electrons are pushed from the left to the right by the light pressure.

frequency shifts, due to the relativistic mass change of the electrons as the wave reaches large amplitudes, and how this mass change gives saturation by shifting the plasma frequency so that the plasma wave falls out of resonance with the beat frequency. In addition to the amplitude limit they calculated the amplitude build-up in time and its decrease after saturation. They worked in Lagrangian coordinates where the equations of motion take the particularly simple form given below:

$$\ddot{\xi} = -\omega_p^2 \left(1 - \frac{3}{2}\frac{\dot{\xi}^2}{c^2}\right)\xi + \frac{e^2 E_1 E_2}{2m^2 \omega_1 \omega_2 k_p} \sin(\Delta k x - \Delta \omega t), \tag{1}$$

where ξ is the displacement of the fluid element from its equilibrium position. The two laser electric fields are given by

$$\mathbf{E}_{1,2} = \mathbf{e}_z E_{1,2} \sin(k_{1,2}x - \omega_{1,2}t). \tag{2}$$

Here, k_p is ω_p/c, and Δk and $\Delta \omega$ are given by

$$\Delta k = k_1 - k_2$$

$$\Delta \omega = \omega_1 - \omega_2, \tag{3}$$

where we may take $\omega_1 > \omega_2$. The $E_1 E_2$ term on the right-hand side of Eq. (1) is the $\mathbf{j} \times \mathbf{B}$ interaction written in terms of the electric fields, and the factor $\left(1 - \frac{3}{2}\dot{\xi}^2/c^2\right)$ gives the lowest order reduction of the plasma frequency due to the relativistic mass increase.

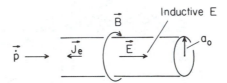

FIGURE 3   As electrons start to move due to momentum transfer $\dot{P}$, a current is set up producing a **B** field and an inductive **E** which opposes the electron drift.

On the other hand, many of the other authors used Eulerian equations of motion and the two time and space scale approach of Bogoliubov and Metropolskii to solve for the motion. This gives the following equation for the electric field:

$$E_x + \omega_p^2 \left(1 - \frac{3}{2} V_x^2 / c^2 \right) E_x = \frac{e E_1 E_2 \omega_p c}{2 m \omega_1 \omega_2} \sin(\Delta k x - \Delta \omega t)$$

$$+ 4\pi e \left[\frac{\partial}{\partial t} (n - n_o) V_x - n_o V_x \frac{\partial V_x}{\partial x} \right] \qquad (4)$$

The second term on the left-hand side of Eq. (4) is related to the relativistic plasma frequency, and the first term on the right-hand side is the beat wave forcing function.

The second term on the right-hand side of Eq. (4) is the controversial one, coming from the requirement that $v_2 = 0$, and leads to the difference in the predicted frequency shift. The two frequency shifts are

$$\Delta \omega = -\frac{3}{16} \epsilon^2 \omega_p \quad \text{(Lagrangian)}$$

$$\Delta \omega = +\frac{5}{16} \epsilon^2 \omega_p \quad \text{(Eulerian)} \qquad (5)$$

where $\epsilon = e E_x / m \omega_p c$.

RESOLUTION OF THE CONTROVERSY

There is a simple physical explanation for the difference. In the Lagrangian treatment the mean motion of the electrons is zero, whereas in the Eulerian treatment the requirement that the second-order velocity is zero is imposed. These are different conditions since the mean second-order flow velocity is $(< n_1 v_1 > + n_o v_2)/n_o$ and not v_2. The Eulerian treatment involves a recoil of the whole plasma due to the energy it absorbs from the light wave. We can calculate this recoil from energy momentum considerations. The plasma wave energy density is given by

$$W = 2n_o m \frac{1}{2} \left(\frac{eE_L}{m\omega_p} \right)^2 = \frac{n_o mc^2 \epsilon^2}{2}. \tag{6}$$

Since the energy is absorbed from the light wave which also carried momentum, there must be momentum transferred to the plasma; the momentum density is

$$P = \frac{W}{c} = \frac{n_o mc^2 \epsilon^2}{2c}. \tag{7}$$

If we assume that this momentum resides in the electrons, then they will have a mean velocity $\bar{v}$ given by

$$n_o m \bar{v} = P = \frac{n_o mc \epsilon^2}{2}. \tag{8}$$

This drift of the plasma will Doppler shift the frequency of the plasma wave by

$$\delta \omega_{ps} = k_p \bar{v} = \frac{\omega_p \epsilon^2}{2}. \tag{9}$$

We see that the following relation holds

$$\text{Doppler Shift} + \text{Lagrange Shift} = \text{Euler Shift}. \tag{10}$$

Thus we understand the difference between the two results, and we might expect on physical grounds that the Eulerian shift is correct. Is this the case? To answer this, we must look deeper into the problem;[2] more details can be found in the papers of Mori[3] and of McKinstrie and Forslund.[4]

If we consider a strictly one-dimensional situation as shown in Fig. 2, then no current (electron drift) can be set up since the plasma will polarize and the resultant electric field will stop the drift. Since the zero current condition is

$$n_1 v_1 + n_o v_2 = 0 \tag{11}$$

and not

$$v_2 = 0 \tag{12}$$

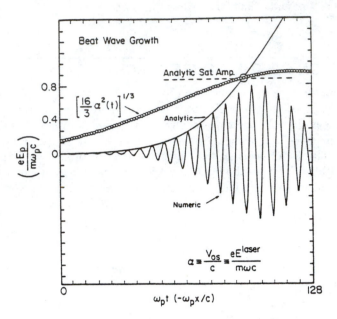

FIGURE 4 Beat wave growth and saturation in a 1-D simulation ("Numeric"), compared with the Rosenbluth-Liu prediction ("Analytic")—modified to include the laser rise time dependence. [From Ref. 5]

as is assumed in the Eulerian treatment, condition (11) should be applied; if this is put in the Eulerian treatment, it leads to the Rosenbluth-Liu result. In this case the momentum is transferred to the ions through the electric field.

For two- or three-dimensional situations (as would occur in real experiments) the polarization can be shorted out by a return current around the outside of the laser column as shown in Fig. 3. In this case you might expect that the electrons could be set in motion and the results of the Euler treatment would be correct. However, if we set up currents, these produce magnetic fields that result in inductive electric fields, which tend to oppose their build-up. This effect can roughly be treated by giving the electrons an effective mass equal to

$$m^* \simeq \frac{a_o^2 \omega_p^2}{c^2} m, \tag{13}$$

where a_o is the radius of the current-carrying channel. From two-dimensional simulations that we have carried out (see Fig. 6), we find typical stable channel sizes of

$$\frac{a_o \omega_p}{c} \simeq (1 \sim 2)\pi. \tag{14}$$

These give effective masses m^*/m of 10 to 50, so we see that the effective drift is very small compared to those predicted by the old Eulerian theory. Thus, the result of Rosenbluth and Liu is substantially correct.

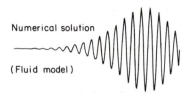

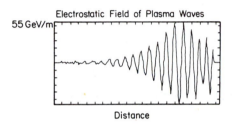

FIGURE 5 Comparison of beat wave potential calculated from fluid model and from particle simulations. [From Ref. 5]

SIMULATION STUDIES

At UCLA we have carried out some analytic and numerical studies of the beat wave build-up. First, according to the calculation of Rosenbluth and Liu, the saturation time and saturation amplitude are given by

$$\tau_s \propto \left(\frac{1}{\alpha_1 \alpha_2^2}\right)^{1/3} \omega_p^{-1} \tag{15}$$

$$E_s \propto (\alpha_1 \alpha_2)^{1/3} \tag{16}$$

where $\alpha_i = eE_i/m\omega_i c$; here, i refers to the ith laser beam. What happens if the laser beams rise in intensity over a finite period of time? If the rise time is short compared to τ_s, then τ_s is the correct answer. Tom Katsouleas has carried out an analysis of this.[5] Figure 4 shows some of his results: plotted are the instantaneous Rosenbluth-Liu saturation limit, $[16\alpha^2(t)/3]^{1/3}$; his analytic calculation of the amplitude, omitting saturation effects; and results from numerically calculating the

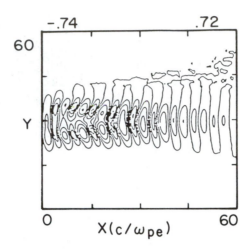

FIGURE 6   Potential contours for a 2-D beat wave simulation.

beat wave build-up, including the relativistic mass effect. We see that one gets very nearly the correct answer by finding the point at which the analytic build-up (omitting saturation effects) crosses the instantaneous Rosenbluth-Liu limit. Figure 5 shows a comparison of the results from the fluid model used by Katsouleas and from one-dimensional particle simulations; the agreement is quite good.

Finally, Fig. 6 shows some results from two-dimensional simulations carried out by Warren Mori.[6] Shown are potential contours for the beat wave. The lasers are of finite width centered at $30c/\omega_p$ and propagating to the right. We see from this figure that the beat wave peaks up first on axis where the laser intensity is highest and then dies down as predicted by theory. The amplitude of the beat wave peaks at later times in the wings of the laser beams where the intensity is lower. This would also be predicted by theory if we could apply the one-dimensional theory locally. It appears that the local approximation is a reasonable one. The observations are in agreement with the theory of Rosenbluth and Liu.

ACKNOWLEDGEMENTS

It is a pleasure for the author to acknowledge the help of Tom Katsouleas and Warren Mori, whose work forms the basis for this paper.

REFERENCES

1. M.N. Rosenbluth and C.S. Liu, *Physical Review Letters* **29**, 701 (1972).
2. The controversy originates from R. Bingham, R.A. Cairns, and R.J. Evans, Rutherford Appleton Laboratory (England), Report No. RAL-84-12 (1984).
3. W.B. Mori, in *Rep. Workshop on Interactions and Transport in Laser Plasmas*, ed. M.G. Haines (CECAM, Orsay, France, 1985), p. 124.
4. C.J. McKinstrie and D.W. Forslund, *Physics of Fluids* **30**, 904 (1987).
5. T. Katsouleas, C. Joshi, J. M. Dawson, F. F. Chen, C. E. Clayton, W. B. Mori, C. Darrow, and D. Umstadter, in *Laser Acceleration of Particles*, AIP Conf. Proc. No. 130, edited by C. Joshi and T. Katsouleas (American Institute of Physics, New York, 1985), p. 63.
6. W.B. Mori, *IEEE Trans. Plasma Sci.* **PS-15**, 88 (1987).

Chuan-Sheng Liu
Department of Physics and Astronomy and
  Laboratory for Plasma and Fusion Energy Studies
University of Maryland
College Park, Maryland 20742

Parametric Instabilities
in Laser Plasma Interaction

Theoretical studies of parametric instabilities in the physics of laser-plasma interactions—in particular, beat wave excitation, stimulated Raman scattering, Raman scattering with a random pump, and nonlinear absolute Raman instabilities at quarter critical density—are reviewed and then compared with experimental observations.

INTRODUCTION

The interaction of light with matter is always a subject of great fascination. With the advent of powerful lasers, its interaction with plasmas becomes a most interesting and important nonlinear problem, because of the richness of collective modes in plasmas and the wide ranging applications, particularly the possibility of laser fusion. I was very fortunate when Marshall Rosenbluth asked me to look into laser plasma interaction physics with him shortly after I arrived at the Institute for Advanced Study in 1971.

To work with Marshall Rosenbluth is like a tour of a grand intellectual landscape, led by a master teacher—to see the height of creativity and the depth of understanding, to witness a most creative mind at work, to share his joy in doing physics, particularly the physics of controlled fusion, and to experience the warmth of his friendship.

The theoretical studies of parametric instabilities that were developed during these years, starting with Rosenbluth's pioneering work on parametric instabilities in inhomogeneous plasmas, have since been extensively investigated analytically, computationally, and experimentally. They have come to form the basis of much of our understanding of laser-plasma interactions in laser fusion experiments. Because a complete review is beyond the scope of this paper, I shall give a brief review of some of the highlights in what follows.

We are quite familiar with the linear dispersion properties of electromagnetic waves in plasmas:

$$k_0^2 c^2 = \omega_0^2 - \omega_p^2 / \left(1 + i\nu/\omega_0\right),\tag{1}$$

where $\omega_p = \left(4\pi n e^2/m\right)^{1/2}$ is the plasma frequency and ν is the electron-ion collision frequency. According to Eq. (1), plasma is transparent to light in the underdense region $\omega_p < \omega_0$ where light propagates, subject only to collisional damping (inverse bremsstrahlung). At the critical density $\omega_p = \omega_0$, light is reflected ($k = 0$). In the overdense region, the light wave is evanescent. Equation (1) is the basis for communication by radio frequency waves via the ionosphere and for determination of the electron densities in the magnetosphere and interstellar medium.

For intense, coherent laser light, its interaction with a plasma is dominated by the collective modes in the plasma, and the plasma becomes a nonlinear, parametric medium.[1] A light wave, when its intensity exceeds certain thresholds for parametric processes, can decay into the collective modes in plasmas—plasma waves, ion waves, or electromagnetic waves if the frequency and wavenumber matching conditions are satisfied:

$$\omega_0 = \omega_1 + \omega_2, \qquad \mathbf{k}_1 = \mathbf{k}_1 + \mathbf{k}_2.\tag{2}$$

Thus, at critical density $\omega_0 = \omega_p$, a light wave can decay into a plasma wave (plasmon) and an ion wave (phonon).[2] At quarter critical density $\omega_0 = 2\omega_p$, a light wave can decay into a pair of plasmons[3] (two plasmon decay) or a plasmon and a photon (Raman scatter).[4,5] Raman scattering can also occur in the underdense region[5,6] $2\left(v_e/c\right)\omega_0 < \omega_p < \omega_0/2$ where the lower limit on density is set by Landau damping. Similarly, a light wave can decay into another light wave and an ion wave[5] (Brillouin scattering) in the underdense region ($\omega_p < \omega_0$). Because these scattering processes have maximum growth rate for backscatter, these parametric instabilities can lead to substantial back and side scatter of laser light. Thus the plasma as a nonlinear, parametric medium behaves qualitatively different from a linear medium. Laser light can be anomalously absorbed at the critical density through parametric decay into plasmon and phonon and, at quarter critical density, through parametric decay into two plasmons. Light can be anomalously scattered by Raman and Brillouin instabilities. When the plasma wave is one of the daughter waves, a few electrons can be accelerated by these plasma waves to become very energetic electrons. These fast electrons can penetrate the core of a pellat, causing the core to preheat, thus adversely affecting its compression by ablation.[7]

BEAT WAVE EXCITATION

The problem we first looked into was the generation of plasma waves by beating two laser lights,[8] for possible heating of tokamak plasmas since $\omega_p \ll \omega_0, \omega_1$ in this case. Because the beat wave has phase velocity near the velocity of light, $\Delta\omega/\Delta k \approx c$, we used Dawson's formalism for plasma waves with Lagrangian displacement[9] ξ_x of the electrons and included the relativistic effect, anticipating that as the oscillating velocity becomes comparable to the speed of light, the nonlinear plasma frequency shift due to the relativistic mass correction may detune the resonance, thereby saturating the growth. The equation for the driven plasma wave is

$$\frac{d}{dt}\left[\frac{\dot{\xi}_x}{\sqrt{1 - \dot{\xi}_x^2/c^2}}\right] = \frac{e}{mc}\left[v_{0y}B_{1x} + v_{1y}B_{0x}\right], \tag{3}$$

where $v_{jy} = (e/2m\omega_j)\{E_j \exp[i(k_j x - \omega_j t)] + \text{c.c.}\}$ with $j = 1$ or 2, is the electron oscillating (quiver) velocity in the laser electric field E_j, and B_k is the magnetic field of laser light k. Writing $\xi_x = A(t)\sin[kx - \omega_p t + \phi(t)]$, we find

$$\dot{a} = \sin\phi, \qquad a\dot{\phi} = \cos\phi + \delta a^3, \tag{4}$$

where $a = A/\lambda_0$, $\delta = 3\omega_p^3 \lambda_0^2/16c^2$, and $\lambda_0 = (v_1 v_2/c^2)(\omega_p/4k_0)$, with $\omega_p = k_0 c$. An exact solution of Eq. (4) gives the relativistic saturation as

$$Ak_0 = \left[\frac{16}{3}\left(\frac{v_1 v_2}{c^2}\right)\right]^{1/3}. \tag{5}$$

This limit has recently been verified in extensive computer simulations. Recent interest in this work is due to the recognition of the very intense electric field produced in the beat plasma wave and its possibility for efficient electron acceleration. This is discussed in more detail in the paper by John Dawson.[10]

STIMULATED RAMAN SCATTERING

Rather than reviewing the whole range of parametric instabilities, which is beyond the scope of this paper, I will discuss simulated Raman scattering in some detail because of the great deal of recent experimental and theoretical interest in this process. After completing the work on beat wave generation theory, we went on to study the parametric instability of Raman scattering by allowing one of the EM wave amplitudes to change as it is driven by the nonlinear current source, which is the product of the density fluctuation δn_p of the plasma wave and the electron oscillating velocity in the pump $\vec{v}_0 : \vec{j}_{NL} = e\delta n_p \vec{v}_0$ with the Maxwell equation,

$$-\nabla \times \nabla \times \vec{E}_1 = 4\pi \dot{\vec{j}}_L/c^2 + 4\pi \dot{\vec{j}}_{NL}/c^2 - \frac{1}{c^2}\frac{\partial^2 E_1}{\partial t^2}.$$

With the scattered EM wave driven by this nonlinear current source, and the plasma wave driven by the ponderomotive force due to the beating of the pump and scattered EM wave, we now have a complete feedback loop for the parametric instability. The two coupled equations for the backscattered wave E_1^* and the plasma wave E_2 in an inhomogeneous plasma are[1,5]

$$\left[\frac{\partial^2}{\partial t^2} + \omega_{pe}^2(x) + \frac{i\nu_1}{\omega_1}\omega_{pe}^2(0) - c^2\frac{\partial^2}{\partial x^2}\right]E_1^* = \frac{e\omega_1}{m\omega_0}E_0^*\frac{\partial E_2}{\partial x} \tag{6}$$

and

$$\left[\frac{\partial^2}{\partial t^2} + \nu_2\frac{\partial}{\partial t} + \omega_{pe}^2(x) - \frac{3}{2}v_e^2\frac{\partial^2}{\partial x^2}\right]E_2 = -\frac{e}{m}\frac{\omega_{pe}^2(0)}{\omega_0\omega_1}\frac{\partial}{\partial x}(E_0 E_1^*). \tag{7}$$

Here $\omega_{pe}(x)$ is the local plasma frequency, $v_e = (2T/m_e)^{1/2}$ is the electron thermal velocity, ν_1 and ν_2 are the damping rates, and

$$\vec{E}_1 = \hat{e}_z E_1(x,t)e^{-i\omega_1 t} + \text{c.c.}$$

$$\vec{E}_2 = \hat{e}_2 E_1(x,t)e^{-i\omega_2 t} + \text{c.c.}$$

and we assume E_1 is parallel to E_0 for simplicity. With the assumption that the amplitude is slowly varying in time, these equations become

$$\left[\frac{\partial}{\partial t} + \frac{\nu_1}{2}\frac{\omega_{pe}^2(0)}{\omega_1^2} + \frac{i\omega_1}{2}\left(1 - \frac{\omega_{pe}^2(x)}{\omega_1^2} + \frac{c^2}{\omega_1^2}\frac{\partial^2}{\partial x^2}\right)\right]E_1^* = -\frac{ieE_0^*}{2m\omega_0}\frac{\partial E_2}{\partial x} \tag{8}$$

and

$$\left[\frac{\partial}{\partial t} + \frac{\nu_2}{2} - \frac{i\omega_2}{2}\left(1 - \frac{\omega_{pe}^2(x)}{\omega_2^2} + \frac{3}{2}\frac{v_e^2}{\omega_2^2}\frac{\partial^2}{\partial x^2}\right)\right]E_2 = \frac{ie\omega_{pe}^2(0)}{2m\omega_0\omega_1\omega_2}\frac{\partial}{\partial x}(E_0 E_1^*). \tag{9}$$

For a density profile $n_0(x) = n_0(1 + x/L)$, we have $\omega_{pe}^2 = \omega_{pe}^2(0)(1 + x/L)$. In a homogeneous plasma, the growth rate with zero damping is

$$\gamma = \frac{k_2 v_0}{2}\left(\frac{\omega_p}{\omega_0 - \omega_p}\right)^{1/2},$$

where $\mathbf{k}_2 = \mathbf{k}_0 - \mathbf{k}_1$. Note that backscattering ($k_2 = 2k_0$) has the maximum growth rate, and forward scattering the minimum. For a laser intensity of 10^{15} watt/cm^2,

$\lambda_0 = 0.35\,\mu$, the growth rate is $\gamma \sim 2 \times 10^{-3}\omega_0$, or the exponential time is about 0.1 ps. Thus the laser light will be backscattered by the Raman parametric instability, with a downshifted frequency.

In laser-fusion experiments, however, there is a steep density gradient with density varying from zero to 100 times the solid density in $100\,\mu m$.[11] The wave vector of the plasma wave therefore must also change rapidly since $\omega^2 = \omega_p^2(x) + 3k^2(x)v_e^2 = \text{const}$ for a stationary medium. The three-wave resonance conditions in Eq. (1) can only be satisfied over a limited zone. Let $\mathbf{K}(x) = \mathbf{k}_0 - \mathbf{k}_1 - \mathbf{k}_p(x) = 0$ at $x = 0$. The width of the resonant zone x_T is then given by

$$K'(x)x_T^2 = \frac{dk_p}{dx}x_T^2 \simeq 1. \tag{11}$$

In a classic paper, Rosenbluth[12] showed that the saturation of the convective amplification of decay waves due to the propagation of the decay waves out of this resonant zone is

$$I = I_0 \exp\left(\pi\gamma_0^2/K'V_1V_2\right), \tag{12}$$

where V_1 and V_2 are the group velocities of the decay waves and γ_0 is the linear growth rate in a homogeneous plasma. From this, one obtains the convective threshold for parametric instability in an inhomogeneous medium as

$$\gamma_0^2/K'V_1V_2 > 1. \tag{13}$$

We also studied the space-time evolution of parametrically excited wave packets in an inhomogeneous medium, including wave damping, to see how the saturation given by Eq. (12) is reached.[13]

Because one of the decay waves in Raman scattering is the light wave with $V_2 = c$, the inhomogeneity imposes a severe threshold condition for Raman backscattering instability:[5,6]

$$\left(\frac{v_0}{c}\right)^2 k_0L > 1, \tag{14}$$

where $v_0 = eE_0/m\omega_0$ in the electron oscillating velocity in the laser field. The threshold intensity $I_{th}\,\left(\text{W}/\text{cm}^2\right) = 4 \times 10^{17}/\lambda_0 L$ (with λ_0 and L in microns). For $\lambda_0 = 0.35\,\mu$ and $L = 10^3\lambda_0$, we have $I_{th} = 3 \times 10^{15}\,\text{watt}/\text{cm}^2$.

At quarter critical density, scattered light wave is near its turning point with reduced group velocity, and the instability becomes absolute with a reduced threshold condition which can be obtained from Eqs. (8) and (9):[1,4,5]

$$\left(\frac{v_0}{c}\right)^{3/2} k_0L > 1. \tag{15}$$

The threshold intensity $I_{ts}\,\left(\text{W}/\text{cm}^2\right) = 5 \times 10^{16} L^{-4/3}\lambda_0^{-2/3}$ (with L and λ_0 in units of microns). For $\lambda_0 = 0.35\,\mu$ and $L = 10^3\lambda_0$, we have $I_{ts} = 5 \times 10^{13}\,\text{W}/\text{cm}^2$, nearly two orders of magnitude down from Raman backscattering.

The side scattered light is also near its turning point in the more underdense region, and the threshold is even more reduced for an infinite slab in the y-direction (side direction):[1,5]

$$\left(\frac{v_0}{c}\right)^{3/2} k_0 L > \frac{1}{4} \left(\frac{\omega_p}{\omega_0}\right)^{1/2}. \tag{16}$$

In a slab with finite length L_y, the additional condition

$$\left(\frac{v_0}{c}\right) k_0 L_y \left(\frac{\omega_p}{\omega_0}\right) > 1 \tag{17}$$

must also be satisfied for light to be sufficiently amplified before propagating out in the y-direction.

In a more realistic geometry of a spherical plasma with radius of curvature R and density scale length L, the Raman sidescatter threshold is found to be[14]

$$\left(\frac{v_0}{c}\right)^2 = 0.24 \left(\frac{c}{\omega_0 L}\right)^{4/3} \left(\frac{\omega_p}{\omega_0}\right)^{2/3} \left(1 - \frac{\omega_p^2}{\omega_0^2}\right)^{1/2} \frac{\left[1 + 2\frac{L}{R}\left(\frac{\omega_0^2}{\omega_p^2} - \frac{2\omega_0}{\omega_p}\right)\right]^{1/3}}{\left(2 - 2\frac{\omega_p}{\omega_0} - \frac{\omega_p^2}{\omega_0^2}\right)}, \tag{18}$$

where the third factor on the right-hand side is due to the local swelling of the pump (laser) wave field, and refraction of the laser light is taken into account.

Over the last two decades, many theoretical papers[15] further elaborated and refined on earlier calculations. Particularly interesting is the role of fast electrons in affecting both the threshold and the observed spectra of Raman scattering.[16]

RAMAN SCATTERING WITH PUMP OF RANDOM PHASE AND AMPLITUDE

The finite bandwidth of the laser frequency has an important stabilizing effect on parametric instabilities. Rosenbluth, Laval, Pellat, Pesme, Williams, and Romani[17] have developed an elegant theory for this effect. In addition, density fluctuations can cause a convective parametric instability to become absolute, as was found numerically by Nicholson and Kaufman.[18]

To study the effects of a pump with random amplitude on stimulated Raman scattering instabilities in an inhomogeneous plasma near the quarter critical density, where decay waves are near their turning point, we solved Eqs. (8) and (9) numerically with the following model for a random pump:[19]

$$E_0 = E_{00} \left\{ 1 + 2 \sum_{n=1}^{N} e^{-\alpha \varepsilon_n^2} \cos\left[\varepsilon_n \Delta\omega \left(t - \frac{x}{v_g}\right)\right] \right\},$$

where ε_n is a random number, $0 \le \varepsilon_n \le 1$. The effective bandwidth is $\Delta\omega_{\text{eff}} = \Delta\omega/2\alpha$ (at $\varepsilon_n = 1/2\alpha$, where the intensity in the frequency spectrum falls to $1/e$

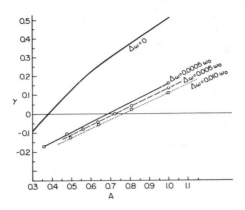

FIGURE 1 Growth rate of absolute Raman backscattering instability in units of $x_0\omega_p/2L$ as a function of laser amplitude $A = eE_0k_0x^2/mc^2$ for different $\Delta\omega/\omega$.

of its peak value). The results are then compared with a coherent pump with the same intensity,

$$E_0^2 = E_{00}^2 \left\{ 1 + 2 \sum_{n=1}^{N} e^{-2\alpha\varepsilon_n^2} \right\}.$$

Figure 1 shows the growth rate of the waves $\left(|E_1|^2 \right)^{1/2}$, $\left(|E_2|^2 \right)^{1/2}$ as a function of the pump intensity given as $A = eE_0k_0x_0^2/mc^2$, where $x_0 = \left[c^2 L/\omega_p(0) \right]^{1/3}$ for different values of $\Delta\omega_{\text{eff}}/\omega_0$. The growth rate is in units of $\left(x_0\omega_p/2L \right)$. Note that even for very small bandwidth $\Delta\omega/\omega_0 \sim 0.05\%$, there is already a factor of two reduction of the growth rate. The threshold intensity for the parametric instability is independent of $\Delta\omega/\omega$ for $\Delta\omega/\omega \lesssim 2\%$ and then increases with $\Delta\omega/\omega$, as shown in Fig. 2. In the latter case the threshold is

$$\left(\frac{v_0}{c} \right)^2 (k_0 L)^{2/3} > \frac{2^{2/5} 3^{1/3}}{20} (\Delta\omega/\omega_0).$$

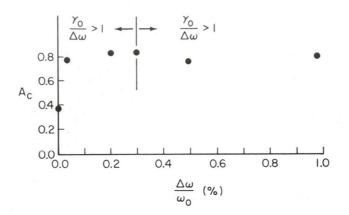

FIGURE 2 Threshold amplitude as a function of $\Delta\omega/\omega$.

NONLINEAR THEORY OF ABSOLUTE RAMAN INSTABILITIES AT QUARTER CRITICAL DENSITY

In an inhomogeneous plasma, the absolute Raman instability is localized in the resonant zone near the quarter critical density. When the daughter waves grow to large amplitude, the ponderomotive pressure due to such localized wave fields would expel the plasma in the region of high field intensity, creating a density cavity.

The magnitude of the density depletion δn in the cavity can be estimated by balancing the thermal pressure with the ponderomotive pressure:

$$\frac{\delta n}{n_0} = -\frac{Z\left(|E_1|^2 + |E_2|^2\right)}{8\pi n\left(ZT_e + T_i\right)}, \tag{19}$$

where Z is the average ionic charge, and E_1 and E_2 are the amplitudes of the scattered EM wave and plasma waves, respectively. The nonlinear frequency shift of the plasma wave is due to the density depletion: $\delta\omega = (1/2)(\delta n/n)\omega_p$. When the nonlinear frequency shift reaches the same order of magnitude as the growth rate of the Raman instability at quarter critical density, one expects the growth due to the instability to saturate:[20]

$$\frac{\delta\omega}{\omega_p} \approx \frac{\delta n}{n_0} \sim \frac{Z\,|E_1|^2}{8\pi n\left(ZT_e + T_i\right)} \sim \frac{Z\,|E_2|^2}{8\pi n\left(ZT_e + T_i\right)} \sim \frac{\gamma_0}{\omega_p} \sim \frac{v_0}{c}. \tag{20}$$

To test these ideas, the coupled equations (8) and (9) were numerically solved[21] with

$$\omega_p^2(x) = \omega_p^2(0)\left[1 + \frac{x}{L} + \frac{\delta n}{n}\right], \tag{21}$$

where $\delta n/n$ is the nonlinear density modification due to the ponderomotive force and is given by Eq. (19). The parameters used in the numerical solution were those for $1\,\mu$m laser light impinging on a plasma with electron temperature $T_e = 600\,\text{eV}$, plasma scale length $300\,\mu$m, and effective ion charge $Z = 6$. The threshold for Raman instability at $(1/4)n_c$ due to inhomogeneity is $I_0 = 1.1 \times 10^{14}\,\text{watt/cm}^2$.

The time evolution of the absolute Raman instability shows three types of characteristic behavior depending on the laser intensity:

1. For laser intensity just above threshold $I \geq I_0$, the parametric instability saturates at a low level and evolves into a coherent nonlinear state (a fixed point in the phase space of the real and imaginary parts of the amplitude E). The level of the density fluctuation at saturation is approximately given by Eq. (20), at which point the nonlinear plasma frequency shift is comparable to the growth rate. In this case the scattering is so feeble that reflectivity by Raman is negligible, i.e., less than 10^{-4}.

2. For laser intensity well above the linear threshold I_0, but still below the chaotic threshold I_{ch}, the Raman absolute mode grows up to large amplitude at the quarter critical density, and not only significantly modifies the local density profile, but also propagates down to regions of much lower density to affect the density profile there. The turbulence, together with the density modification, destabilizes the Raman mode at lower density values that were initially below the threshold. This is reminiscent of observations of the simultaneous occurrence of $(1/4)n_c$ absolute and underdense Raman scattering in many experiments. This nonlocal, turbulent destabilization produces intense fluctuations which, remarkably, after some duration τ, diminish and settle down to a coherent stable state again, not too different from Case 1. This nonlinear turbulent state is termed the chaotic transient, during which there is strong backscattering and intense plasma waves.

3. For $I \gtrsim I_{ch}$, the duration of the chaotic transient effectively becomes infinite when the laser intensity approaches I_{ch} (Fig. 3). Large backscatter and plasma waves persist throughout, with scintillation of Raman backscatter as seen in the experiments. This is indeed chaos, quite similar to the crisis route to chaos discovered by Grebogi, Ott, and Yorke[22] in studying nonlinear dynamics by mappings in which the system makes a transition from a stable fixed point through a chaotic transient to chaos.

EXPERIMENTAL OBSERVATIONS OF PARAMETRIC INSTABILITIES

In recent years, there have been extensive experimental observations of various parametric decay processes, particularly Raman scatter and two plasmon decay.[23-26]

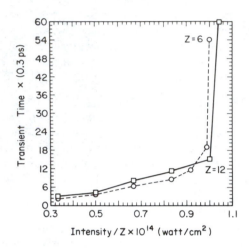

FIGURE 3 Duration of the turbulent transient as a function of laser intensity/atomic number.

In both of these cases, intense plasmons are produced which can accelerate electrons to very high energy, causing preheat of the core plasma and rendering it more difficult to compress to the desired density for laser fusion. Furthermore, for laser wavelengths longer than $1\,\mu m$, Raman scattering also causes substantial reflection, thereby reducing the effective absorption.

Experimental observation of parametric instabilities has been greatly aided by the use of Thomson scattering to detect the density fluctuation of the plasma waves.[23,24] With Thomson scattering, experimentalists have been able to measure both the threshold and the growth rate of parametric processes with great accuracy.

High power laser-plasma interaction experiments carried out at Livermore by Drake, Turner, et al.[25,26] showed significant Raman back and side scatter for long scale length plasmas: up to 10% of the laser energy can be Raman scattered. In addition, the x-ray spectrum shows the existence of hot electrons with temperature 20–40 keV (i.e., speeds comparable to the plasma velocity of plasma waves), whose function is well-correlated with the fraction of Raman scattered light.

Another interesting experimental observation of Raman scattering is that, in a very underdense plasma, Raman backscatter seems to occur below the convective threshold for the average intensity.[27] One possible explanation is the self-focusing and filamentation of laser light into hot spots. This has been analyzed theoretically by Tripathi and myself.[28] The ponderomotive force of the intense light in the hot spot causes a cavity in the plasma density which, in turn, localizes the plasma wave to a few Debye lengths. This localized plasma wave limits the zone of parametric instability and reduces the growth rate in the hot spot. Thus, Raman scattering growth is only marginally affected by modulational instability. A hot spot in the laser, however, may still enhance the local growth. The more likely possibility is that the absolute Raman scattering at quarter critical density, which has a lower threshold than convective threshold, may modify the density profile in the more underdense region with substantial turbulence, which can trigger parametric Raman scattering in the underdense region, as shown in Sec. V.

ACKNOWLEDGEMENTS

This work has been supported by the Office of Naval Research and the Naval Research Laboratory.

REFERENCES

1. C. S. Liu, in *Advances in Plasma Physics*, Vol. 6 (John Wiley, New York, 1976), p. 121.
2. F. W. Perkins and J. Flick, *Physics of Fluids* **14**, 2012 (1971).
3. C. S. Liu and M. N. Rosenbluth, *Physics of Fluids* **19**, 967 (1976).
4. J. F. Drake, Y. C. Lee, P. K. Kaw, G. Schmidt, C. S. Liu, and M. N. Rosenbluth, *Physics of Fluids* **17**, 776 (1974); J. F. Drake and Y. C. Lee, *Physical Review Letters* **31**, 1197 (1973).
5. C. S. Liu, M. N. Rosenbluth, and R. B. White, *Physics of Fluids* **17**, 1211 (1974).
6. C. S. Liu and M. N. Rosenbluth, Institute for Advanced Study Report C00-3237-11 (1972).
7. W. L. Kruer in *Laser Plasma Interaction III*, edited by M. B. Hooper, SUSSP Publication, Edinburgh (1986); K. Estabrook and W. L. Kruer, *Physics of Fluids* **26**, 1892 (1983); W. L. Kruer, K. Estabrook, B. F. Lashinski, and A. B. Langdon, *Physics of Fluids* **23**, 1326 (1980) and **26**, 1982 (1983).
8. M. N. Rosenbluth and C. S. Liu, *Physical Review Letters* **29**, 701 (1972).
9. J. M. Dawson, *Physical Review* **113**, 383 (1959).
10. J. M. Dawson, this book, Chapter 8.
11. C. E. Max, in *Interaction of Laser-Plasma*, Les Houches School XXXIV, (North Holland, Amsterdam, 1982).
12. M. N. Rosenbluth, *Physical Review Letters* **29**, 5651 (1972).
13. M. N. Rosenbluth, R. B. White, and C. S. Liu, *Physical Review Letters* **31**, 1140 (1973).

14. C. S. Liu, M. N. Rosenbluth, and R. B. White, in *Plasma Physics and Controlled Nuclear Fusion Research 1974* (IAEA, Vienna, 1975), Vol. II, p. 515.
15. B. B. Afeyan and E. A. Williams, *Physics of Fluids* **28**, 3379 (1985).
16. A. Simon and R. W. Short, *Physical Review Letters* **53**, 1912 (1984).
17. G. Laval, R. Pellat, D. Pesme, A. Ramani, M. N. Rosenbluth, and E. A. Williams, *Physics of Fluids* **20**, 2049 (1977).
18. D. R. Nicholson and A. N. Kaufman, *Physical Review Letters* **33**, 1207 (1974).
19. P. Guzdar, Y. C. Lee, J. F. Drake, and C. S. Liu, "Stimulated Raman Scattering with Random Pump at Quarter Critical Density," University of Maryland Report No. LPF-86-047 (June, 1986).
20. H. H. Chen, C. S. Liu, C. Grebogi, and V. K. Tripathi, in *Plasma Physics and Controlled Nuclear Fusion Research 1978* (IAEA, Vienna, 1979), Vol. III, p. 181.
21. P. Guzdar, C. S. Liu, W. Shuy, and Y. C. Lee, "Nonlinear Raman Instabilities in Inhomogeneous Plasmas," manuscript in preparation.
22. C. Grebogi, E. Ott, and J. Yorke, *Physica* 7D, 181–200 (1983).
23. H. A. Baldis and C. J. Walsh, *Physical Review Letters* **47**, 1658 (1981).
24. G. McIntosh, J. Meyer, and Y. Zhang, *Physics of Fluids* **29**, 3451 (1986); W. Rozus, A. A. Offenberger, and R. Felosejers, *Physics of Fluids* **26**, 1071 (1983).
25. R. E. Turner, D. W. Phillion, E. M. Campbell, and K. G. Estabrook, *Physics of Fluids* **26**, (2) 579 (1983).
26. R. E. Turner *et al.*, *Physical Review Letters* **57**, 1725 (1986).
27. W. Seka, E. A. Williams, K. S. Craxton, L. M. Goldman, R. W. Short, and K. Tanaka, *Physics of Fluids* **27**, 2181 (1984).
28. C. S. Liu and V. K. Tripathi, *Physics of Fluids* **29**, 4188 (1986).

Norman M. Kroll
Department of Physics
University of California, San Diego
La Jolla, California 92093

Free Electron Lasers

The subject of this paper will be Marshall Rosenbluth's contributions to and impact on the development of the free electron laser.

BACKGROUND

As far as I know, Marshall Rosenbluth's first involvement in the free electron laser problem took place at a JASON summer study in 1978. Some very exciting experimental results had recently been obtained, which set the stage for this summer meeting. Although for a few years prior to this meeting there had been a good deal of theoretical discussion about free electron lasers, they were not being taken too seriously outside of the community of people directly working on them. However, the appearance of two papers changed the situation dramatically, which shows the importance of experimental results to confirm theoretical conjectures. The first was the observation of the amplification of CO_2 radiation.[1] The second was the first operation of a self-excited free electron laser oscillator.[2] The significant feature of that oscillator was its 3.4 micron wavelength—which was a very impressive three-order-of-magnitude decrease in wavelength compared to what had ever been achieved with a traveling wave tube (which is what a free electron laser really is).

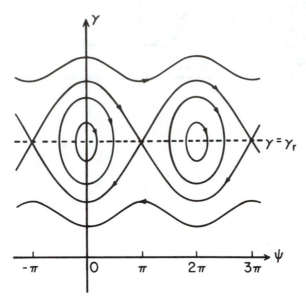

FIGURE 1   Particle orbits in the phase space of energy γ and phase ψ.

Also, that first device produced 7 kilowatts of peak power, which was also quite impressive. The efficiency, however, was not very impressive: only one hundredth of one percent.

The origin of that low efficiency was the central issue at the time of the summer study meeting. If one thinks of the free electron laser as a type of traveling wave tube, the efficiency is naturally low, because all traveling wave tubes operate by having a velocity resonance through which particles interact with some sort of electromagnetic wave. Thus, if the particles were to give up a significant fraction of their energy, they would fall out of resonance with the wave and no longer interact with it effectively. However, that is not actually the way in which things happen, because the resonant particles do not even lose enough energy to become non-resonant.

The actual situation is as follows. In a free electron laser, there is a wiggler magnetic field that combines with the signal field (the electromagnetic wave) to produce a ponderomotive potential wave, which travels down the laser with a characteristic velocity. The resonant electrons move with the velocity of this ponderomotive wave. Their energy is specified by the quantity $\gamma_r mc^2$, where γ_r is given by

$$\gamma_r = \left\{ \frac{k_s}{2k_w} \left[1 + \left(\frac{EB_w}{mc^2 k_\omega} \right)^2 \right] \right\}^{1/2}. \tag{1}$$

Note that the value of γ_r depends only on the choice of the signal frequency and on the parameters of the wiggler field structure. It is convenient to use the moving ponderomotive wave as the frame of reference. In this frame, the phase variable

$$\psi(z) = (k_w + k_s)z - \omega_s t(z) \qquad (2)$$

measures the position of the electron relative to the ponderomotive potential wells. The quantity ψ assumes the values of zero (as well as $\pm 2\pi$, $\pm 4\pi$, etc.) at the minima of those wells—all moving along with the velocity that corresponds to the energy γ_r. Shown in Fig. 1 is a phase space plot of energy γ versus ψ for the particle motion: it looks like the phase space plot for the motion of a simple pendulum, since it describes motion in a sinusoidal well that corresponds to the potential. The energy of individual particles moving in this potential oscillates back and forth. The untrapped orbits with $\gamma > \gamma_r$ correspond to particles that are running faster than the well, so they just zoom over the top. With higher and higher energies, their energy oscillates less and less. The untrapped orbits with $\gamma < \gamma_r$ correspond to particles that are running slower than the wells. The closed orbits correspond to particles that are trapped in the wells. All of the particles experience rather small energy excursions. From a simplistic consideration of these orbits, it is not obvious why one would expect any amplification at all.

The way that amplification does occur is illustrated in Fig. 2. This is a picture of only the separatrix of the phase space plots of Fig. 1, which defines the limits of a ponderomotive well within which electrons are trapped. For this case shown in Fig. 2, γ_r had the value $\gamma_r = 200$. In order to be able to extract energy from this device, the electrons must be introduced at an energy $\gamma_{initial}$ somewhat above γ_r, as shown in Fig. 2(a). The electrons arrive at arbitrary initial values of their phase variable, i.e., at arbitrary positions with respect to the well. Those that are outside the separatrix move along in a path that takes them over the top. However, those that are inside the separatrix start following confined orbits. The electrons all move with different velocities, but after a certain amount of time has passed, the situation becomes as shown in Fig. 2(b). The average energy has, in fact, decreased. However, if the interaction is not stopped at this time, the particles will move back up again and the energy will increase, as shown in Fig. 2(c). Indeed, on average, there is a small increase in the energy of the electrons that are not inside the well. Those inside the well have their average energy simply reduced to γ_r. One can do a little better than that by choosing where the interaction is terminated. Nevertheless, the efficiency of a device like this is intrinsically low.

The Stanford group, in their original paper on the oscillator,[2] suggested that efficiencies of the order of 20% might be achieved by recirculating the electrons in a storage ring. The small amount of energy lost by the electrons in passing through the wiggler could then be restored by passing them through an RF cavity, just as one replaces the energy lost to synchrotron radiation. It was noted in those days that a 1 GeV storage ring circulating one ampere would yield an average power of one megawatt if only one-tenth of a percent of the circulating power could be skimmed off.

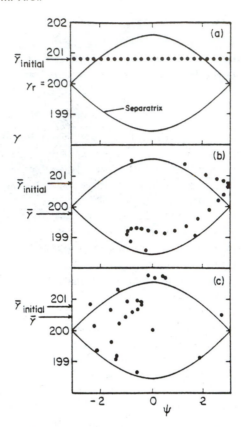

FIGURE 2 Simulation of the motion of particles injected with an initial energy, $\gamma_{initial}$, above $\gamma_r = 200$: (a) corresponds to the initial time, (b) and (c) to successively later times. [From Ref. 5]

It was this background that led to our being asked at that summer study to evaluate the potential of the free electron laser for the efficient generation of high average power in the visible region of the spectrum. By the time that meeting took place, it was already clear that the storage ring idea was in a bit of trouble. The difficulty can be seen from Fig. 2. Clearly, the energy has decreased by an amount slightly larger than $\gamma_{initial} - \gamma_r$. However, after the electrons have passed through the wiggler, their energy spread is on the order of the "width" of the pondermotive well. Hence the energy spread of the electrons that emerge is larger—in this case by a factor of two—than the mean energy loss. Every time the electrons pass through the wiggler, they are being heated but they are also losing some energy. There is a mechanism in storage rings for cooling electrons, viz., synchrotron radiation. Unfortunately, the synchrotron radiation energy necessary to overcome the energy spread acquired is much larger than the energy extracted by the FEL process.

Indeed, a pioneering analysis by Renieri[3] of the uniform wiggler in a storage ring had shown that the free electron laser output is of the order of the fractional energy acceptance of the storage ring, divided by the energy of the storage ring, times the synchrotron radiation rate. This is a number that is of the order of a percent or less. Of course, the synchrotron radiation rate must be paid for by the replacement of the synchrotron radiation with RF power, and in turn the RF power has to be generated. Hence there was no reason to think that a uniform wiggler would operate more efficiently in a storage ring than as a single pass device. Storage rings are a convenient means for obtaining current pulses, and this result did not imply that storage rings are not of interest for free electron lasers; but it did make it appear that high average power and high efficiency would not be easily obtained from them.

Nevertheless, during the early days of the summer study, a number of ideas were already being circulated about how to design wigglers with clever axial variation which would evade the Renieri limit. The study group that was put together at JASON also spent a good bit of time trying to think of some such schemes. However, there was a famous gain-spread theorem, proved by John Madey,[4] which indicated that an axially varying wiggler cannot be designed which, in the linear regime, will not produce energy spread. The formula for this theorem states that in the linear regime the energy gain is directly related to the derivative of the energy spread with respect to the energy. Thus, it seemed clear that all of the claims for designs for fancy wiggler fields that could beat this result and produce gain without energy spread were the result of incorrect approximations. The form of the Madey theorem actually implies that something close to the Renieri limit will apply to any quasilinear regime. During that summer, Marshall Rosenbluth produced a nice proof, based upon the Fokker-Planck equation, that one would indeed encounter this type of limit. Like many of his back-of-the-envelope calculations, it was never published. There appeared to be only one idea in this area which held out any hope of solving the storage ring problem, the transverse gradient wiggler, another invention of the Madey group. It evaded the gain spread relation by introducing transverse variations in the wiggler. At the time it appeared complicated and difficult to deal with analytically, but well worth further study. Since we won't have time to return to this topic later, I will mention that we did in fact study it over the next four summers and found, finally, that it had interesting and possibly useful properties, but that it also fails to evade the Renieri limit.

Before leaving the notion of the storage ring, let me mention that there was another idea at that time for improving the overall efficiency. According to that idea electrons would not be re-used, as is done in a storage ring, but rather their energy would be recovered after they had been sent through the wiggler. That idea has progressed far, and a number of concepts being pursued today make use of it. It is most effective, however, when combined with efficient first passes.

TAPERED WIGGLER

The idea of tapered wigglers was something that we discussed over coffee, while spending most of our time worrying about these storage ring problems. Eventually, however, we—Phil Morton, Marshall Rosenbluth, and myself—individually started doing preliminary calculations on this idea and almost immediately it was realized that this could be quite an attractive scheme. That was the origin of our work on the tapered wiggler and the beginning of Marshall Rosenbluth's serious involvement in the free electron laser problem.[5,6]

The idea for the tapered wiggler can be explained with the use of Figs. 1 and 2, which were described earlier. Once electrons are trapped in the potential wells, one can imagine manipulating them in such a way that they can be made to give up energy. For example, the simplest thing to do is simply to change the parameters of the wiggler field so as to reduce γ_r. If this can be accomplished in such a way that the trapped electrons stay trapped, then they will have to lose energy as γ_r decreases. The only time-dependent field that is present is the signal field, which contributes to the formation of the well, and therefore the electrons must transfer their energy to the signal field. This is the basic idea behind the tapered wiggler.

Actually, it is not necessary to change γ_r. Another method that can be used is to try to accelerate the electrons by putting them in an electric field. This could be a DC field or an RF field that is synchronized, so that on average it always tries to accelerate the electrons. The electric field tries to accelerate the electrons, but because they are trapped in the wells, they cannot move faster than the wells. Nevertheless, since this whole picture is moving to the right at the speed of the ponderomotive potential well, the electric field is still doing work on the electrons. Since the work done has to go somewhere, it goes into the signal field. This is the basic physics for this alternate idea.

Figure 3 illustrates how the tapered wiggler operates like a decelerating bucket. Initially, electrons in the free electron laser fill a certain region (the shaded rectangle) in $\gamma - \psi$ phase space. As γ_r is reduced, the separatrix (the tear-drop shaped curve) moves down, holding trapped electrons. From these decelerated trapped electrons, one can extract a substantial amount of energy. By starting this procedure in a gentle fashion, one can arrange to trap a fairly large fraction of the electrons.

Another way of representing the results, which has some interesting features, is given in Fig. 4, which shows the energy distribution of the electrons after they emerge from the wiggler as a function of the phase variable with which they enter. Notice, first of all, that the electrons that are not trapped actually gain a little energy. The ones that are trapped throughout the process lie on the bottom of the wells; according to the theory, there should be a large peak of energies at the bottom. The sloping gradient of the right-hand side comes about because the potential wells shrink (depending upon the design) as particles go through the wiggler. The shrinkage leads to detrapping of particles, so that a continuum of energies will be associated with the side of these wells. Notice, too, the small wiggles on the upper right side of the curves; the presence of these little scallops corresponds to peaks

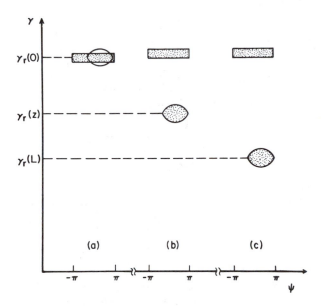

FIGURE 3 Schematic of the separatrix as a decelerating bucket: (a) refers to the initial situation, (b) and (c) to later times. [From Ref. 6]

in the electron distribution. These features are all confirmed in the experimental results.

Another feature in Fig. 4 is the difference between the three curves labeled (a), (b), and (c). The three curves correspond, respectively, to better extraction rates and, hence, to higher efficiencies: curve (a) indicates an efficiency of 18%, curve (b) an efficiency of 31%, and curve (c) an efficiency of 45%. However, the reason why one does not always design to achieve a high efficiency like 45% is that the circulating power in the optical resonator must be extremely high. Actually, in all three of the cases in Fig. 4, the power is quite high. The one with the lowest power, curve (c), still requires a circulating power of approximately 20 gigawatts (peak power). The result is expressed in terms of the power, rather than the power density, because it was assumed that the length of the device is limited by diffraction effects. Under that assumption, high powers will be involved.

The results shown in Fig. 4 apply to cases involving a high Q oscillator and in which the field does not change very much across the oscillator. If the field does change considerably — the simplest example of which would be an amplifier —

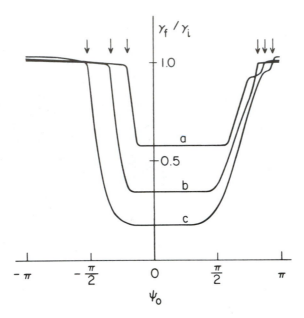

FIGURE 4   Energy spectrum of the emerging electrons, $\gamma_{\text{final}}/\gamma_{\text{initial}}$, as a function of the entry phase ψ_0. All electrons enter with energy equal to $\gamma_r(0)$. The three curves refer to minimum power configurations, with various efficiencies: (a) $P_{\text{cir}} = 21$ GW and $\eta_e = 18.1\%$, (b) $P_{\text{cir}} = 210$ GW and $\eta_e = 31.4\%$, (c) $P_{\text{cir}} = 1326$ GW and $\eta_e = 44.8\%$. The vertical dashed lines above the respective curves denote the positions of the trapping boundaries ψ_1 and ψ_2. [From Ref. 5]

then different results are obtained, although the basic idea is still the same. That is, the theory must be modified to describe buckets that grow. Figure 5 shows the results for an amplifier design that was published in the original paper. Again, note the very high values for the peak electron beam power. The input signal power is 640 megawatts. The gain in this amplifier is quite substantial, of the order of a factor of 100 in the power gain. With sufficient current to drive it, the device was designed for an input beam power corresponding to case (b), for which an efficiency of 20% was predicted. Increasing the input power by increasing the beam current leads to still higher efficiency because the buckets grow fast enough to capture some electrons which were not trapped initially. However, without enough beam power to drive this particular device, it peters out. At a certain point, the wells are no longer able to keep the electrons trapped; and when they detrap, the full extraction that is designed into the device cannot be attained.

The issue of high power is noteworthy, because the free electron device at Livermore has been able to improve over these results. The reason that device has

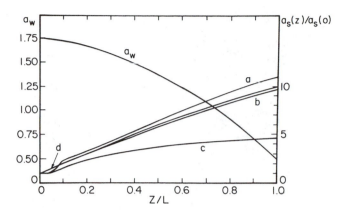

FIGURE 5   Simulated performance of a matched profile amplifier for various peak electron beam powers P_b and electronic efficiencies η_e: (a) $P_b = 350$ GW and $\eta_e = 21.4\%$, (b) $P_b = 295$ GW and $\eta_e = 20.3\%$, (c) $P_b = 145$ GW and $\eta_e = 9.1\%$. Curve (d) shows the wiggler field profile, $a_s(z)/a_s(0)$. The input signal power is 640 MW. [From Ref. 5]

done better is that its length is not limited by diffraction effects. It is a microwave device in a waveguide. Hence, one can obtain higher extraction figures with smaller input powers simply by making the device long compared to the diffraction length.

An interesting question is whether optical methods, which might be more convenient than a waveguide, could be used for the same purpose. The technique of optical trapping is one such notion and has been developed in a paper by Scharlemann, Sessler, and Wurtele.[7] The underlying idea is that the index of refraction along the laser beam is slightly larger than unity. Therefore, in principle, it may be possible for the electron beam to act as a light pipe for the laser light. (In our original paper there was a brief suggestion to the same effect, although we thought that the focusing effect was more likely to cause trouble than do any good. Others, however, took the idea much more seriously.) Figure 6, taken from the paper of Scharlemann et al.,[7] is a three-dimensional plot of laser intensity versus the distance along the wiggler, z, and the transverse distance, r. This result was obtained for a uniform wiggler, so it shows the trapping with an unamplified beam. Nevertheless, the beam is trapped, and its intensity is maintained over the entire length, even though it fluctuates some. That length is 60 Rayleigh lengths, which is sixty times longer than we would have assumed in our original power estimates. This may have very important implications either for the power levels that can be attained or, with practical power levels, for the extraction efficiencies that can be achieved.

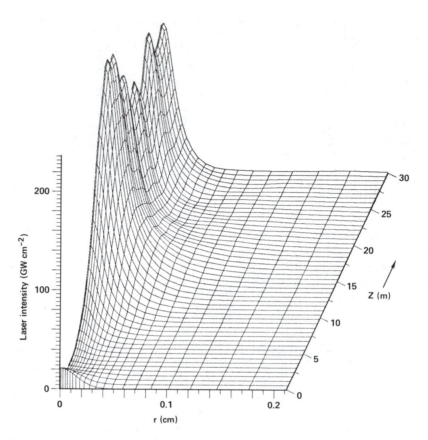

FIGURE 6 A three-dimensional plot of laser intensity versus the distance along the wiggler, z, and the transverse distance, r. [From Ref. 7]

SIDEBAND INSTABILITY

After we had the idea for the tapered wiggler and thought it looked rather attractive, we naturally began to worry about what might go wrong. It was quite in character for Marshall Rosenbluth to worry about instabilities. Indeed, shortly after our euphoria over this device, he brought up the issue of the sideband instability. As it has turned out, the free electron laser sideband instability is actually identical with the Kruer-Dawson-Sudan instability,[8] transferred from electrostatic waves to the ponderomotive potential of the free electron laser. Had we been aware of that work at the time, we might have avoided some imperfections in our original work on

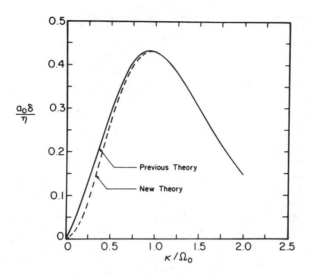

FIGURE 7 Sideband gain as a function of the frequency: κ is the sideband frequency and Ω_0 is the bounce frequency at the bottom of the well. [From Ref. 9]

the subject. The sideband instability is quite a knotty problem, and its theoretical study is still an active ongoing pursuit.

The origin of the sideband effect comes from the fact that the electrons trapped in the wells can slosh back and forth in them at a frequency that is often referred to as the synchrotron frequency because of the analogy of free electron lasers with accelerators. Plasma physicists would call it the bounce frequency. The particles do not all bounce back and forth at the same frequency because the wells are deep and the particles are not all trapped at the bottom. Thus, a wide range of frequencies is available. The oscillation is nonlinear for the same reason, and therefore harmonics also play a role. So, while the effect is loosely peaked at the synchrotron frequency, it is quite broad, as is illustrated in Fig. 7, which shows the sideband gain as a function of the frequency. Our original theory produced the curve labeled "previous theory," and some later improvements yielded the other curve.[9] The two curves are not very different. Sideband instability effects are also seen in numerical simulations of FEL operation.[9]

TABLE 1 Design characteristics for the Los Alamos free electron laser experiment.
[From Ref. 10]

Optical	
Wavelength	10.6 μm
Design Optical Power	500 MW
Strehl Ratio	0.5
Rayleigh Length	400 mm
Radius of Focus	0.16 cm
Electron Beam	
Peak Electron Current	20 amp
Electron Energy	20 MeV
Energy Spread	$\pm$0.5%
Emittance	2π mm·mrad
Wiggler	
Wiggler Length	100 cm
Length of Exit/Entrance Regions	5.0 cm
Taper in Wavelength	12%
Max./Min. Wavelength	2.7/2.4 cm
Number of Wiggler Cycles	40
Total Number of Magnets	314
Magnet Size	$0.5 \times 0.5 \times 3.5$ cm^3
Gap Between Magnets	0.88 cm
Peak Magnetic Field	0.31 T
Performance (Theoretical)	
Design Energy Extraction Efficiency	2.8%
Design Optical Gain (@ 500 MW)	2.4%
Small Signal Gain	3.1%

EXPERIMENTAL RESULTS

At this point, we will discuss experimental results that tend to confirm these theoretical ideas. A fairly large set of experiments over a broad range of wavelengths have been stimulated by these suggestions. Although many groups have made contributions, the discussion here will be limited to the work that has been done by the groups at Los Alamos and Livermore. The characteristics of the components that were used for the Los Alamos experiments[10] are listed in Table 1. The first experiments done at Los Alamos were designed to test the trapping and electron energy loss theory. Because it provides a large amount of input power, a CO_2 laser

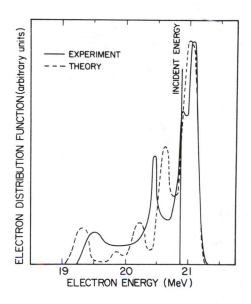

FIGURE 8 Energy distribution of the emerging electrons. The vertical line indicates the incident energy of the electrons. A bucket appears for an optical power near 0.2 GW. [From Ref. 10]

was used. The work was done at 20 MeV, with a wiggler whose length was 1 meter. The experiment required a laser with the kind of power needed to do the trapping. In keeping with the power that was available with a CO_2 laser, this device was designed to achieve an extraction efficiency that was rather modest compared to the large numbers discussed earlier. Nevertheless, it was sufficient to illustrate quite well the basic features of the theory.

Figure 8, a plot of the energy of the electrons that come out of the device, shows the first onset of trapping and energy degradation. Here the input power was 200 MW, which was just sufficient to begin to get some significant trapping. The peak at the lowest energy corresponds to those electrons which remain trapped in the bucket throughout the wiggler. The electrons which lose an intermediate amount of energy are those that detrap due to bucket shrinkage. The intermediate peak in the electron distribution correspond to the little shoulder pointed out in Fig. 4. As the buckets shrink, the particles loaded in buckets are swinging back and forth at the synchrotron oscillation frequency, and they tend to slosh over the top where

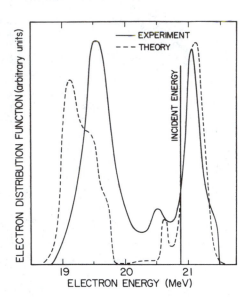

FIGURE 9 Same as Fig. 8, but for an optical power of 0.7 GW. More than 50% of the electrons are trapped. [From Ref. 10]

the phase ψ reaches its maximum value. The peak indicates that some bunching in ψ takes place after capture. The peak above the initial energy corresponds to particles which were never trapped. At higher input powers more complete trapping and hence more effective energy extraction is achieved. Figure 9 shows the results obtained with input power of 750 MW. The large peak at the lower energy shows that more particles are trapped, and the smaller number between the peaks shows that detrapping is less significant. The comparison with theory exhibited in Figs. 8 and 9 shows good qualitative agreement. The differences have been attributed to fluctuations in optical power, to fluctuations in initial energy, and to the neglect of lateral displacement in the electron motion.

Figure 10 shows how the amount of extraction efficiency that can be attained goes up as the optical power is increased. At about 900 MW, the extraction efficiency is up to a little under 4%. Again, the theoretical curve and the experimental

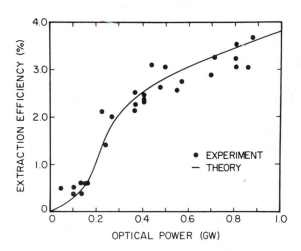

FIGURE 10 Extraction efficiency versus optical power. Saturation of the efficiency indicates the onset of electron trapping. [From Ref. 10]

results agree quite well and provide confirmation of the basic theory and of the practical consequences.

The Los Alamos device has since been operated as an oscillator, and there are some preliminary results. The 2% efficiency associated with the results shown in Fig. 11 is the highest efficiency that has been obtained on any self-excited oscillator. Although less than what is hoped for, it is still a great improvement over the original efficiency of one-hundredth of a percent. The onset of trapping can be seen in Fig. 11. The trapping is rather poor because the oscillator has not yet been operated up to the power for which it is designed. The problem is that the mirrors begin to burn at high powers. It is expected that the efficiency of this oscillator will improve to 4% when there is available a mirror design able to withstand the power that develops.

The existence of the sidebands has also been experimentally demonstrated. Spectral analysis of the output from the Los Alamos oscillators shows the appearance of the sideband, mostly on the longer wavelength side. Eventually the sideband becomes a dominant feature of the spectrum. Therefore, it is a very real effect. It may be possible to do something about the sideband problem by means of the introduction of some type of optical discrimination in the optical cavity.

TABLE 2 Features and parameters for the Electron Laser Facility (ELF) at Livermore.

Operates in an amplifier mode:
    $\lambda_s = 8.6$ mm
    P_{in} (interaction region) ~ 50 kW
    P_{in} (amplification mode) ~ 5 kW
Pulsed electromagnetic wiggler:
    $\lambda_w = 9.8$ cm
    $L_w = 3$ m
    $B_{w,max} = 5$ kG
    Linear polarized
    No axial magnetic guide field (horizontal focusing provided by external,
    horizontally focusing quadrupoles)
    Each two periods independently controlled
Interaction region:
    Oversized waveguide (3 cm × 10 cm)
    Fundamental excitation mode: TE_{01}
Electron beam characteristics:
    Beam energy $\cong 3.5$ MeV
    Field emission cathode
    30 ns pulse length
    1 Hz prf
    Accelerator current ~ 4 kA
    Wiggler current ~ 850 A
    Normalized beam brightness: 2×10^4 A/(cm-rad)^2

    The free electron laser experiment at Livermore, named ELF, has also produced some very interesting amplifier results. This is a microwave device, which operates at 8.6 mm with the use of an induction accelerator. Its principal features are listed in Table 2. Because the wiggler field is provided by a set of electromagnets, it is "tunable": i.e., it can be more or less tapered at will, in place. This has proven to be a useful design feature. The signal frequency corresponds to a wiggler wavelength of $\lambda_w = 9.8$ cm, with the length of the wiggler being $L_w = 3$ m. Although the use of a high γ approximation is common in free electron theory, in the Livermore experiment γ is not very high, due to the rather low beam energy of 3.5 MeV. Hence it has been necessary for the Livermore group to extend the theory by taking into account three-dimensional effects, electrostatic effects, and so forth.

    Figure 12 shows results from the Livermore amplifier experiments.[11] The input signal for the amplifier had a power level of 50 kW. The microwave signal experiences exponential amplification for the first 1.3 m of the wiggler, after which it saturates if the wiggler is uniform (i.e., not tapered). The wiggler field was then tapered, beginning from the 1.3 m point, where the signal is about 180 MW. This

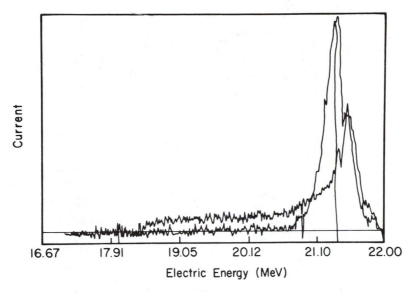

FIGURE 11   Current versus electric energy as obtained with a self-excited oscillator. The efficiency exceeds 2%.

means that instead of the 640 MW that was mentioned earlier, the effective input to the tapered wiggler is really 180 MW. Nonetheless, an efficiency of 35% was achieved, compared to the 6% efficiency achieved without taper. Since there is a waveguide, the device can be made to be many Rayleigh ranges in length. In fact, these results could be extended, since all the available length was not utilized. The degree of taper that was possible in this experiment was limited; there is no further tapering after 2.4 m. In the future when the taper is increased, it will be interesting to see how much more efficient the device becomes and how much more power can be obtained.

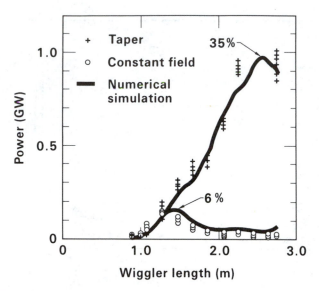

FIGURE 12   Amplified signal output as a function of wiggler length for a uniform wiggler (0) and for a tapered wiggler (+). [Adapted from Ref. 11]

CONCLUSION

Although in this paper I have been able to describe only a few, Marshall Rosenbluth has made many important contributions to the free electron laser problem. A fuller list of his contributions would include the following:

- Tapered wigglers
- Sideband instability
- Resonant coupling between betatron and synchrotron oscillations
- Analytic theory of transverse gradient wiggler
- The gain-spread-transverse excitation theorem
- Phase area displacement
- Pulse build-up
- Two-dimensional effects

Many collaborators were involved in all of this work, and many others have also worked on these subjects throughout the historical development of the free electron laser. Finally, another of his crucial contributions has been his long-term advisory role in the U.S. free electron laser program. Not only his work, but also his interactions with all the groups working on free electron lasers, have had — and, I expect, will continue to have — an enormous impact in this field.

Parenthetically, it may noted that the free electron laser has relevance to plasma heating. There is now a project at Livermore, in which a slightly upgraded version of the ELF, having somewhat higher power output and more frequent pulses, would be used as an electron cyclotron heating device for a tokamak plasma.[12,13] Electron cyclotron heating has potentially great advantages in being highly tunable, which allows the power to be deposited in the profile exactly where desirable and also allows for feedback applications. With these free electron lasers, the necessary power for heating is now available at reasonable cost.

ACKNOWLEDGEMENTS

The assistance of Dr. James Van Dam and Dr. Vernon Wong in the conversion of my talk into a manuscript is gratefully acknowledged. This work was supported by the U.S. Department of Energy (Contract: DE-AS03-81ER40029).

REFERENCES

1. L.R. Elias, W.M. Fairbank, J.M.J. Madey, H.A. Schwettman, and T.I. Smith, *Physical Review Letters* **36**, 717 (1976).
2. D.A.G. Deacon, L.R. Elias, J.M.J. Madey, G.J. Ramian, H.A. Schwettman, and T.I. Smith, *Physical Review Letters* **38**, 892 (1977).
3. A. Renieri, *Il Nuovo Cimento* **53B**, 11 (1979).
4. J.M.J. Madey, *Il Nuovo Cimento* **50B**, 64 (1979).
5. N.M. Kroll, P.L. Morton, and M.N. Rosenbluth, *IEEE Journal of Quantum Electronics* **QE-17**, 1436 (1981).
6. N.M. Kroll, P.L. Morton, and M.N. Rosenbluth, in *Free-Electron Generators of Coherent Radiation* (Physics of Quantum Electronics, Vol. 7), edited by S.F. Jacobs, H.S. Pilloff, M. Sargent, III, M.O. Scully, and R. Spitzer, (Addison-Wesley, Reading, MA, 1979), p. 113.
7. E.T. Scharlemann, A.M. Sessler, and J.S. Wurtele, *Physical Review Letters* **54**, 1925 (1985).
8. W.L. Kruer, J.M. Dawson, and R.N. Sudan, *Physical Review Letters* **23**, 838 (1969).
9. M.N. Rosenbluth, H.V. Wong, and B.N. Moore, ARA Report No. I-ARA-83-U-62 (NTIS No. AD-136333) (November, 1983).

10. R.W. Warren, B.E. Newnam, J.G. Winston, W.E. Stein, L.M. Young, and C.A. Brau, *IEEE Journal of Quantum Electronics* **QE-19**, 391 (1983).
11. T.J. Orzechowski, B.R. Anderson, J.C. Clark, W.M. Fawley, A.C. Paul, D. Prosnitz, E.T. Scharlemann, S.M. Yarema, D.B. Hopkins, A.M. Sessler, and J.S. Wurtele, *Physical Review Letters* **57**, 2172 (1986).
12. K.I. Thomassen *et al.*, *Free-Electron Laser Experiments in Alcator-C* (U.S. Government Printing Office, Washington, D.C., 1986).
13. W.M. Nevins, T.D. Rognlien, and B.I. Cohen, *Physical Review Letters* **59**, 60 (1987).

Guy Laval
Centre de Physique Theorique
Ecole Polytechnique
91128 Palaiseau Cedex, France

Plasma Turbulence: The Statistical Approach

Three important works by M.N. Rosenbluth on plasma turbulence are reviewed. The first one concerns the study of parametric instabilities in a plasma with time independent fluctuations. The second one deals with the nonlinear theory of drift waves in a sheared magnetic field. Lastly, we discuss the methods for computing thermal transport in a magnetic field with destroyed magnetic surfaces in the collisional case.

INTRODUCTION

This paper is not a review of plasma turbulence theory even restricted to the statistical approach. Its purpose is only to describe a few important problems of plasma turbulence in which M.N. Rosenbluth took a prominent part. These problems do not all deal with nonlinear dynamics, but rather with the statistical description of stochastic phenomena induced by turbulence.

First, we consider the excitation of parametric instabilities in a medium with random inhomogeneities. This is a rather standard case of linear propagation in random media. The randomness is entirely extrinsic and we could expect that the usual methods of statistical theory would apply. We shall show why these methods were unable to provide a complete answer and how M.N. Rosenbluth, with his co-workers, found a very elegant solution by using a new original scheme.

The second case is a very complicated turbulence problem where one tries to compute the linear response of a turbulent plasma. The turbulence is supposed to result from the excitation of drift waves in a sheared magnetic field. In this case, the stochastic instability plays an important role, and self-consistent effects cannot be ignored. However, the equations of the Direct Interaction Approximation were solved and gave results in accordance with what would be expected from rough estimates.

Lastly, we shall discuss the computation of thermal conductivity for a collisional plasma in a magnetic field with destroyed magnetic surfaces. Here the role of the stochastic instability does not reduce to resonance broadening. Its detailed mechanism is needed to describe the transport process. It seems that, in such a case, the statistical approach is not of great help, so that only heuristic estimates or sophisticated numerical computations are available.

PARAMETRIC INSTABILITIES IN A RANDOM MEDIUM

Parametric instabilities in the presence of space-time fluctuations are of particular importance in plasma physics. Such fluctuations can arise from finite bandwidth effects of the pump wave or they can be induced by wave propagation in a plasma with fluctuating macroscopic parameters. The interest of this problem lies in the possibility of a strong modification of thresholds and saturated states of the instabilities.

Usually, the parametric instabilities can be described with envelope equations. For studying the linear thresholds, pump depletion is neglected and the equations are written

$$D_1 a_1 = \gamma_0 S(t - x/V_0)a_2 \tag{1}$$

$$D_2 a_2 = \gamma_0 S^*(t - x/V_0)a_1, \tag{2}$$

where

$$D_1 = \partial/\partial t + V_1 \partial/\partial x \tag{3}$$

$$D_2 = \partial/\partial t + V_2 \partial/\partial x. \tag{4}$$

Here, a_1 and a_2 are the complex amplitudes of the product waves; γ_o is the growth rate of the coherent case; V_1 and V_2 are the group velocities of the decay waves; and $S(t)$ describes a stochastic process such that $\langle S(t)S^*(t) \rangle = 1$.

A closed set of approximate equations can be obtained as discussed in Ref. 1. It has been found that whenever the product $V_1 V_2$ is positive, for short correlation times of $S(t)$ one can easily obtain equations for the correlation functions that reduce to the usual random phase approximation (RPA). This property results essentially from the fact that time ordering is not destroyed by wave propagation, so that the usual arguments for using the Bourret approximation can be applied. On the other hand, for $V_1 V_2 < 0$, one has to use an additional constraint, namely that the correlation times of the product waves a_1 and a_2 are small, if we want to be allowed to use these random phase approximations.

The case $V_1 V_2 < 0$ is important since it corresponds to situations where absolute instabilities are possible. Then, we expect that in a plasma of finite length, unstable normal modes will exist. Such modes can be studied with the RPA equations, but obviously it will not be possible to determine the instability threshold since we need many unstable modes to build daughter waves a_1, a_2 with short correlation times.

In Ref. 2, the problem has been solved for the case of wave propagation in a stationary random medium. For a normal mode with an exponential time dependence $\exp(-i\omega t)$, the complex amplitudes were written

$$a_1(x,t) = |V_2/V_1|^{1/4} b_1 \exp[-i\omega(t - x/V_1] \tag{5}$$

$$a_2(x,t) = |V_1/V_2|^{1/4} b_2 \exp[-i\omega(t - x/V_2)]. \tag{6}$$

The boundary conditions are then $b_1(0) = 0$ and $b_2(L) = 0$ for a plasma extending from $x = 0$ to $x = L$. The differential equations for $b_1(x,\omega)$ and $b_2(x,\omega)$ being supposed to be solved with $b_1(0,\omega) = 0$ and $b_2(0,\omega) = 1$, the eigenvalue condition takes the form

$$b_2(L,\omega) = 0. \tag{7}$$

Setting $b_2(L,\omega) = r \exp(i\psi)$ and using the Nyquist theorem, we find that the average number of zeros of b_2 with $\text{Im}\,\omega > \sigma$ is given by

$$N(L,\sigma) = \frac{1}{2\pi} \int_{-\infty+i\sigma}^{+\infty+i\sigma} \left\langle \frac{\partial \psi}{\partial \omega} \right\rangle d\omega. \tag{8}$$

This number provides the average number of unstable modes with growth rates larger than σ. The authors were then able to obtain a Fokker-Planck equation for the probability distribution of $b_1(x,\omega), b_2(x,\omega)$. They solved it in the asymptotic limit of a short correlation length ℓ_c of the fluctuations. Then they were able to obtain both the distribution of unstable normal modes and the instability threshold length L_t, which was given by

$$L_t = \left[(\pi^2 \Delta k/8) / \ln |\Delta k| \right] \frac{|V_1 V_2|^{1/2}}{\gamma_0} \tag{9}$$

where $\Delta k = |V_1 V_2|^{1/2}/(\gamma_0 \ell_c)$, with $\Delta k \gg 1$. The previous RPA equations would have given

$$L_t^{\text{RPA}} \sim L_t \ln |\Delta k|. \tag{10}$$

As expected, far from threshold, both methods give the same average growth rate.

RENORMALIZED DIELECTRIC RESPONSE FOR ELECTROSTATIC DRIFT WAVE TURBULENCE

Drift wave turbulence is one of the central topics of anomalous transport theory. The simplest model of present physical situations in tokamaks takes into account magnetic shear, but the plasma is assumed to be an inhomogeneous plane slab. The linear theory is now well known. In a pioneering work, Hirshman and Molvig[3] noticed that the linear response could be strongly modified by a very low level of turbulence. In the linear theory of drift waves in a sheared magnetic field, a special role is played by electrons in regions where $k_\parallel V_e < \omega$, where $k_\parallel$ is the wave number along the magnetic lines, V_e is the electron thermal velocity, and ω is the wave frequency. Such electrons have a non-adiabatic response to the electric potential, and consequently they stabilize the normal modes. In Ref. 3, it was shown that such electrons would no longer exist as soon as turbulent diffusion was strong enough to induce large fluctuations of $k_\parallel$ during a wave period. This condition can be written

$$\left[D(k_\parallel' V_e)^2/3 \right]^{1/3} = \omega_c \gtrsim \omega, \qquad (11)$$

where $k_\parallel' = dk_\parallel/dx$, x being the radial coordinate and D the turbulent diffusion coefficient. A destabilizing effect is then expected. Hirshman and Molvig used a simple theory in which resonance broadening was applied without taking into account self-consistent effects. They found that turbulence at low level was destabilizing and that saturation resulted from the usual shear damping at large enough amplitudes.

In Ref. 4, Diamond and Rosenbluth tried to apply a systematic method to this interesting and very complicated problem. They used the Direct Interaction Approximation (DIA) to compute the linear response equations. Then self-consistency was taken into account and the energy was conserved in a natural way. They succeeded in solving the equations in the whole range of variation of ω_c. For $\omega_c < \omega$, turbulence was found to stabilize the mode, whereas destabilization was recovered for $\omega_c > \omega$ as predicted by Hirshman and Molvig. The interest of this case lies in the fact that resonance broadening has here a dramatic effect on the dynamics of the system and DIA is able to take it properly into account, which shows that it is able to describe at least orbit diffusion. Another interest of this work has been to show that the DIA equations could be solved even in such a complicated physical case.

ELECTRON HEAT TRANSPORT IN A CONFIGURATION WITH DESTROYED MAGNETIC SURFACES

The good confinement properties of a toroidal configuration are based upon the heat conductivity across field lines being very low when compared with the conductivity along field lines. It assumes that the field lines are themselves confined inside a closed region and, more precisely, that they remain very close to an ensemble of nested tori. Small perturbations induced by field irregularities or by self-consistent fluctuations of the plasma current density can destroy the field line confinement. The field lines behave in a chaotic way and diffuse across the plasma to the edge.

When many overlapping resonances are responsible for such diffusion, the diffusion equation can be obtained by quasilinear theory as shown by Rosenbluth, Sagdeev, Taylor, and Zaslavski[5] in 1966. The resulting collisionless diffusion coefficient for electrons is easily obtained by assuming that particles move along field lines with a constant velocity v.

For a displacement L along field lines, the average squared radial displacement across the unperturbed field is given by

$$\langle \Delta r^2 \rangle = 2 D_{st} L, \tag{12}$$

where D_{st} is the field line diffusion coefficient. The collisionless electron thermal diffusion coefficient is then $\chi_c = v D_{st}$.

If collisions are taken into account, the particle dynamics must be described more carefully. First, collisions lead to a diffusive motion along field lines. When this process is taken into account, the diffusion coefficient for particles vanishes, since $\langle \Delta r^2 \rangle$ behaves like $t^{1/2}$. Collisions also induce a small transverse diffusion coefficient $\chi_\perp$ so that particles do not follow exactly the magnetic field lines. Rechester and Rosenbluth[6] noticed that this small transverse motion is amplified by the stochastic instability of the chaotic lines. They heuristically derived a diffusion coefficient, taking into account these processes.

First, they showed that the microscopic scale of transverse diffusion is no longer the Larmor radius, but rather a scale δ resulting from a balance between collisional diffusion and the stochastic instability. They found

$$\delta = L_c \left(\frac{\chi_\perp}{\chi_\parallel} \right)^{1/2}, \tag{13}$$

where L_c^{-1} is the averaged spatial growth rate of the stochastic instability for field lines and $\chi_\parallel$ is the collisional diffusion coefficient along the field lines. Moreover, that small scale is amplified by the stochastic instability, so that after a distance $L_{c\delta}$, the particle is on a field line decorrelated from the initial one, with

$$L_{c\delta} = L_c \ln \left[(k_\perp \delta)^{-1} \right], \tag{14}$$

where $k_\perp^{-1}$ is the transverse correlation length of the field perturbations. The resultant overall radial thermal diffusion coefficient χ_r is then given by

$$\chi_r = D_{st} \chi_\parallel / L_{c\delta}. \tag{15}$$

For $L_{c\delta}$ larger than the collisional mean-free-path, they recovered the collisionless diffusion coefficient χ_c.

Since this derivation involves many steps, it would be useful to obtain the result by the tools of the statistical theory of random media. Quasilinear theory yields

$$\chi_r = D_{st}\sqrt{k_\perp^2 \chi_\perp \chi_\parallel}, \tag{16}$$

which is several orders of magnitude smaller in practical applications. Moreover, it is readily shown, by computing next-order terms, that the quasilinear result is meaningless as soon as the Rechester-Rosenbluth result reaches its domain of validity. The reason for the failure of quasilinear theory lies in the fact that diffusion along field lines induces a back and forth motion so that time ordering is no longer equivalent to space ordering and decorrelation does not take place for large time intervals in the perturbation series expansions. Krommes, Kleva, and Oberman[7] found that the usual DIA did not give a better result than quasilinear theory. They showed also that by taking into account vertex renormalization, the effect of the stochastic instability could be included.

However, no precise analytical or numerical check of the Rechester and Rosenbluth formula has ever been reported. Therefore, in this paper, we outline a method that makes it possible to compute this diffusion coefficient in a rather simple way.

A SIMPLE MODEL FOR DIFFUSION IN DESTROYED MAGNETIC SURFACES

The heat transport equation for the temperature T may be written

$$\frac{\partial T}{\partial t} = \chi_\parallel \left(\hat{b} \cdot \nabla\right)^2 T + \chi_\perp \Delta_\perp T, \tag{17}$$

where $\hat{b}$ is the unit vector along a field line and

$$\Delta_\perp T = \nabla \cdot (\nabla_\perp T) \tag{18}$$

with $\nabla_\perp T = \nabla T - \hat{b}(\hat{b} \cdot \nabla)T$.

With the assumption that the unperturbed field with unit vector $\hat{b}_0$ is perpendicular to the x-direction, it can be shown that the diffusion coefficient χ_x resulting from the fluctuation $\hat{b} = \hat{b}_0 + \tilde{b}(\mathbf{r})$ is obtained in a variational form as

$$\chi_x = \min\{\langle \chi_\parallel \phi^2 + \chi_\perp (\nabla_\perp n)^2 \rangle\}, \tag{19}$$

where ϕ and n are related by the constraint

$$\phi = \hat{b} \cdot \nabla n + \tilde{b}_x. \tag{20}$$

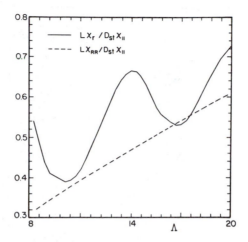

FIGURE 1   The solid curve line shows $L\chi_r/D_{st}\chi_{\parallel}$ versus $\Lambda = KkL^2/L_s$ for $\beta^2 = 10^{-2}$. The dashed line shows $L\chi_{RR}/D_{st}\chi_{\parallel}$ versus Λ, where χ_{RR} is given by the Rechester and Rosenbluth formula.

The brackets mean either an averaging over the small scale length of $\tilde{b}$ or a statistical averaging.

As a simple example, we have considered the case where

$$\hat{b}_o = \hat{e}_z + \frac{x}{L_s}\hat{e}_y \tag{21}$$

$$\tilde{b} = \hat{e}_x \sum_{p=-\infty}^{+\infty} KL \cos(ky)\, \delta(z - pL). \tag{22}$$

This corresponds to the well-known case in which the intersections of a given field line with planes $z = z_0 + pL$, where p are successive integers, are given by the standard map. For $z \neq pL$, the minimization leads to a simple equation in the unperturbed field, which can be integrated. We have only to write jump conditions at $z = pL$. As trial functions, we take for $n(x, y, 0)$ a truncated Fourier series such that

$$n(x, y, 0) = \sum_{m,q} n_{m,q} \left[\exp i \left(m\, k_y + q\, k_x \frac{L}{L_s} \right) \right], \tag{23}$$

m and q being integers with $|m| < M$ and $|q| < Q$. Minimization of the variational expression for χ_r provides a set of coupled algebraic equations for $n_{m,q}$, which can be solved numerically.

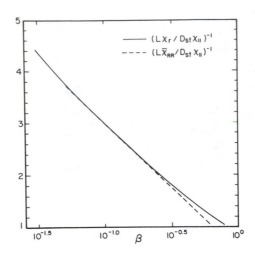

FIGURE 2 The solid curve shows $[L\chi_r/D_{st}\chi_\parallel]^{-1}$ versus $\beta = (\chi_\perp/\chi_\parallel)^{1/2}kL$, for $\Lambda = 10$. The dashed line shows $[L\bar{\chi}_{RR}/D_{st}\chi_\parallel]^{-1}$ versus β, for $\Lambda = 10$, with $\bar{\chi}_{RR} = \frac{1}{2}\chi_{RR}$.

The results are easily expressed in terms of the stochasticity parameter Λ of the standard map, where $\Lambda = KkL^2/L_s$, and then compared with the Rechester-Rosenbluth theory. With $\beta^2 = k^2L^2\chi_\perp\chi_\parallel \ll 1$, the result of Rechester and Rosenbluth can be written as

$$\chi_{RR} = \frac{D_{st}\chi_\parallel}{L}\left[\frac{\ln\left(\frac{\Lambda}{2}\right)}{\ln\left(\beta^{-1}\ln\left(\frac{\Lambda}{2}\right)\right)}\right], \tag{24}$$

where D_{st} is the diffusion coefficient for the field lines as it has been computed in Ref. 8 for $\Lambda \gg 1$. Figure 1 shows the numerical solution for χ_r (normalized to $D_{st}\chi_\parallel/L$) as a function of Λ for fixed β, compared to the Rechester-Rosenbluth result χ_{RR}. We have found for $10 < \Lambda < 20$ and $10^{-2} < \beta < 10^{-1}$ that χ_r behaves approximately as predicted by Rechester and Rosenbluth. The precise dependence with respect to Λ is hidden by the oscillations of D_{st}, which are a residue of long correlation effects that remain in the standard map,[9] but the linear dependence on $\ln\beta$ is quite accurate for a given value of Λ. Figure 2 shows χ_r^{-1} and χ_{RR}^{-1} plotted as functions of β for fixed Λ; within a numerical factor of about 2, there is very good agreement with the Rechester-Rosenbluth formula.

CONCLUSION

It appears that the usual statistical methods like the quasilinear approximation, the random phase approximation, or the direct interaction approximation have indeed a restricted range of applicability. They seem to be unable to take into account such effects as propagation in opposite directions, diffusion, or the detailed process of stochastic instabilities. Therefore, in particular for the case of heat conductivity in a turbulent magnetic field, it seems necessary to use new statistical tools, as was done by the authors of Ref. 2 in the case of parametric instabilities.

REFERENCES

1. G. Laval, R. Pellat, D. Pesme, A. Ramani, M.N. Rosenbluth, and E.A. Williams, *Physics of Fluids* **20**, 2049 (1977).
2. E.A. Williams, J.R. Albritton, and M.N. Rosenbluth, *Physics of Fluids* **22**, 139 (1979).
3. S.P. Hirshman and K. Molvig, *Physical Review Letters* **42**, 648 (1979).
4. P.H. Diamond and M.N. Rosenbluth, *Physics of Fluids* **24**, 1641 (1981).
5. M.N. Rosenbluth, R.Z. Sagdeev, J.B. Taylor, and G.M. Zaslavski, *Nuclear Fusion* **6**, 297 (1966).
6. A.B. Rechester and M.N. Rosenbluth, *Physical Review Letters* **40**, 38 (1978).
7. J.A. Krommes, R.G. Kleva, and C. Oberman, *Journal of Plasma Physics* **30**, 11 (1983).
8. A.B. Rechester, M.N. Rosenbluth, and R.B. White, *Physical Review A* **23**, 2664 (1981).
9. A.B. Rechester and R.B. White, *Physical Review Letters* **44**, 1586 (1980).

Tsung-Dao Lee
Department of Physics
Columbia University
New York, New York 10027

Soliton Stars and Black Holes

The possibility of a new type of cold stable stellar configuration called a soliton star is described, based on the extension of the nontopological soliton solution in general relativity to include gravity.

INTRODUCTION

My first publication in physics was a collaboration with Marshall Rosenbluth and C.N. Yang.[1] The text of that paper covers less than a page; in fact it is the shortest paper I have ever authored. In it we introduced the concept of the universal Fermi interaction and examined the idea of the intermediate boson. This paper was written when we were all students at the University of Chicago. At that time, the Chicago faculty was packed with superstars, such as Fermi, Chandrasekhar, Mayer and Mayer, Mullikan, Teller, and Urey. The student group was also a strong one, with such talents as Chamberlain, Chew, Garwin, Goldberger, Wolfenstein, and Steinberger. Looking back, even by University of Chicago standards at that time, this short paper has not fared badly. But that is because I had the good fortune to share an office with Marshall Rosenbluth.

That was forty years ago and I had just come fresh from China. I found Chicago strange and exotic. Marshall was the native I turned to for guidance and advice. Even then, Marshall already knew everything. I was impressed, not only by the depth of his physics, but also by the wide range of his knowledge in general.

During the war years, automobiles were extremely rare in China. In the wartime capital of Chungking there were a few military trucks. The drivers were called "pilots" because driving a truck, like piloting an airplane, was considered such a difficult skill. I was quite surprised by the abundance of cars in Chicago. I felt I had to learn how to drive. One day I asked Marshall if he could teach me. As expected, he immediately said yes. We borrowed a friend's car. Marshall explained, first abstractly, the function of the steering wheel, the brake, the clutch, and the gear shift. Next, he demonstrated some fancy maneuvering at very high speed on Chicago's Lake Shore Drive. Then he stopped and said, "You take over." This was the one and only driving lesson I have ever had in my life, and it worked like a charm. For years, I thought that was the proper way to learn to drive. As in physics, all you need is to first understand the basic principles and then the rest follows. Decades later, Marshall told me he himself only got his first driving lesson the night before he gave me mine. This demonstrates Marshall's superb confidence and his enormous generosity in sharing his knowledge with others. Of course, it also reconfirms my belief about how driving should be taught, since both Marshall and I learned it the same way.

I am very happy to be here for the celebration of Marshall Rosenbluth's sixtieth birthday. This talk on soliton stars[2-5] may remind him of our early Chicago days when *everything* was field theory. In this context, it may be appropriate to begin my talk by reviewing the well-known Chandrasekhar limit.

THE CHANDRASEKHAR LIMIT

Consider a white dwarf, or a neutron star, of radius R, mass M, and fermion number N. The gravitational force is balanced by the Fermi pressure. From the equipartition of energy we expect, for the equilibrium state, the magnitude of the gravitational energy to be comparable to that of the kinetic energy. For ultra-relativistic fermions, we have

$$\frac{GM^2}{R} \sim \frac{N^{4/3}}{R} \tag{1}$$

where G is Newton's constant. Let m be the effective mass, defined by

$$N = \frac{M}{m}. \tag{2}$$

For a neutron star, the fermions are neutrons and m is the neutron mass m_N; for a white dwarf, they are the electrons and $m = 2m_N$, since there are two nucleons per each electron. Combining (1) and (2), one sees that a critical mass M_c exists:

$$M_c \sim \frac{1}{G^{3/2}m^2}.$$

Relating (in units $\hbar = c = 1$)

$$G = \ell_P^2, \tag{3}$$

where ℓ_P is the Planck length, given by

$$\ell_P \cong 10^{-33}\text{cm},$$

we find (because $m_N^{-1} \sim 10^{-14}\text{cm}$)

$$M_c \sim \frac{m}{\ell_P^3 m^3} \sim 10^{57} m_N \sim M_\odot, \tag{4}$$

with $M_\odot$ the solar mass. For the white dwarf, this is the well-known Chandrasekhar limit, which is about

$$1.4 M_\odot.$$

For M bigger than $1.4 M_\odot$ but less than M_c of the neutron star, white dwarfs cease to exist; instead, one has a neutron star. For the neutron star, because of general relativity[6] and nuclear forces, M_c is somewhat smaller than 4 times the white dwarf limit, as would be indicated by Eq. (4); it is commonly accepted as $\lesssim 5M_\odot$, depending on the physical assumptions.[7,8]

For M bigger than M_c of the neutron star, the solution becomes singular ($R = 0$). The star collapses into a black hole. This relatively low critical mass $M_c \lesssim 5M_\odot$ has been used as a criterion for the observation of black holes.

This critical mass M_c for stellar collapse is relatively insensitive to the assumption of the equation of state of matter. For any cold matter (temperature $T = 0$) and with the assumption of the usual thermodynamical limit, the pressure p must be a function of the density $\sim M/R^3$. Take for example

$$p \propto \left(\frac{M}{R^3}\right)^\gamma.$$

By balancing the gravitational force with the force exerted by the pressure, we have, in place of Eq. (1),

$$\frac{GM^2}{R^2} \sim pR^2 \propto \left(\frac{M}{R^3}\right)^\gamma R^2;$$

i.e.,

$$GM^{2-\gamma} \propto R^{4-3\gamma}. \tag{5}$$

To estimate gravitational collapse we may set R to be the Schwarzschild radius $2GM$. Substituting that into (5), we find

$$G^{-3+3\gamma} \propto M^{2-2\gamma}$$

and therefore the critical mass M_c is always proportional to $G^{-3/2}$, which leads again to Eq. (4), independent of γ. (For a relativistic Fermi gas, $\gamma = 4/3$.)

Nevertheless, we would like to ask: Can a cold stable star exist with $M > 5M_\odot$, without becoming a black hole? In the following we shall introduce the notion of "soliton stars" whose critical mass M_c can be much larger than that indicated by Eq. (4).

NONTOPOLOGICAL SOLITON

To illustrate the basic mechanism, consider the following example of a nontopological soliton,[9,10] first without gravity. The theory contains an additive quantum number N (like the baryon number) carried by either a spin $\frac{1}{2}$ field ψ, or a spin 0 complex field ϕ, with its elementary field quantum having $N = \pm 1$. (If one wishes, one may think of ψ as the quark field.) In addition, there is a scalar field σ. Take, as a first example, the self-interaction of σ to be the typical degenerate vacuum form:

$$U(\sigma) = \frac{1}{2}m^2\sigma^2 \left(1 - \frac{\sigma}{\sigma_o}\right)^2 . \tag{6}$$

We may assign $\sigma = 0$ to be the normal vacuum state, and $\sigma = \sigma_o$ the (abnormal) degenerate vacuum state. (Theories of this type have been studied in the literature, e.g., in connection with the spontaneous T violation,[11,12] the abnormal nuclear model,[9] the bag model,[13,14] and the Higgs mechanism.[15]) The soliton contains an interior in which $\sigma \cong \sigma_o$, a shell of width $\sim m^{-1}$, over which σ changes from σ_o to 0, and an exterior that is essentially the vacuum. The N-carrying field ψ, or ϕ, is confined to the interior; this produces a kinetic energy E_k (assuming for simplicity that the mass of ψ, or ϕ, is zero when $\sigma = \sigma_o$, but nonzero when $\sigma = 0$):

$$E_k \sim \begin{cases} \dfrac{N^{4/3}}{R} & \text{for fermions} \\[2mm] \dfrac{N}{R} & \text{for bosons.} \end{cases} \tag{7}$$

In the simplest case of a scalar boson field ϕ, N conservation requires ϕ to be complex and vary as $e^{-i\omega t}$. Because the σ field changes from σ_o in the interior to 0 outside, there is also a surface energy

$$E_s = sR^2,$$

where s is the surface tension, related to σ_o and σ-mass m by

$$s \sim m\sigma_o^2.$$

The radius R can be calculated by minimizing the total energy $E = E_k + E_s$. Setting $\partial E/\partial R = 0$, we have the equipartition

$$E_k = 2E_s.$$

Hence, the soliton mass M (which is the minimum of E) can be written as

$$M = 3E_s = 3sR^2, \tag{8}$$

and the total conserved particle number N is related to M by

$$M \propto \begin{cases} N^{8/9} & \text{for fermions} \\ N^{2/3} & \text{for bosons.} \end{cases} \tag{9}$$

Very little is known about the mass of such Higgs-like scalars, except that their masses are probably > 30 GeV. For m and $\sigma_o \gg 1$ GeV and for a normal nucleus, the above soliton mass M would be much larger than N (the quark number) times $\frac{1}{3}$ of the nucleon mass; therefore, the soliton configuration is unstable when N is small. But because the exponent of N in Eq. (9) is < 1, when N is sufficiently large the soliton mass is always less than that of the free particle solution, and that insures its stability against decaying into N free particles (or $\frac{1}{3}N$ free, or nearly free, nucleons, in the case of N quarks). But, is there a limit to this stability of very large N?

SOLITON STARS

To find the upper limit, we must include the gravitational field. Gravity becomes important when the soliton radius R becomes of the same order as $2GM$. Thus, the critical mass M_c may be estimated by simply setting

$$R \sim 2GM_c,$$

which, because of Eq. (8), leads to

$$M_c \sim \frac{1}{G^2 s} = \frac{1}{\ell_P^4 s}.$$

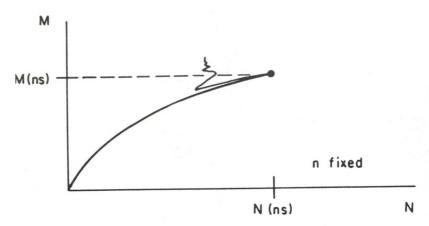

FIGURE 1   A schematic drawing of mass M versus particle number N for a soliton (or mini-soliton) star; n denotes the number of nodes of the scalar field.

A typical Higgs-like field σ may have $\sigma_o \sim m$ (with m^{-1} much less than ℓ_P); we estimate

$$M_c \sim (\ell_P m)^{-4} m. \qquad (10)$$

For example, if m is 300 GeV, we have $M_c \sim 10^{12} M_\odot$ and $R \sim 1$ light month. Hence we find the answer to our question: depending on the physical theory, the critical mass for a hadron star to become a black hole can be much greater than $5 M_\odot$. Such cold stable stellar configurations are called *soliton stars*. The estimate Eq. (10) holds for both fermion and boson soliton stars.

We note that because of the existence of the surface energy, which leads to Eq. (9), the system, though large, does not have a thermodynamical limit (i.e., M is not proportional to N). This is why M_c can have a power dependence on the Planck length, different from -3, as shown by Eq. (4).

Let $n - 1$ denote the number of nodes (or the number of minima at finite radii) of the σ-field. The lowest energy state for a soliton star is always $1s$. An interesting feature is that for a given n, the relation between M and N exhibits a curious behavior, as shown schematically in Figure 1. This zigzag feature is independent of the statistics of the particles that carry the quantum number N, but is characteristic of the type of nonlinear equations with which we have to deal. For each given n, there is a maximum $N = N(ns)$, with a corresponding mass $M(ns)$, beyond which there is no solution. For $N < N(ns)$, depending on N we may have one solution, or two or three or an infinite number of solutions. Furthermore,[16]

$$M(ns) \cong \frac{1}{2n - 1} M(1s)$$

and

(11)

$$N(ns) \cong \left(\frac{1}{2n-1}\right)^2 N(1s).$$

Thus, only the lowest $1s$ solution is stable. For $M > M(1s)$ there is no (spherically symmetric) solution. The critical mass M_c is therefore $M(1s)$.

Another interesting feature is that when $M = M(1s)$, the radius R is $\sim (M_c/s)^{\frac{1}{2}} \neq 0$. Unlike the case of the Chandrasekhar limit, the solution is not singular. Consequently, it is possible to find non-singular solutions for the black holes when $M > M(1s)$ (and $N > N(1s)$).

SOLITON BLACK HOLES

It is most convenient to adopt the isotropic coordinates

$$ds^2 = -e^{2u}dt^2 + e^{2v}(dr^2 + r^2 d\theta^2 + r^2 \sin^2\theta d\phi^2),$$
(12)

where, for the spherically symmetric solution, u and v depend only on the radius r. Let $2\pi\rho$ be the circumference of a two-sphere. From Eq. (12), we see that

$$\rho = re^v.$$
(13)

The dependence of ρ on r is plotted schematically in Fig. 2; the shaded region refers to the star (with nonzero matter density and Higgs field $\sigma \cong \sigma_o$, the false vacuum). At $r = R + O(m^{-1})$, the σ-field changes from σ_o to 0 over a distance $\sim m^{-1}$. Outside the surface, $r > R + O(m^{-1})$, the matter density becomes zero, and the metric is determined by the Schwarzschild solution:

$$e^u = \frac{r-a}{r+a}, \qquad e^v = \left(\frac{r+a}{r}\right)^2$$
(14)

where $a = \frac{1}{2}GM$ is the "Schwarzschild" radius in the isotropic coordinates.

For a soliton star, R is $> a$, and ρ is a monotonic function of r, shown in the top drawing in Fig. 2.

For a soliton black hole, R is $< a$. Although $d\rho/dr$ is > 0 inside the star, outside the star ρ is no longer a monotonic function. From Eqs. (13) and (14), we see that outside the star,

$$\rho = a\left(\sqrt{\frac{r}{a}} + \sqrt{\frac{a}{r}}\right)^2;$$
(15)

hence, $d\rho/dr$ is negative for $R < r < a$ (inside the horizon) and positive for $r > a$ (outside the horizon). The horizon is located at $r = a$ (i.e., $\rho = 2GM$). Notice

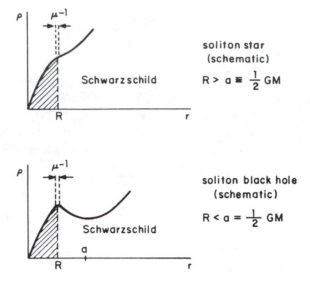

FIGURE 2   Schematic drawings of the circumference $2\pi p$ of a two-sphere versus the radius r in the isotropic coordinates for a soliton star and a soliton black hole. The shaded region denotes matter in the star, and $a = \frac{1}{2}GM$ is the Schwarzschild radius.

that the inside and outside regions of the horizon are now related "space-like" to each other (since the metric is time-independent). This is illustrated by the lower drawing in Fig. 2.

The lowest $1s$ configuration of a soliton star—for N sufficiently large, but less than the critical value of Eq. (10)—is stable.

The soliton black hole is not. The star decays very slowly by sending matter towards the horizon; at the initial stage, the decay (measured by the total particle number N of the star) obeys the usual exponential dependence in time t:

$$N \propto e^{-\Gamma t} \tag{16}$$

with a decay rate

$$\Gamma \cong (Rx_{\mathrm{in}})^{-1} 8\pi G \sigma_o^2 m\omega (a + R)^2 \exp\left[-2m\left(\frac{a^2 - R^2}{R} + 2a\ \ln\frac{a}{R}\right)\right], \tag{17}$$

where ω is the rate of t-dependence in the internal symmetry space, given by Eq. (1), and x_{in} is a parameter $O(1)$, determined by the solution. Because this is a black hole solution, R is less than a. For $\sigma_o \sim m \sim 300$ GeV, since R and a are ~ 1 light month, we find[17]

$$\Gamma \sim (ma)^{\frac{1}{2}} m \exp[-10^{34}]. \tag{18}$$

This very slow time dependence changes only the solution within the horizon ($r < a$). Outside the horizon ($r > a$), the Schwarzschild solution holds at all t.

LATITUDE OF THE MODEL

If we take, instead of Eq. (6), an MIT-bag-like potential which gives the false vacuum a higher potential than the real vacuum, in the absence of the gravity the soliton mass would be given, instead of by Eqs. (7) and (8), as

$$M = sR^2 + pR^3 + \begin{cases} N^{4/3}/R & \text{for fermions} \\ N/R & \text{for bosons.} \end{cases} \tag{19}$$

Because $\partial M / \partial R = 0$, we have

$$2sR^2 + 3pR^3 = \begin{cases} N^{4/3}/R \\ N/R \end{cases},$$

and therefore

$$M = 3sR^2 + 4pR^3.$$

Next, we turn on the gravitational field. The critical mass M_c can again be estimated by setting $R \sim GM_c$. This gives

$$1 \sim 3sG^2 M_c + 4pG^3 M_c^2,$$

or

$$M_c \sim \frac{1}{3sG^2} \frac{2}{1 + \sqrt{1 + \xi^2}}, \tag{20}$$

where

$$\xi = \frac{4}{3s} \sqrt{\frac{p}{G}}.$$

Hence, when $p = 0$ and $s \neq 0$, we have $\xi = 0$, and $M_c \sim 1/sG^2$ as given before by Eq. (10). If $s = 0$ but $p \neq 0$, then $M_c \sim 1/p^{1/2}G^{3/2}$, which has the same power of G as the Chandrasekhar limit of Eq. (4). Consequently, there is an enormous latitude of M_c for a soliton star, which can vary from a galactic mass to a solar mass.

What happens when p and s are both zero? As we shall see, in that case there is an equilibrium solution for any mass (at least classically).

MINI-SOLITON STAR

Consider the simple theory consisting of only a "free" spin 0 *complex* field ϕ of mass $m \neq 0$, plus gravity. There is again a conserved additive quantum number N. As shown in Eq. (1), for $N \neq 0$, ϕ must be time-dependent:

$$\phi = \sigma(r)e^{i\omega t}.$$

Although this and related problems have been studied in the literature,[18,19] there are two new results found in Ref. 3:

1. Absence of a critical mass for gravitational collapse for the equilibrium solution. Physically, this can be seen as follows:

Set N relativistic particles in the same orbit in a zero-node s-state of wavelength $\sim R$. Equipartition gives the balance between the kinetic and gravitational energies:

$$\frac{N}{R} \sim \frac{GM^2}{R}. \tag{21}$$

Write $N \cong M/m$ where m is the mass of the free particle; one sees that there is an upper bound $M(1s)$ for the stellar mass of

$$M(1s) \sim \frac{1}{Gm}.$$

2. For N relativistic particles in the ns orbit (number of nodes $= n - 1$), the wavelength is $\sim R/n$, and therefore Eq. (1) is replaced by

$$\frac{nN}{R} \sim \frac{GM^2}{R},$$

which gives an upper bound

$$M(ns) \sim \frac{n}{Gm}. \tag{22}$$

The corresponding upper bound on the particle number N is

$$N(ns) \sim \frac{n}{Gm^2}. \tag{23}$$

In Ref. 2, it is shown that this linear dependence of $M(ns)$ on n is quite accurate; it holds to a few parts in 10^4. When the node number n is increased, there is no overall upper bound in the stellar mass M for the equilibrium solution (at least classically). More recently, J.J. van der Bij and M. Gleiser[19] have derived a generalization of this result to boson stars with a non-minimal energy-momentum tensor.

At any fixed node number n, the M versus N curve has again a zigzag behavior, as in Fig. 1. (Here, in contrast to Eq. (11), $M(ns)$ and $N(ns)$ both increase with n.) Thus, for a given $N < N(ns)$, there can be more than one solution. The lowest mass branch is very close to the Newtonian approximation. For $N > N(ns)$ [but $< N((n+1)s)$], the equilibrium solution for the ns state ceases to exist and is replaced by the $(n+1)s$ state. In the example of $m \sim 300$ GeV, $M(1s)$ is $\sim 10^9$ kg, the radius is $\sim 6 \times 10^{-17}$ cm, and the corresponding density is extremely high, $\sim 10^{43}$ times that of a neutron star! Because of the smallness of its size, we call such a configuration a mini-soliton star.

CONCLUSION

Nonlinear field theories have been found to be of importance in all elementary particle interactions: QCD, the electroweak theory, GUT, etc. Many of their physical properties are still in the developing stage. For stellar configurations, although the nonlinearity of gravitation is fully recognized through general relativity, that of the matter field is far from adequately explored. The simple examples given here are only meant to illustrate the physical richness of this exciting new domain.

ACKNOWLEDGEMENT

This research was supported in part by the U.S. Department of Energy.

REFERENCES

1. T.D. Lee, M.N. Rosenbluth and C.N. Yang, *Physical Review* **75**, 905 (1949).
2. T.D. Lee, *Physical Review D* **35**, 3637 (1987).
3. R. Friedberg, T.D. Lee, and Y. Pang, *Physical Review D* **35**, 3658 (1987).
4. T.D. Lee and Y. Pang, *Physical Review D* **35**, 3678 (1987).
5. R. Friedberg, T.D. Lee, and Y. Pang, *Physical Review D* **35**, 3640 (1987).
6. J.R. Oppenheimer and R. Serber, *Physical Review* **54**, 540 (1938); J.R. Oppenheimer and G.M. Volkoff, *Physical Review* **55**, 374 (1939).
7. J.B. Hartle,*Physics Reports* **46C**, 201 (1978).
8. R.M. Wald, *General Relativity* (University of Chicago Press, Chicago, 1984); S. Weinberg, *Gravitation and Cosmology* (Wiley-Interscience, 1972); C.W. Misner, Kip S. Thorne, and J.A. Wheeler, *Gravitation* (W.H. Freeman, 1973); S. Chandrasekhar, *The Mathematical Theory of Black Holes* (Oxford University Press, 1983). See also the references quoted therein.
9. T.D. Lee and G.C. Wick, *Physical Review D* **9**, 2291 (1974).
10. R. Friedberg, T.D. Lee, and A. Sirlin, *Physical Review D* **13**, 2739 (1976); *Nuclear Physics B* **115**, 1, 32 (1976).
11. T.D. Lee, *Physical Review D* **8**, 1226 (1973); *Physics Reports* **9C**, 143 (1974).
12. S. Weinberg, *Physical Review Letters* **37**, 657 (1976).
13. A. Chodos, R.J. Jaffe, K. Johnson, C.B. Thorn, and V.F. Weisskopf, *Physical Review D* **9**, 3471 (1974); W. A. Bardeen, M.S. Chanowitz, S.D. Drell, M. Weinstein, and T.M. Yan, *Physical Review D* **11**, 1094 (1975).
14. R. Friedberg and T.D. Lee, *Physical Review D* **15**, 1694, and *Physical Review D* **16**, 1096 (1977).

15. P.W. Higgs, *Physics Letters* **12**, 132 (1964); *Physical Review Letters* **13**, 321 (1964); *Physical Review* **145**, 1156 (1966); F. Englert and R. Brout, *Physical Review Letters* **13**, 321 (1964); G.S. Guralnik, C.R. Hagen, and T.W.B. Kibble, *Physical Review Letters* **13**, 585 (1964); T.W.B. Kibble, *Physical Review* **155**, 1554 (1967).
16. Y. Pang, to be published.
17. R. Friedberg, T.D. Lee, and Y. Pang, to be published.
18. R. Ruffini and S. Bonazzola [*Physical Review* **187**, 1767 (1969)] and W. Thirring [*Physics Letters* **127**B, 27 (1983)] have examined the lowest branch of the 1s solution. Similar analysis has been extended by M. Colpi, S.L. Shapiro, and I. Wasserman [*Physical Review Letters* **57**, 2485 (1986)] to include a repulsive quartic interaction.
19. J.J. van der Bij and M. Gleiser (Fermilab preprint) recently examined boson stars, which are like mini-soliton stars, but with a non-minimal stress tensor.

Roald Z. Sagdeev
Space Research Institute
USSR Academy of Sciences
Moscow, USSR

Encounter with Comet Halley

In March of 1986, an international armada of six spacecraft encountered
the comet Halley and performed *in situ* measurements. These encounters
led to the discovery of a number of cometary plasma physics phenomena.
Another important result was that a value for the average density of the
cometary nucleus could be estimated, which is found to be compatible with
"snow ball" models for the nucleus.

INTRODUCTION

Fusion plasmas cover many orders of magnitude in density, with laser plasmas at the
high end of the scale. If the very low density plasmas found in space are included,
this range extends over twenty orders of magnitude. Throughout this wide range,
the same physical behavior is found.

Here I will focus on a mini-object in space, a comet. A comet is a body that
is roughly a hundred billion tons. It is rather small: for instance, the nucleus of
Halley's comet has an elongated shape, a kind of irregularly shaped ellipsoid, about
15 kilometers in size (major axis). While traveling through space, this tiny object

is able to influence a very large environment around itself. As it approaches the Sun, it controls the behavior of plasma out to a distance of twenty to thirty million kilometers. The reason for such a strong influence is rather simple. The solar wind plasma is extremely rarified, and therefore even the small amount of material that is being continuously evaporated from the surface of a cometary nucleus and then ionized is sufficient to dominate the solar wind environment. In so doing, it creates a number of interesting plasma physics phenomena,[1] which hark back to the old days when I worked in fusion. In the solar wind plasma near comets, one can find almost every kind of plasma instability that has ever been discovered—many by Marshall Rosenbluth. For example, as the Vega-1 spacecraft was approaching Halley's comet, it was able to produce direct measurements that indicated the solar magnetic field was undergoing tearing, which is a phenomenon to whose explanation Marshall Rosenbluth contributed a great deal.

COMETARY PLASMA PHYSICS PHENOMENA

In March of 1986, an armada of six space probes from four space agencies approached Halley's comet and carried out *in situ* measurements (see Fig. 1). Three of these probes came very close to the comet's nucleus, crossing many physically important areas—e.g., the bow shock produced around the comet. I will mainly refer to results obtained by the two USSR probes, Vega-1 and Vega-2, which approached the comet's nucleus to within 9,000 and 8,000 kilometers, respectively, within three days of each other. The closest approach was by the Giotto spacecraft, which came within 600 kilometers a week later; at that distance, the spacecraft, with its relative speed of 60 km/sec, was almost destroyed by dust particles.

At large distances from the comet, the spacecraft identified solar wind protons and alpha particles as the main ion constituents of the plasma. Closer to the nucleus, there appeared a new component in the plasma, of cometary origin. The solar wind component did not penetrate into the inner region of the plasma corona of the comet. Detailed analysis of this region showed what kinds of phenomena develop there.

It is worthwhile to mention that one of the most important phenomena observed throughout this million kilometer scale area was that of Alfvén waves produced by a rather unusual type of cyclotron instability. Newly born ions, which are created by the photo-ionization of neutral molecules or by the evaporation of atoms from the comet, rapidly become highly energetic through their $E \times B$ immersion in the solar wind magnetic and electric fields. Eventually these ions contribute the major part of the thermal energy of this mixed plasma, and there is a sufficient amount of excess free energy to produce cyclotron wave instability. Alfvén wave MHD turbulence finally builds up throughout the entire region. This particular turbulence is very important in controlling dissipative processes within the shock front. It is also

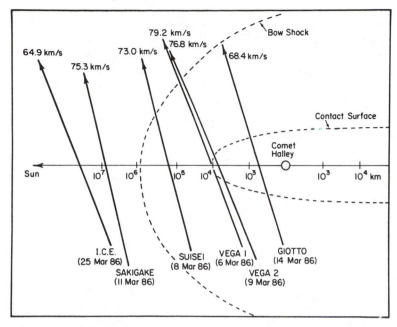

FIGURE 1 Rendezvous with Comet Halley: shown for each flyby is the distance between the satellite and the comet on the date of closest approach.

important in producing a Fermi acceleration mechanism in which Alfvén waves play the role of those magnetized clouds invented by Fermi to accelerate cosmic rays. This cometary environment could be viewed as a small-scale model prototype for galactic cosmic ray shocks. Many cosmic ray physicists consider the shocks produced by supernova explosions to be the main source of the accelerated cosmic ray particles.

Far from the nucleus of the comet, where the number of these newly born contaminating ions is small, the energy source of cyclotron turbulence is sufficiently weak that only very mild fluctuations in the magnetic field were measured at frequencies corresponding to the heavy ion cyclotron resonances. Approaching the nucleus, the spacecraft detected more and more heavy ions, and the amplitude of the MHD turbulence increased. It is quite remarkable that quasilinear theory, without any special tricks, was able to provide a satisfactory explanation of this phenomenon. It explained the spatial variation of the amplitude of the turbulence, and it also gave a reasonable explanation for the Fourier spectra of the MHD turbulence.

On the other hand, developing the analogy with cosmic ray acceleration, we can consider the small number of contaminating heavy ions as an energetic component that acts like "quasi" cosmic ray particles. Fortunately, because the spacecraft made *in situ* measurements, we were able to obtain both the energies of the particles and the spectrum of the turbulence. Hence, this was a unique case in which one could make a direct check of the Fermi theory, using modern plasma theory tools such as the quasilinear equations, and then compare with the data for the measured energy distribution of the accelerated ions. Of course, these energies are not in the energy range for actual cosmic rays; however, with an appropriate scaling law, one can extrapolate these mechanisms to actual cosmic rays. Even in this case, nonlinear plasma theory explains the energy spectra of these energetic particles rather well.

The bow shock wave at a comet is quite different from typical solar wind shocks at a planetary magnetosphere. The effective size of the obstacle, which is responsible for bow shock formation, is much larger for a comet than for a planet. The effective size for a comet is the spatial scale for contamination, or mass loading, of the solar wind by the cometary plasma. The measured bow shock stand-off distance is consistent with theoretical estimates that are based on solar wind mass loading arguments.

The problem of cometary bow shock structure has great importance for the physics of collisionless shocks in a plasma, the first theory of which emerged in the late 1950's. The shock registered by the Vega-1 spacecraft was identified by a sharp enhancement of extremely-low-frequency (ELF) waves and by a jump in the magnetic field strength. Thus, it is similar to the strong quasi-perpendicular bow shock of the Earth. The only difference is that the role of the ions that are reflected from the shock front is played here by the cometary ions, which leak easily from the shocked to the unperturbed solar wind, where they are accelerated in a self-consistent electric field. The inbound crossing of the bow shock by Vega-2 was characterized by a smooth increase in plasma waves and plasma heating. This is consistent with an almost field-aligned solar wind plasma flow, in which case the cometary shock is similar to the cosmic ray shock and the energy dissipation is caused by the thermal conductivity of cometary ions.

The solar wind loading by cometary ions that continues behind the shock results in a gradual deceleration of the plasma flow and a magnetic field build-up. Both Vega-1 and Vega-2 reached the outer edge of a thick magnetic barrier separating loaded solar wind flow from unmagnetized outflow of cometary plasma.

NATURE OF THE COMETARY NUCLEUS

The Vega missions, together with the Giotto encounter with Halley's comet and the earlier ICE flyby of the Giacobini-Zinner comet, were important not only from the point of view of plasma physics, but also because they produced other kinds of significant physics results. In March of 1986, Halley's comet lost its privileged status as a purely astronomical object, as had Giacobini-Zinner's comet earlier. Now, after the excitement over the space encounters has faded and the data has been digested for over a year, we may compare the findings of the space probes with the results that had been known from remote sensing. As brightly as comets can shine in the sky as they approach the Sun, the remotely observable pattern still represents only the outflow of the comet, i.e., the expansion of gases and dust in interplanetary space. The gases and dust must have been released by a source, which, largely obscured by them, is an astronomically unresolved point-like object. At any given moment, the outflow that is observed through scattered and re-radiated solar light accounts for only a tiny fraction (less than 10^{-3}) of the mass of the source.

Several decades ago, the astrophysicist F. Whipple invented the notion of the cometary nucleus: a hypothetical icy body capable of sublimation due to solar radiation heating and thus serving as the source. Qualitative and even quantitative speculations about such a nucleus used to be based on remote sensing chemistry of the outflowing gases, with water vapor being the major constituent, accompanied by an ever-growing list of other species. That list has now been completed by mass spectrometry measurements onboard the space probes. The overall experimental arrangement involved a few dozen different instruments aboard several spacecraft. In addition to a brief description of the plasma experiments, I will now describe how we finally made an estimate of the mass and the density of the cometary nucleus.

To find the density, we needed the mass and the volume of the nucleus. The volume was evaluated by an analysis of the imaging data from the Vega-1 and Vega-2 probes,[2-5] and hence the problem was reduced to the determination of the mass. The traditional way to do this is to measure the gravitational effect of the celestial body being studied on the motion of some other body. With a comet, this method is impossible because the mass of the cometary nucleus is too small. To measure a comet's mass, one must consider a different effect, one that is nongravitational. The flow of the substance ejected by the nucleus creates a force influencing the cometary orbit. In this sense, therefore, a comet is like a rocket, and the mass of its nucleus can be determined, if the force and acceleration are known.[6]

The presence of this "jet propulsion" force provides a systematic difference of $\Delta P \cong 4$ days between the observed and the "gravitationally" predicted times for the perihelion passage of comet Halley. This force is equal to

$$F = K \dot{M} \bar{u}, \tag{1}$$

where $\dot{M}$ is the mass production rate, $\bar{u}$ is the flow velocity, and K is the factor describing the degree of asymmetry of the flow ($K = 0$ for spherically symmetric flow and $K = 1$ for one-dimensional flow). Let us suppose that the vector for the force F lies in the orbital plane. Then the orbital integral of the energy leads to a simple relationship for the mass:

$$M = \frac{3aP}{\mu_\odot \Delta P} \int_{t_1}^{t_2} K \dot{M} \bar{u} v \cos(\beta - \alpha) dt, \tag{2}$$

where a is the semimajor axis, P is the orbital period, $\mu_\odot$ is the solar gravitational constant ($GM_\odot$), v is the absolute value of the velocity of the comet, β is the angle between $\vec{v}$ and the Sun-to-comet direction, α is the angle between the direction of the flow and the comet-to-Sun direction, and t_1 and t_2 are the times for the turn-on and cut-off of cometary activity, respectively.

At first glance, Eqs. (1) and (2) may appear oversimplified, since a real mass flow involves different components with different values of K, $\dot{M}$, u, and α. However, the dust velocity near the surface is practically zero, because the dust particles need time to be accelerated by the gas. Thus, we can take into account the gas only. The main constituent of the cometary gas is H_2O. The mass production for the gas is quite well understood.

The asymmetry factor K was evaluated as follows. Images obtained by the Vega and Giotto probes showed that only the illuminated side of the nucleus is the source of the cometary dust. Near-infrared spectrometry on Vega also directly demonstrated a strong asymmetry of the H_2O molecule flux. Note that $K = 2/3$ if the local flow is proportional to the solar illumination (for a spherical shape), with the maximal possible value being $K = 1$. For the nominal model, we assumed the mean of these two limits, viz., $K = 5/6$, and for the other two models, we took $K = 1$ and $K = 2/3$.

As for the angle α, in principle it cannot be zero, owing to the rotation of the nucleus and the thermal inertial of its surface layer. Measurements of the infrared thermal emission of the nucleus by Vega-1 indicated that the temperature maximum lags by approximately $10°$. However, the effective direction of the gas flow does not necessarily coincide with the thermal maximum. Note that with $\alpha = 0$, the jet propulsion effect would have no influence on the period if the mass production function $\dot{M}(\vec{r})$ is symmetric in shape. However, the values for $\dot{M}(\vec{r})$ based on the continued observations of Halley's comet with space telescopes show a pronounced asymmetry, so that $\Delta P \neq 0$ even with $\alpha = 0$. We chose $\alpha = 5°$ as the upper limit, combining this value with the maximal model for $\dot{M}(\vec{r})$. For the minimal model we used $\alpha = 0°$ and for the nominal model, $\alpha = 2°$.

Table 1 gives values for the mass and the density calculated for all three models.[7] The volume is assumed to be $V = 5 \times 10^{11} m^3$, corresponding to the $16\,km \times 18\,km \times 8\,km$ ellipsoid that was used to approximate the irregular shape of the nucleus obtained from the Vega images.

TABLE 1 Mass, density, relative mass loss $\Delta M/M$, and characteristic decay time t_0 from three models of comet Halley

Model	Mass $(10^{11}$ ton)	Density $(g\,cm^{-3})$	$\Delta M/M\,(\%)$	$t_0(yr)$
Nominal	2.9	0.6	0.14	54,000
Minimal	1.1	0.2	0.50	15,000
Maximal	7.6	1.5	0.05	150,000

The nominal model corresponds to an average density of $\rho = 0.6\,\mathrm{g\,cm^{-3}}$. For the minimal model, it is approximately three times smaller, and for the maximal model approximately three times larger. The relative mass loss $\Delta M/M$ per revolution is also given in Table 1, along with the characteristic time t_0 that corresponds to exponential decay of the cometary mass with constant $\Delta M/M$ and P. The real lifetime of the comet is much longer than t_0, due to gradual weakening of the activity and decreasing $\Delta M/M$. There is the possibility that the eventual fate of at least some comets is that they will "decay" into asteriods after the termination of their activity.

This evaluation of the density has physical implications with regard to determining the nature of the cometary nucleus. A compact structure of ice, mixed with stony material, could be acceptable if the density were nearer to that of the maximal model.

CONCLUSION

Eventually the sad moment came when we were no longer able to see Halley's comet and we had to say good-bye to it for seventy-six more years. I am reminded of the statement (quoted yesterday by Bruno Coppi) from Freeman Dyson's book,[8] that the universe somehow probably knew that we were coming. In view of the seventy-six year interval until the reappearance of Halley's comet, I would like to ask if the universe knows whether we are staying.

That type of question is, of course, much more difficult to answer than most of the problems in physics for which we were trained as theoretical physicists. I consider myself to be a student of Landau and a few other Soviet physicists. I am also proud to consider myself a student of Marshall Rosenbluth. When I met him for the first time I was only twenty-five years old. He was probably thirty years old at the time and already an experienced teacher.

In conclusion, let me tell a true story. After the encounter with Halley's comet, a group of scientists was invited by General Secretary Gorbachev to brief him about the comet Halley and the Vega project. The very first question he asked was, "How could you manage to encounter Halley's comet on the last day of the party congress?" My answer was, "We have the perfect alibi: the orbits of the comets are controlled by God." He then said, "That means that God is with us." (This part of our conversation has been published in the magazine called *The Communist*, which is printed by the Central Committee.) Not long afterwards I received a proof of the essence of this story, namely, that "He is with us." A few months after this conversation, a larger group of scientists—including Europeans, Americans, and Japanese—was invited to the Vatican by the Pope to describe the exploration of Halley's comet. Representatives from the various space agencies had the opportunity to give five-minute lectures each about the results from their respective projects. All of the results had been conveniently compiled in one book,[9] with Giotto di Bondone's depiction of the comet Halley on its title page.[10] Each speaker at the appropriate time had only to invite the Pope to open the book to the particular page indicated by a marker of a certain color. The page on which I presented comments concerned the Vega observations of Halley's comet. It was a great pleasure for me to invite the Pope, "Your Holiness, please open to the page indicated by the red (!) marker." Later, when I left the audience, I also realized that it was the seventh of November, which is exactly the day on which the October Revolution occured. This all proves that "God is with us."

Just as the first copy of this book on the encounter with Halley's comet was presented to the Pope, I now take special pleasure in delivering another copy to Marshall Rosenbluth, the "godfather" of plasma physicists.

REFERENCES

1. A description of plasma phenomena observed by various cometary space probes may be found in A. A. Galeev, *Astronomy and Astrophysics* **187**, 12–20 (1987). [The contributions to this special issue on Halley's comet have also been published, together with several review papers, in the book *Exploration of Halley's Comet*, edited by M. Grewing, F. Praderie, and R. Reinhard (Springer-Verlag, Berlin, 1987).]
2. R. Z. Sagdeev *et al.*, *Nature* **321**, 262–269 (1986).
3. R. Z. Sagdeev *et al.*, in *Proceedings of the 20th ESLAB Symposium: Exploration of Halley's Comet* (ESA SP-250, Paris, 1986), Vol. II, pp. 307–316.
4. H. U. Keller *et al.*, *Nature* **321**, 320–326 (1986).
5. H. U. Keller *et al.*, in *Proceedings of the 20th ESLAB Symposium: Exploration of Halley's Comet* (ESA SP-250, Paris, 1986), Vol. II, pp. 347–350.
6. F. Whipple, *Astrophysical Journal* **111**, 375–394 (1950).
7. R. Z. Sagdeev, P. E. Elyasberg, and V. I. Moroz, *Nature* **331**, 240–242 (1987).
8. Freeman Dyson, *Disturbing the Universe* (Harper and Row, New York, 1979), p. 250.
9. *Encounter '86: An International Rendezvous with Halley's Comet*, edited by N. Langdon (European Space Agency, Paris, 1986, ESA-BR-27).
10. The European Space Agency named its spacecraft "Giotto" because it is thought that the Italian painter Giotto di Bondone depicted the A.D. 1301 apparition of the comet Halley in place of the usual Star of Bethlehem in a fresco, *The Adoration of the Magi*, in the Scrovegni Chapel in Padua. [See R. J. M. Olson, *Scientific American* **240**, 160–170 (1979), and R. J. M. Olson and J. M. Pasachoff, *Astronomy and Astrophysics* **187**, 1–11 (1987).]

Charles F. Kennel*
Department of Physics and
  Institute of Geophysics and Planetary Physics
University of California, Los Angeles
Los Angeles, California 90024

Cosmic Ray Acceleration:
A Plasma Physicist's Perspective

The present theory of acceleration of cosmic rays by supernova shocks will be reviewed impressionistically. The astrophysical arguments underlying a "standard model" of shock acceleration will be summarized, but not criticized. The problems of plasma physics posed by the standard model will be discussed and major unsolved issues will be emphasized. Observations of shocks in the solar system and numerical simulations of the scattering of energetic particles from finite amplitude MHD turbulence, both of which may be pertinent to the standard model, will be milked for their possible significance.

INTRODUCTION

That I should accept the invitation to speak about astrophysical plasma physics to this symposium was obvious from the beginning. To be asked to participate in

*On sabbatical leave at California Institute of Technology, Pasadena, California

honoring Marshall Rosenbluth is a significant honor in itself. But what astrophysical topic should I choose? That was the problem. What topic would suit this audience, and Marshall, who are so used to rigorous and good plasma physics? Ultimately I chose to discuss cosmic ray acceleration with you. Let me tell you why.

I have it on good authority that an august panel of the National Academy of Sciences, in a receptive mood, once asked Lyman Spitzer whether the Academy ought to "do something for plasma astrophysics." Spitzer is reputed to have said no. Since it will never be possible to measure the microscopic plasma processes that regulate the transport (and sometimes even the macroscopic configurations) of astrophysical plasma physical objects, the subject was inherently speculative, and nothing organized need be contemplated.

That was some years ago. Since then we have developed two strategies—numerical simulations and *in situ* measurements of analogous plasma configurations in the solar system—that offset (somewhat) what is the fundamental disability of astrophysical plasma physics: our inability to measure the important plasma microprocesses.

I chose cosmic ray acceleration for two reasons. One is that it is built on the solid foundation of the Alfvén Ion Cyclotron instability, which our honoree, Marshall Rosenbluth, and Roald Sagdeev constructed in the 1950's. The second reason is that a rather tight chain of reasoning put together by my colleagues in the astrophysics and cosmic ray communities has reduced cosmic ray acceleration to a well-posed plasma problem. This problem, which has been addressed by numerical simulations and by solar system observations, is probably the most successful example to date—perhaps the only one—in which the interactions between plasma physicists, space physicists, and astrophysicists have brought a major problem into clear focus.

I hope you will forgive me—maybe you will even thank me—for skipping over most of the technical details. I will, for example, outline some of the astrophysical arguments that pose what I have called the "'standard model" of cosmic ray acceleration, but I will not discuss some of the astrophysicists' doubts about the standard model. Our main interest will be the plasma physics implied by the standard model. Here, I will concentrate on the interesting and largely unsolved questions of plasma physics more than I will on the various detailed answers that have been proposed.

SOME BASIC FACTS ABOUT COSMIC RAYS

Here I remind you of some basic facts about cosmic rays that have been developed over decades by the cosmic ray and astrophysical research communities. My presentation draws largely on the excellent and up-to-date review articles by Roger Blandford[1] and by Blandford and David Eichler.[2]

The cosmic ray energy density is comparable with that of the interstellar plasma and probably exceeds that of the interstellar magnetic field. The galactic cosmic rays with energies above 3 GeV, which is sufficiently high that they escape modulation by the solar wind, are remarkably isotropic, to one part in 10^4 for protons. This suggests that the cosmic rays do not propagate directly to us from their sources, but diffuse with scattering mean free paths so short that their flux is isotropic to very high order. The energy spectrum of galactic cosmic rays is remarkably smooth and uniform over a remarkably wide range of energies:

$$I(E) \sim E^{-2.7} \text{p/cm}^2 \cdot \sec \cdot \text{sr for 3 GeV} < E < 10^5 \text{GeV}. \tag{1}$$

Small wrinkles in the spectrum near 10^5 GeV can be interpreted as signalling the beginning of the change to a predominantly extragalactic origin for the highest energy cosmic rays.

By comparing the energy dependences of the fluxes of secondary cosmic rays, such as Li, Be, and B, which are produced by spallation reactions of the primary cosmic ray flux with the interstellar medium, it is possible to infer the mean lifetime of cosmic rays in the galactic disk. This lifetime is the order of 20 million years at GeV energies and decreases approximately as $E^{-0.5}$. Thus, the primary acceleration spectrum is approximately

$$I_o(E) \sim E^{-2.2}. \tag{2}$$

The experts will argue about the second significant figure in the above power law index, but for us, it is sufficient to say that the primary goals of cosmic ray acceleration theory are, first, to account for the overall cosmic ray intensity and, second, to explain a power law index of about 2.2 that extends over several orders of magnitude in energy.

Knowing the flux we observe and the cosmic ray lifetime, it is possible to estimate the rate at which cosmic ray energy leaves the disk of the galaxy to enter the halo, possibly to escape to the intergalactic medium. This rate is about 3×10^{40} ergs/sec, give or take a factor of 3—a most impressive rate, indeed. As we will argue shortly, there are compelling reasons to believe that the cosmic rays are accelerated out of the interstellar medium. Supernovae constitute by far the largest energy input to the interstellar medium, and even they must convert about 3% of the energy released in each explosion (exclusive of neutrinos) to cosmic rays, in order to account for the above galactic energy loss rate. Since any other mechanism would have to be even more efficient, it is certainly simplest and reasonable to assume that galactic cosmic rays are accelerated by supernovae. There is some suggestive circumstantial evidence: we can observe the synchrotron radiation from electrons accelerated by the interaction of supernova remnants with the interstellar medium. The spectral indices of the synchrotron photons (a typical value might be $\alpha = 0.6$) imply a primary electron energy spectral index of 2.2, about that inferred for the spectral index for freshly accelerated cosmic ray ions.

The cosmic rays are probably not accelerated in the explosion of the star itself. Any promptly accelerated particles ejected by the explosion would lose energy to the subsequent adiabatic explosion of the supernova remnant, since the energetic particles are tightly bound to the fluid by Alfvén wave turbulence. Besides, there is precious little evidence of the nuclear and isotopic abundances in the cosmic rays that would signal the fact that they originated in the nuclear cataclysm of a stellar explosion. If anything, the cosmic ray abundances appear to be closer to those expected for the interstellar medium—a medium whose composition has been enriched in iron (mainly) relative to solar composition by the stellar nucleo-genesis cycle. Moreover, it appears that the cosmic rays are accelerated out of the interstellar medium in one event. For example, there cannot be a significant pool of pre-accelerated particles with energies < 1 GeV awaiting the next supernova event to be kicked to higher energies; these would lose energy to the interstellar medium so rapidly by ionization that they would make the energy constraints on the acceleration efficiency even more difficult to satisfy.

In sum, most theorists assume as a general working model that the cosmic rays are accelerated to energies well above the GeV range with an $E^{-2.2}$ spectrum in one supernova event and that the cosmic rays from various blended sources diffuse to earth where we observe them.

The above arguments suggest that it is a long-lived feature of the interaction of supernova remnants with the interstellar medium that accelerates the cosmic rays. The shock wave that must precede every expanding remnant into the undisturbed interstellar medium immediately comes to mind. Radio observations tell us that such shocks certainly accelerate relativistic electrons, because the synchrotron radio intensity increases suddenly at the outer boundary of the remnant, about where measurements of bremsstrahlung X-rays indicate that few KeV electrons are being heated downstream of a sharp discontinuity.

We are therefore led to investigate how ions are accelerated by the interaction of supernova shocks with the interstellar medium. Such shocks can last some tens of thousands of years before they weaken into insignificance, and the hope is that, as we shall see, particles have enough time to diffuse back and forth across the shock front to achieve the high energies that are observed. The supernova remnant and its associated shock decelerate as the shock sweeps up interstellar matter, so that for most of its lifetime, the Mach number of the shock is relatively low— comparable with those we encounter in the interplanetary medium. Moreover, most of the volume of the interstellar medium into which the supernova shock propagates is in the so-called "hot low density" phase, a fully ionized plasma phase, whose temperature is comparable to that of the solar wind (since it comes from the blended winds of stars), but whose density is smaller than that of the solar wind. It is difficult to measure the magnetic field in the hot low density phase, but it is thought that β, the ratio of thermal to magnetic energy density, probably exceeds ten.

The above arguments suggest that it will be profitable, despite the known strong dependence of collisionless shock structure upon plasma parameters, to examine how shocks in the solar systems accelerate energetic particles, not only for its own sake, but also to see what the solar system tells us about cosmic ray acceleration. Certainly, solar system shocks are the best hope we have for studying the microphysics of particle acceleration by shocks.

THE "STANDARD MODEL" OF PARTICLE ACCELERATION BY SHOCKS

Fermi was the first to suggest that scattering from moving magnetic irregularities would produce the power law distribution characteristic of cosmic rays. The properties of the scattering magnetic clouds have to be carefully tailored to produce the observed spectral index, and nowadays Fermi's magnetic clouds have been replaced by large amplitude MHD turbulence—Alfvén waves. The waves, which act as scattering centers, are set in systematic relative motion one to another over a large spatial scale by a collisionless shock. In the shock frame of reference, the waves upstream are blown towards the shock at highly super-Alfvénic speeds and then transmitted through the shock into the downstream sub-Alfvénic flow, where they propagate slowly. Therefore the upstream and downstream wave distributions are in relative motion, and a particle which scatters from the waves back and forth across the shock systematically gains energy. Its net energy gain depends on the number of times it crosses the shock and not on the wave amplitudes upstream and downstream, so long as the amplitudes are large enough over a large enough distance to reflect a particle back towards the shock. The MHD turbulence can be present in the interstellar medium before the shock sweeps over it (and indeed for the longest wavelengths required it may well have to be) or it can be generated by the energetic particles themselves via the Alfvén ion cyclotron instability.

There has evolved a relatively simple picture of Fermi accelerating shocks that in broad outline fits our observations of quasi-parallel interplanetary shocks and planetary bowshocks in the solar system—those shocks that propagate sufficiently parallel to the upstream magnetic field that shock-heated particles can escape freely into the upstream region from downstream to start the Fermi acceleration process.[3] The overall shock consists of a thin subshock (in which the dissipation is due to the "conventional" microturbulence responsible for "ordinary" collisionless shock structure), a broad foreshock upstream, and a postshock region in which energetic particles interact with MHD turbulence. We emphasize that although many people would call the subshock the shock itself, the entire shock includes the subshock, the foreshock, and the postshock region.

The energetic particles, which are approximately isotropic in the frame of the shock, have a pitch angle anisotropy of the firehose sense ($p_\parallel > p_\perp$) in the frame of the upstream foreshock plasma and so destabilize Alfvén waves, which amplify as they are carried by the super-Alfvénic flow towards the shock. The linear instability theory firmly states that circularly polarized waves propagating parallel to the upstream magnetic field grow the fastest. This is illustrated in Figs. 1(a) and 1(b).

Figure 1(a) shows a schematic of the Fermi acceleration process. The top panel shows the flow velocity profile, from upstream to downstream, of an accelerating shock. Far upstream, the flow velocity is much larger than the Alfvén speed; far downstream, it is somewhat less than the Alfvén speed. The overall shock consists of a broad foreshock, in which the upstream flow is gradually decelerated by its interaction with the energetic particle pressure gradient, and a thin subshock,

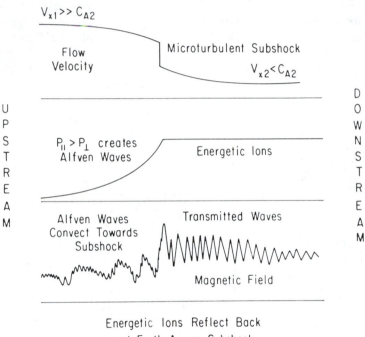

FIGURE 1(a)   Fermi acceleration schematic: flow velocity profile of an accelerating shock, from upstream to downstream (top); profile of the energetic ion pressure (middle); Alfvén waves (bottom).

across which the thermal plasma is heated by microturbulence. The middle panel of Fig. 1(a) shows the profile of the energetic ion pressure. It increases exponentially in the foreshock and is constant downstream. The ion pressure, which is approximately isotropic in the plane of the shock, has a $P_\parallel > P_\perp$ anisotropy in the frame of the upstream flow. This anisotropy destabilizes circularly polarized Alfvén waves, which grow as they are carried by the super-Alfvénic flow toward the shock. The bottom panel of Fig. 1(a) sketches the Alfvén waves, which grow in amplitude as they approach the subshock. As they are transmitted through the subshock, they

FERMI ACCELERATION

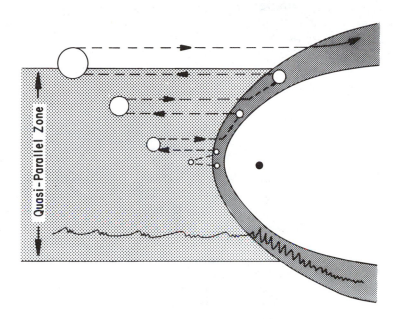

FIGURE 1(b) Fermi acceleration at the earth's bow shock.

couple to a variety of MHD modes, and most of this wave energy propagates downstream. If the large amplitude waves upstream and downstream can scatter particles leaving the shock (in either direction) back towards the subshock and the scattered particles can cross the subshock, the cycle of upstream and downstream scattering results in a net energy gain.

Figure 1(b) sketches our current understanding of how Fermi acceleration might operate at the earth's bow shock. Shown is the special and optimum case for which the solar wind magnetic field is parallel to the flow velocity. In this case, the magnetic field lines connected to the portion of the bow shock where it is quasi-parallel occupy the lightly shaded region. The more heavily shaded region is the postshock flow downstream of the shock. The shaded regions are observed to contain intense MHD turbulence. A shock-heated superthermal proton starts the process off by wandering upstream. As it scatters back toward the shock from the MHD turbulence, its energy increases, as is indicated by the sizes of the white dots in Fig. 1(b). As it scatters, it also diffuses laterally, and it eventually escapes the Fermi acceleration zone. This shows that Fermi acceleration becomes more efficient the larger and the longer lived the shock system. Many of the energetic particles observed in the vicinity of the earth may have been accelerated by processes inside the magnetosphere as well, and so the best way to evaluate Fermi acceleration theory is to use interplanetary shocks, which are very much larger than the earth's bow shock.

If we now assume that energetic particles scatter back and forth across the shock, and that the scattering Alfvén wave distribution has a mean phase speed of $V_1 \gg C_{A1}$ upstream (in the shock frame) and a speed $V_2 \approx C_{A2}$ downstream, the resulting momentum distribution will be a power law whose spectral index is related to the shock compression ratio, i.e.,

$$f(p) \sim p - \beta \tag{3a}$$

$$1/\beta = \frac{1}{3}[1 - V_2/V_1] \tag{3b}$$

Since $V_2/V_1 \leq 1/4$ for all shocks, it is immediately clear that the spectral index in Eq. (3b) expected from shock acceleration is comparable to what is needed for cosmic rays. This fact, which was discovered virtually simultaneously by Axford et al.,[4] Bell,[5,6] Blandford and Ostriker,[7] and Krymsky,[8] has encouraged people to work in earnest on the plasma physics of shock acceleration ever since.

Of course, many questions leap into mind as soon as the significance of Eq. (3b) registers. For example, not all shocks have a compression ratio of 4, which applies to the strong shock limit; the observed spectral index requires an MHD Mach number of 2-3; how do the cosmic rays "pick out" these and only these shocks? The cosmic ray pressure gradient itself decelerates the incoming fluid, so that the velocity of the scattering centers does not jump suddenly from V_1 to V_2 at the subshock, but decreases slowly in the foreshock as shown in Fig. 1. This self-consistent effect will modify the spectrum of accelerated particles somewhat. The models that trace out the evolution of the foreshock-subshock system as a supernova remnant expands are being developed now.

These models are necessarily oversimplified. The simplest picture would have a regular, spherical shock expanding into a uniformly magnetized, fully ionized region. The top panel in Fig. 2 shows the ideal supernova remnant (SNR) interaction. The supernova creates a spherical blast wave which propagates into a uniformly magnetized, fully ionized interstellar medium (ISM). Our present understanding indicates that Fermi acceleration would be most efficient where the shock is quasi-parallel.

In fact, the shock fronts in supernova remnants are observed to be highly irregular and nonsymmetric, and the interstellar magnetic field undoubtedly has structure that we cannot observe. The supernova shocks can sweep over cold clouds. This more realistic picture of the interaction of supernova remnants with the interstellar medium is shown in the bottom panel of Fig. 2. The observed supernova remnants

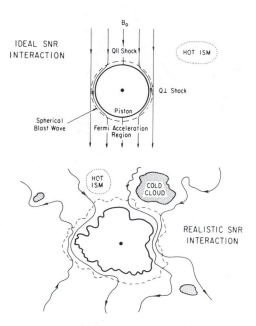

FIGURE 2   An idealized picture (top) and a more realistic picture (bottom) of the interaction of supernova remnants with the interstellar medium.

have highly irregular surfaces and are at best quasi-spherical overall. The magnetic field upstream of the expanding blast wave is essentially unknowable, but it would be naive to assume that it is simple. Finally, the interstellar medium has a two phase structure, and some SNR's overrun cold, partially ionized interstellar clouds in the course of their expansion. All these effects complicate the cosmic ray acceleration problem. The complex magnetic field topology, however, could conceivably help by confining accelerated particles near the shock front.

It is possible that we will encounter insuperable astrophysical difficulties, but in my opinion, the issues of basic plasma physics are equally central, and it is to these that I now turn my attention.

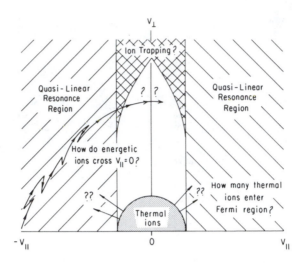

FIGURE 3   Plasma physics of the standard model of cosmic ray acceleration.

KEY QUESTIONS OF PLASMA PHYSICS ASSOCIATED WITH THE STANDARD MODEL OF COSMIC RAY ACCELERATION

Figure 3 schematically illustrates some of the important issues of plasma physics associated with our present picture of cosmic ray acceleration. This figure attempts to illustrate questions associated with the standard model, by showing a $(V_\perp, V_\parallel)$ map of velocity space, supposedly downstream of the subshock. The dark region shows shock-heated thermal ions diffusing to energies where they can interact with MHD turbulence and can migrate back and forth across the subshock easily. The regions where energetic particles can interact with Alfvén turbulence via first-order ion cyclotron resonance are separated by a region where no resonant waves are expected and where the quasilinear pitch angle diffusion coefficient should be zero. How, then, do particles diffuse across $V_\parallel = 0$? The final question is: How do particles diffuse off the page, i.e., to high energies?

The first issue associated with the standard model is the so-called "seed-particle" or efficiency problem. Since the cosmic rays appear to be accelerated directly out of the interstellar medium, we must ask how an originally cold plasma particle begins to participate in the Fermi process. We expect the thermal interstellar plasma to be heated by the collisionless dissipation in the subshock, and that some of these subshock-heated particles will acquire sufficient energy to stream along field lines, catch, and pass through the subshock from behind. When these seed particles get upstream, there must be enough of them to stimulate the growth of the Alfvén waves resonant with them to amplitudes sufficient to reflect them back to the shock from upstream. They then are subject to Fermi acceleration thereafter.

Only one step in the above chain of processes is reasonably well understood. A subshock-heated ion with two or three times the downstream thermal speed can easily catch the subshock, provided the subshock propagates at angles less than about 45° to the upstream magnetic field direction—i.e., the subshock is "quasi-parallel."[3] The remaining issues cannot be settled until we have a clearer picture of the microdissipation in the subshock. The rate at which this microdissipation creates a superthermal tail in the downstream ion distribution determines how many seed particles are produced and determines the overall acceleration efficiency. Thus the rate of diffusion across the boundary in phase space between the thermal and Fermi particle regions acts as a valve which regulates the number of particles that can be subject to Fermi acceleration.

The second issue is the "reflection problem." Linear theory predicts that a proton with a component of velocity, $V_{\parallel}$, parallel to the magnetic field will resonate with circularly polarized Alfvén waves according to the cyclotron resonance condition

$$V_{\parallel} = \frac{\omega - \Omega_0/\gamma}{k_{\parallel}} \qquad (4)$$

where Ω_o is the ion cyclotron frequency and γ is the proton Lorentz factor. In addition, the linear instability theory indicates that the waves with $\omega \simeq \Omega_o/\gamma$ that would resonate with small $|V_{\parallel}|$ particles are absent from the spectrum. The quasilinear theory of ion pitch angle scattering, which retains the linear resonance condition above, predicts that the pitch angle diffusion coefficient will be zero at 90° pitch angle. Thus there is no way in quasi-linear theory to reverse the particle's parallel velocity, as the Fermi mechanism requires.

In short, shock Fermi-acceleration is necessarily a strong turbulence process. Simple energy balance arguments indicate that the MHD waves will reach order unity amplitude near the subshock, so that we do not expect quasilinear theory to apply anyhow. Most theorists simply assume that ion reflection occurs and move on, but precisely how this comes about is not well understood. One possibility, sketched in Fig. 2, is that gyro-phase trapping in finite amplitude waves will broaden the wave particle resonance at high perpendicular momentum, as indicated, so that there is indeed a non-zero pitch angle diffusion coefficient at small $p_{\parallel}$. Another possibility is that particles mirror from local peaks in the magnitude of the turbulent magnetic field.

The third issue, and the one perhaps most difficult to solve in the long run, concerns how particles diffuse to very high energies. For example, using a quasi-linear estimate for the scattering rate (assuming, however, that particles diffuse across $p_{\parallel} = 0$), Galeev et al.[9] have found that the exponential scalelength of the energetic particles in the foreshock becomes comparable to the radius of curvature of typical supernova remnants at about 3×10^{11} eV, energy, well below 10^{14} eV. Furthermore, there are not enough 10^{14} eV particles to destabilize the very long wavelength waves needed to scatter them. There are two points of view about this.

The first is to say that a few times 10^{11} eV is all one gets from supernova shocks and to look for other shocks of galactic scale to boost particles to higher energies. At the other extreme, one can argue that the uniformity of the spectrum suggests that all the particles to 10^{14} eV are accelerated by the same process in a self-similar way. This alternative presents by far the more challenging theoretical problem. It implies that the scattering rate at high energies has to be much faster than quasilinear and is not governed by the linear resonance condition.

The high energy limit will be very difficult to attack using either numerical simulations or measurements in interplanetary space because of the very large range of spatial scales involved in scattering particles with a spread of 10^5 in Larmor radius.

After we have discussed observations of shocks in the solar wind, we will propose another issue: whether the MHD waves are circularly polarized and propagate parallel to the magnetic field, as the linear growth rate would predict, or whether they refract to become highly steepened, elliptically polarized waves, as some solar system observations suggest.

SOLAR SYSTEM OBSERVATIONS PERTINENT TO THE STANDARD MODEL

Before we go much further, we should ask whether anything that looks like the standard model even occurs in the solar system. Here I will be guilty of describing only work with which I have been involved; I hope my space physics colleagues who happen to read this article will forgive my failure to cite their contributions here. I have tried to do them justice, however, in a detailed review article on solar system shocks.[10] Moreover, I do not wish to imply that Fermi acceleration is the only way that particles in the solar system are accelerated, or even the only way that solar system shocks accelerate particles. I am going to concentrate on solar system tests of Fermi-acceleration theory, because of their pertinence to the standard model.

It now seems clear that quasi-parallel interplanetary shocks do have extended foreshocks that contain energetic protons and finite amplitude MHD turbulence. Moreover, planetary bowshocks also have foreshocks upstream of those portions of the curved shock surface where the shock is locally quasi-parallel. However, it is much easier to study the full spatial extent of the foreshock in the interplanetary shock case, because the measurements are not aliased by finite radius of curvature effects.

The interplanetary shock which was encountered by the ISEE-3 spacecraft on November 11-12, 1978, was an especially good accelerator of energetic protons: the energy density of 35-56 KeV protons became comparable to that of the solar wind magnetic field just upstream of the shock. (We must remember that we cannot expect particles to become highly relativistic as they do in the supernova case, because the solar system shocks have only existed 2-3 days, as opposed to thousands of years, when we observe them. Therefore, we can only study the acceleration of nonrelativistic to mildly relativistic particles in interplanetary space). Several years ago, we began to assemble as many plasma and energetic particle diagnostics of this shock as possible, with the ultimate goal of testing Lee's detailed version of the particle acceleration theory.[11] The final paper in our series[12-14] has recently been published.

Figure 4 presents an overview of our measurements of the November 11–12 shock. It shows approximately 6 hours of data surrounding the subshock passage: data concerning the amplitudes of low frequency MHD turbulence (bottom panel), ion acoustic waves (middle panel), and the energetic particle fluxes in various energy ranges, the lowest energy range being 35-56 Kev (top curve in top panel).

The top panel in Fig. 4 shows the time profiles of energetic particles associated with the passage of a quasi-parallel interplanetary shock of fast Mach number 2.6 over the ISEE-3 spacecraft. Concentrate on the curves labelled (1) and (2) for 35-56 KeV and 56-91 KeV protons. Somewhat before 2230 UT on Nov. 11 (line labelled "bubble"), ISEE-3 entered a closed magnetic field region connected at both ends to the oncoming shock. This region contained enhanced fluxes of energetic protons and MHD waves (bottom panel). These fluxes were bound to the fluid by the waves, as is indicated by the decrease in proton anisotropy (third panel from bottom), whereas earlier they were streaming away from the shock. Other diagnostics shown include the spectral indices deduced from the ratios of fluxes in the energy channels indicated, and the amplitude of 3.16 KHZ electrostatic waves, thought to be in the ion acoustic mode. The proton fluxes in channels 1, 2, and 3 started to increase about 30-45 minutes before shock encounter, and they levelled off downstream, consistent with the predictions of Fermi acceleration theory. [For a detailed description of the figure, and a more precise discussion of all the panels, see Ref. 12.]

The November 11-12 shock was actually quite a complicated event. When ISEE-3 observed it, the shock was propagating in a closed magnetic island, and the activity near 2230 UT on November 11 was associated with the passage of the island over the spacecraft. Only later, at about 2345 UT, did the energetic proton fluxes below 200 KeV begin an exponential increase that leveled off at the subshock. Since this behavior is precisely that predicted by the standard model, we subjected the interval between 2345 UT and the shock encounter at 0028 UT to detailed study. Given the measured shock speed of 640 km/s, this time interval corresponded to a region 1.6 million kilometers in thickness ahead of the subshock in which the particle fluxes increased exponentially. The foreshock was therefore several thousand energetic proton Larmor radii thick.

The top panel of Fig. 5 shows the plasma velocity profile, and the bottom panel shows the energetic proton flux profiles in three different energy ranges that were measured around the time the subshock passed over the spacecraft. The proton flux in the foreshock increased exponentially in each energy range and then leveled off at the subshock. The upstream flux scalelength increased with increasing energy. All these features agree with Lee's detailed theory. Indeed, all our particle measurements were in good agreement with theory. The energetic protons had a power law

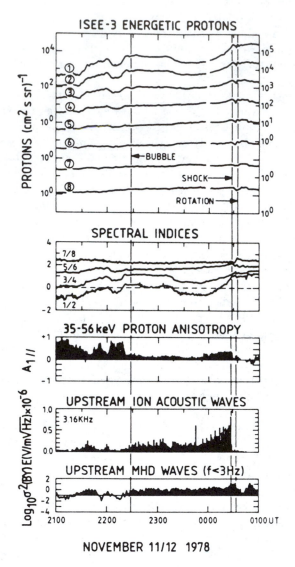

FIGURE 4   The November 11-12 interplanetary shock [From Ref. 12]

energy distribution whose spectral index agreed with the measured shock compression ratio. The energy dependence of the scalelength and the measured pitch angle anisotropies agreed with theory, as did certain details of the α-particle flux profiles.

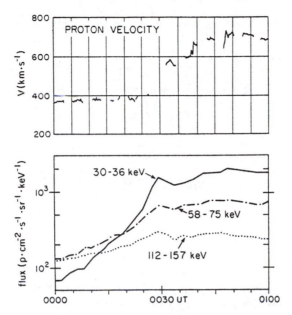

FIGURE 5 Particle fluxes and associated fluid velocity: The top panel shows the plasma fluid velocity measured as the subshock passed over ISEE-3. Since the shock speed was 640 km/s, the fluid velocity in the shock frame decreased from 260 km/s upstream to about 70 km/s immediately downstream of the subshock (0028 UT). The energetic proton fluxes increased exponentially upstream and levelled off downstream. The exponential scale length increased with particle energy, in quantitative agreement with theory. [From Ref. 14]

If Lee's theory—which used quasilinear scaling for the pitch angle diffusion rate but lost the precise correspondence relating protons of a given $V_\parallel$ with Alfvén waves of a given $k_\parallel$ embodied in the linear resonance condition—did well with the energetic particles, it fared poorly with the waves that scatter the particles. Figure 6 shows the magnetic field power spectral density measured in three time intervals, the first about one particle scalelength upstream of the subshock, and the last just before the subshock passed over the spacecraft. For all three time intervals, the spectrum was flat out to 100 mHz frequency and then fell off in a high frequency tail. The waves with frequencies below 100 mHz can resonate in linear theory with the observed energetic particles, whereas those above 100 mHz cannot.

Lee's theory[11] predicted an $f^{-7/4}$ frequency spectrum below 100 mHz, whereas a flat spectrum was observed. The flat portion of the spectrum did increase in amplitude as the subshock approached and reached an integrated amplitude $\delta B \sim (1/4)B_o$ at the shock. Thus the waves were far from the small amplitude regime. The trapping frequency based upon $\delta B/B_o \sim 1/4$ was about the width of the flat portion of the spectrum, so it is possible that wave growth saturated by gyro-phase trapping.

The high frequency tail of the wave distribution increased only in the last measurement just before the subshock passed over the spacecraft; this component can only have been created nonlinearly and should tell us something about how the MHD waves saturated.

A NUMERICAL EXPERIMENT PERTINENT TO THE STANDARD MODEL

Numerical simulations of the interactions between finite amplitude Alfvén wave turbulence and energetic particles are difficult indeed, because the wavelengths are much longer than a thermal ion Larmor radius and extend over a broad range, so it is not surprising that there have not yet been many simulations. Nonetheless, one is in progress by Zachary, Max, Arons, and Cohen,[15] using the massive computer facilities of the Lawrence Livermore National Laboratory. These authors have kindly permitted me to discuss their work in advance of publication. Any errors in interpreting their results are entirely mine.

The simulations were initialized with a power law flux of energetic protons of a given energy density, and the subsequent evolution of the particle and wave spectrum was followed. Since the simulations have one space dimension, only parallel propagating, circularly polarized Alfvén waves were generated. As we indicated earlier, the linear instability theory indicates that parallel propagating waves should grow the fastest. The power spectrum obtained after about one hundred gyroperiods had elapsed (Fig. 7) was similar to that of the November 11-12, 1978, interplanetary shock. The numerical spectrum contained a flat region of low frequencies and a high frequency tail. The waves observed at the November 11-12 shock were circularly polarized, just as in the numerical simulation. Thus, at least at the level of this simple comparison, the simulations have a degree of verisimilitude.

Zachary et al.[15] tested the quasilinear theory of pitch angle scattering by comparing two runs: one that arrived at a small $\delta B/B \sim 0.04$, and one that achieved $\delta B/B \sim 0.06$, about twice that reached by the November 11-12, 1978, shock. The diffusion rates measured in the low amplitude run essentially agreed with quasilinear theory, whereas the large amplitude case differed substantially. These results can be visualized by following the quantity μ, the cosine of the pitch angle, as a function of time for selected test particles. Several such μ-trajectories are shown in Fig. 8, whose left- and right-hand panels contain the small and large amplitude cases, respectively. Non-resonant particles oscillated in the wave-fields without diffusing in either case (top panels). The middle left panel shows a resonant particle that diffused in pitch angle in agreement with the quasilinear rate, and the bottom panel shows a particle that gradually diffused across $\mu = 0$ and so reversed the sign of its parallel velocity. Such particles were relatively rare, and the overall diffusion

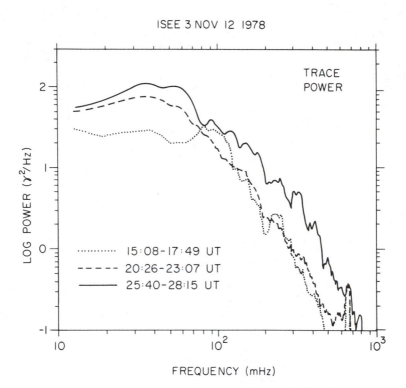

FIGURE 6   Development of wave spectrum: Shown are three magnetic field power spectra measured as the shock approached ISEE-3. The times these spectra were taken may be referred to the proton flux profiles in Fig. 5. The spectrum was flat at low frequencies (where the energetic ions have first-order cyclotron resonance) and fell off at high frequencies, where there is no resonant interaction. The high frequency tail increased its amplitude without changing its spectral distribution as the shock approached. [From Ref. 14]

rate was quasilinear. The large amplitude case (bottom two panels on the right) was profoundly different: the resonant test particles seem to diffuse slowly and then to take single large steps in μ. The overall diffusion rates are strongly influenced by the large "super-steps" in μ and are faster than quasilinear. Exactly what happens at a super-step is presently under investigation.

While the above results are preliminary and subject to revision, they promise to be an important help in understanding cosmic ray acceleration.

TIME EVOLUTION OF WAVE SPECTRUM
(Zachary et al, 1987)

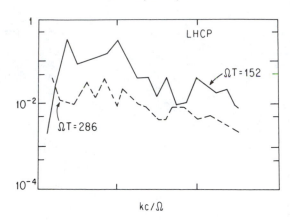

FIGURE 7   Numerical simulation of wave spectrum: This shows the spectral density of the left-hand circularly polarized component of the waves generated by energetic particles in the simulations of Zachary *et al.*[15] It is broadly similar to that in Fig. 6. [From Ref. 15]

WAVEFORMS ASSOCIATED WITH STRONG SHOCK-RELATED MHD TURBULENCE IN THE SOLAR SYSTEM

Figure 9 shows the time series of one vector component and the magnitude of the magnetic field taken of two times upstream of the November 11-12, 1978, interplanetary shock. The left-hand panel shows waveforms measured near 2345 UT at the leading edge of the energetic proton foreshock, and the right hand panel shows the waveform taken just ahead of the subshock. In both cases, the wave polarizations were circular, as is indicated by the large variations in the vector component and small variations in the magnitude. Far upstream, the vector amplitude was already large, but the waveform was relatively sinusoidal. Near the subshock, the waves, while they did not achieve a significantly larger amplitude, were apparently driven very strongly nonlinearly, as is evidenced by the rapid variations in phase and the associated modulations in the magnitude of the magnetic field. This behavior is reminiscent of the modulational envelopes expected for nonlinear Alfvén waves,[16] and is characteristic of parallel propagating, circularly polarized waves.

Now let us discuss the waveforms observed in the MHD turbulence associated with smaller shock systems. Figure 10 shows two kinds of waveforms found in the foreshock of the earth's bow shock. One waveform (top panel) circularly polarized, is

PITCH ANGLES OF SELECTED TEST PARTICLES

$n_{CR}/n_0 = 4 \times 10^{-3}$, $< \frac{\delta B}{B} > = .06$ $n_{CR}/n_0 = 4 \times 10^{-2}$, $< \frac{\delta B}{B} > = 0.6$

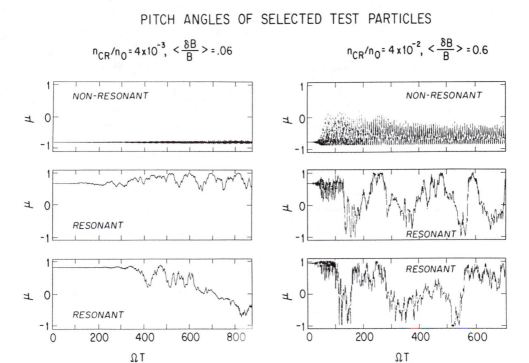

FIGURE 8   Pitch angle diffusion of test particles: The left- and right-hand columns show results for weak and strong turbulence, respectively. The top panels in each column plot the pitch-angle versus time for test particles that are far from linear first-order cyclotron resonance. The bottom two panels show resonant particle interactions. Note the large excursions in the strong turbulence case. [From Ref. 15]

hardly steepened although its amplitude is order unity, and only slightly modulates the magnetic field. These are the parallel Alfvén waves that the linear instability theory predicts. The other waveform is elliptically polarized and strongly steepened, and strongly modulates the field magnitude. The linear theory indicates that such obliquely propagating MHD waves cannot be amplified to the observed levels by the energetic protons in the foreshock.[17]

The key to understanding why there are two types of waveforms is to recognize where the two types are observed. The circularly polarized waves are found at the leading edge of the earth's foreshock (Fig. 11), where they are amplified. We believe that as they are blown farther downstream in the earth's curved foreshock, they

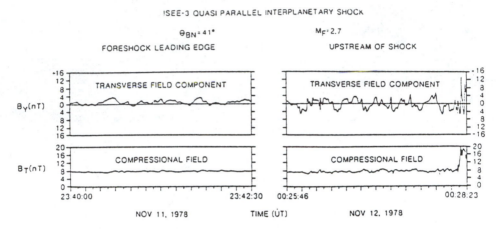

FIGURE 9   Waveforms observed upstream of the November 11-12, 1978, interplanetary shock: The left-hand panel shows one component (B_Y) and the magnitude of the magnetic field measured by ISEE-3 at the leading edge of the foreshock. The right panel shows the same magnetic field parameters just ahead of the subshock.

are refracted away from parallel propagation. As soon as they propagate obliquely, they develop a density compression which causes them to steepen until they achieve ion inertial (c/ω_{pi}) scalelengths, at which point they break and damp their energy into thermal solar wind ions.[17]

Which type of wave should the cosmic ray scattering theory use: the parallel propagating circularly polarized waves predicted by linear theory, or the obliquely propagating waves that might be generated by refraction? At the linear level, the difference between the two cases may be slight, but at the nonlinear strong turbulence level at which shock acceleration theory must be formulated, the difference seems profound. For example, consider the lowest order nonlinear model equations which describe the two cases. Parallel Alfvén waves are described by the so-called "derivative nonlinear Schrodinger equation,"[18] which has a cubic nonlinearity and second-order dispersive term and is characterized by modulational envelope solitons. On the other hand, oblique MHD waves in the fast mode are governed by the Korteweg-DeVries equation, which has a quadratic nonlinearity and a third-order dispersion and is characterized by ordinary solitons. Because their nonlinearity is stronger and their dispersion is weaker, oblique MHD waves can obviously steepen much more. Such differences are bound to affect the nonlinear wave-particle scattering rates.

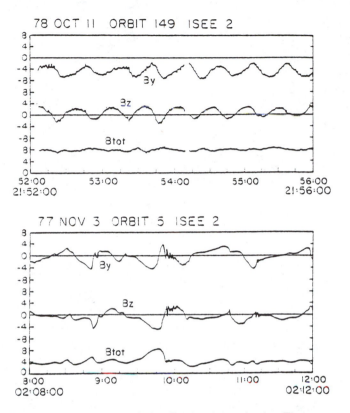

FIGURE 10   Waveforms upstream of the Earth's bow shock: The top panel shows a finite amplitude, circularly polarized, unsteepened wave observed at the leading edge of the ISEE-1 spacecraft. The bottom panel shows a highly steepened elliptical waveform, typically observed deep in the earth's foreshock (see Fig. 11). [From Ref. 17]

Moreover, we can argue that parallel propagating waves generated in supernova foreshocks may well refract. The possibility that refraction of the type found in the earth's foreshock (Fig. 11) can also occur in interstellar shocks is illustrated in Fig. 12. Two arguments suggest that it could. First of all, the foreshock is part of the shock, so the magnetic field gradually changes direction in the foreshock, in the sense predicted by the Rankine-Hugoniot relations. The dependence of the change in field direction in the foreshock has been computed by Kennel et al.[10] Moreover, the cosmic rays will increase in β in the foreshock, and their pressure is anisotropic. When β is large enough, the anisotropy affects the Alfvén wave dispersion relation. Both the changes in magnetic field and particle pressure in the foreshock suggest that Alfvén waves might refract as they are convected towards the subshock. If they are originally generated propagating parallel to the magnetic field, refraction will

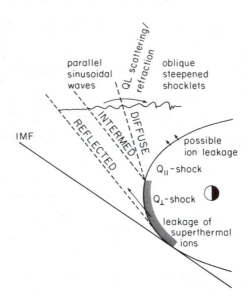

FIGURE 11   Schematic of the Earth's foreshock: The field lines downstream of the tangent field line (IMF) define the earth's foreshock, since shock-heated particles have access to this region. Ions escaping from the quasi-perpendicular zone of the subshock are found in the region labelled "reflected;" these destabilize circularly polarized waves, which refract as they are convected further downstream by the solar wind. As they refract, they steepen to form wave envelopes like the one shown in the bottom panel of Fig. 10. [From Ref. 17]

make them oblique. At this point, they might steepen into shocklets. This suggests that wave refraction may also come from the energetic particle spatial gradient in interstellar foreshocks.

CONCLUDING REMARKS

I have given an impressionistic review of the current status of the theory of cosmic ray acceleration, emphasizing four issues of plasma physics pertinent to standard model: the seed particle problem, the reflection problem, the high energy problem, and, for short, the polarization-steepening problem. None of these is a problem in weak turbulence. The physics of finite-amplitude dispersive MHD turbulence and

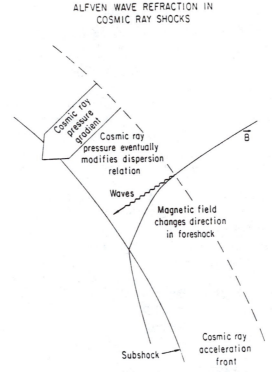

FIGURE 12   Possible refraction of waves in supernova foreshocks: The D.C. magnetic field in the foreshock must change its direction upstream of the subshock, so that the wave normal vector may gradually deviate from the magnetic field direction. Also, the energetic particles in the foreshock quickly achieve a beta of order unity, at which point their anisotropy can modify the Alfvén wave dispersion relation.

its interactions with energetic particles is at the very core of the standard model's plasma physics.

ACKNOWLEDGEMENTS

I would like to thank R. Blandford, B. Buti, F. Coroniti, T. Hada, R. Kulsrud and R. Pellat for interesting discussions, as well as the Princeton Plasma Physics Laboratory and the California Institute of Technology for their hospitality during the writing of this paper. This research was supported by NASA NGL-05-007-190 and NSF-ATM 85-03434.

REFERENCES

1. R. Blandford, in *Magnetospheric Phenomena in Astrophysics*, edited by R.I. Epstein and W.C. Feldman (Conf. Proc. 144, American Institute of Physics, N.Y., 1986), pp. 1–23.
2. R. Blandford and D. Eichler, *Physics Reports* **154**, 1 (1987).
3. J.P. Edmiston, C.F. Kennel, and D. Eichler, *Geophys. Res. Lett.* **9**, 531 (1982).
4. W.I. Axford, E. Leer, and G. Skadron, *Proc. 15th Int. Conf. Cosmic Rays* **11**, 132 (1977).
5. A.R. Bell, *Mon. Not. R. Astron. Soc.* **182**, 147 (1987a).
6. A.R. Bell, *Mon. Not. R. Astron. Soc.* **182**, 443 (1987b).
7. R. Blandford and J.P. Ostriker, *Astrophysical Journal* **221**, L29 (1978).
8. G.F. Krymsky, *Dokl. Akad. Nauk. SSR* **234**, 1306 (1977).
9. A.A. Galeev, R.Z. Sagdeev, and V.D. Shapiro, *Proc. Joint Varenna-Abastumani Intl. School and Workshop* (European Space Agency, August, 1986), p. 297. (ESA Sp. 251-ISSN 0379-6566)
10. C.F. Kennel, T. Hada, and J.P. Edmiston in *Collisionless Shocks in the Heliosphere*, Geophysical Monographs No. 34 (American Geophysical Union, Washington, D.C., 1985), Vol. 1, pp. 1–36.
11. M.A. Lee, *J. Geophys. Res.* **88**, 6109 (1983).
12. C.F. Kennel *et al.* (16 authors), *J. Geophys. Res.* **89**, 5419 (1984).
13. C.F. Kennel *et al.* (12 authors), *J. Geophys. Res.* **89**, 5436 (1984).
14. C.F. Kennel *et al.* (8 authors), *J. Geophys. Res.* **91**, 11917 (1986).
15. A. Zachary, C.E. Max, J. Arons, and B.I. Cohen, manuscript in preparation; this work was published in preliminary form as part of the Ph.D. thesis by A. Zachary, "Resonant Alfvén Wave Instabilities Driven by Streaming Fast Particles," Lawrence Livermore National Laboratory, University of California Report. No. UCRL-53793 (May 8, 1987).
16. Y.H. Ichikawa, K. Konno, M. Wadati, and H. Sanuki, *J. Phys. Soc. Japan* **48**, 279 (1980).
17. T. Hada, C.F. Kennel, and T. Teresawa, *J. Geophys. Res.* **92**, 4423 (1987).
18. K. Mio, T. Oyino, K. Minomi, and S. Takeda, *J. Phys. Soc. Japan* **41**, 265 (1976).

Harold P. Furth
Plasma Physics Laboratory
Princeton University
Princeton, New Jersey 08544

Progress in Toroidal Confinement and Fusion Research

During the past thirty years, the characteristic value of $T_i n \tau_E$ for toroidal confinement experiments has advanced by more than seven orders of magnitude. Part of this advance has been due to an increase of gross machine parameters. Most of the advance is associated with improvements in the "quality of plasma confinement." The combined evidence of spherator and tokamak research clarifies the role of magnetic field geometry in determining confinement and points to the importance of shielding out plasma edge effects. A true physical understanding of anomalous transport remains to be achieved.

INTRODUCTION

When I first met Marshall Rosenbluth, exactly thirty years ago, the forefront of fusion research had reached approximately the point marked by the asterisk in the Lawson diagram of Fig. 1. The product $T_i(0)n(0)\tau_E$ of central ion temperature, central density, and global energy confinement time stood at about $10^8\,\text{keV}\,\text{cm}^{-3}\,\text{sec}$—disturbingly far from the goal of $5 \times 10^{15}\,\text{keV}\,\text{cm}^{-3}\,\text{sec}$ for an ignited D-T reactor. During the intervening decades, experimental progress has been fairly steady, so

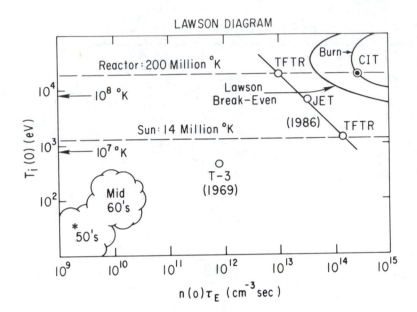

FIGURE 1 Progress of toroidal confinement research in the Lawson diagram.

that the current $T_i(0)n(0)\tau_E$ product (at the end of 1986) stands at 2×10^{14} keV cm^{-3} sec. On this nostalgic occasion, it may be appropriate to re-examine the means whereby such substantial progress was made in toroidal confinement research.

Some favorite fusion reactor candidates of the mid-1950's were the tokamak (invented by A. Sakharov and I. Tamm, and independently by L. Spitzer, Jr.), the stellarator (invented by Spitzer), and the reversed-field pinch (invented by M. Rosenbluth[1]). In the meantime, MHD theory has evolved to include finite resistivity[2] and some other important effects.[3] There have been a number of stylistic refinements in the architecture of tokamaks (cross-sectional shaping and profile optimization), stellarators (reduction of connection lengths and Pfirsch-Schlüter currents), and RFP's (utilization of the dynamo mechanism), but the favorite present-day reactor candidates are still the tokamak, the stellarator, and the RFP. Contrary to a widespread expectation of the 1950's, the evolution of magnetic architecture does not seem to have been the principal key to progress in the Lawson diagram.

The experimental results of the 1950's and early 1960's were mostly discouraging. Stellarators exhibited Bohm diffusion[4]

$$D_{\mathrm{Bohm}} = \left(\frac{1}{16}\right) cT_e/eB \sim 6 \times 10^3 T_e(\mathrm{eV})/B(k\Gamma)\,\mathrm{cm}^2\,\mathrm{sec},\qquad(1)$$

$$\chi_{\mathrm{Bohm}} \sim 3D_{\mathrm{Bohm}},\qquad(2)$$

over a broad range of plasma parameters. For stellarator configurations that obeyed the MHD stability conditions, there was no apparent further dependence of confinement on such architectural features as the rotational transform and shear. (One should note, however, that the available range of effective $\langle B_p \rangle$ was quite limited, corresponding to $\langle B_p \rangle / B_t < 10^{-2}$.) In other types of toroidal configuration, such as the tokamak and the RFP, the observed confinement scaling was not obviously Bohm-like, but the observed magnitudes of anomalous transport seemed to offer no improvement relative to Eqs. (1) and (2).

At this stage of fusion research, the value of $T_i(0)n(0)\tau_E$ stood at about $10^{10}\,\mathrm{keV\ cm^{-3}}$ sec, and reactor extrapolations based on the Bohm formula called for unacceptably large plasma minor radii, in excess of 10^3 cm. Theoretical interpretations of this state of affairs tended to appeal to drift waves and/or small-scale resistive MHD modes, but there was a lack of demonstrated correspondence between experiment and theory.

At this low point in the history of toroidal confinement research, the Ioffe "minimum-B" experiment,[5] first reported in 1961, struck a note of good cheer and served to revitalize the search for superior architecture. The potential advantages for toroidal confinement of a favorable magnetic well were recognized, and the invention of "minimum-average-B" configurations, with and without current-carrying rings floating inside the plasma, proved to be an exhilarating pastime.[6]

The relative architectural merits of some major toroidal confinement options are illustrated in Table 1. If one wishes to eliminate possible drivers of anomalous transport, the ability to operate without any net $J_\|$ is clearly an asset. Some other potentially favorable features are strong shear and a deep average magnetic well, with short connection lengths between regions of good and bad curvature. To reduce the possible threat from trapped-particle modes, there is a further benefit from avoidance of local mirror trapping in regions of bad curvature. In regard to these figures of architectural merit, floating-ring devices[7-10] are clearly most advantageous—but unfortunately they are poorly suited for use in a D-T reactor. Marshall Rosenbluth played a key role in the invention of a powerful nonfloating-ring solution of the stellarator type,[11] currently known as the heliac. Compared with these entries in categories 1 and 2 of Table 1, the tokamak configuration has few and feeble architectural merits—except for the virtue of simplicity.

TABLE 1 Search for an Ideal Architectural Solution

	No $J_\parallel$	Strong Shear	Magnetic Well	No Unfavorable Trapping
1. Best Possible Configuration (incompatible with D-T)				
Multipole	✓	—	✓	✓
Multipole with B_t	✓	✓	✓	✓
Spherator	✓	✓	✓	✓
2. Ideal Reactor-Compatible Solution				
Heliac	✓	✓	✓	—
3. Nonideal but Simple Architecture				
Tokamak	—	—	✓	—

The next section reviews the experimental data obtained during the late 1960's and early 1970's in architecturally optimal configurations of the spherator type (Fig. 2). The confinement results are related to the early stellarator experience and to the major advances introduced by the tokamak approach.[12] Following sections briefly review the present state of tokamak research; compare anomalous transport phenomena in spherators and tokamaks; and attempt to infer a general model for toroidal confinement.

SPHERATOR EXPERIMENTS

During the early 1960's, simple "levitron" experiments[9] were carried out with transiently free-floating copper rings, in order to test the ideal MHD stability theory. Later the emphasis switched to the study of near-vacuum-field low-β confinement, and a number of superconducting-ring devices were built. The present discussion will make particular use of the results of the FM-1 device[13] of Fig. 3, which was operated by S. Yoshikawa *et al.* during the early 1970's.

The FM-1 contained a superconducting ring of 75 cm major radius, capable of carrying several hundred kA of ring current and remaining afloat in its own B_p field for an eight-hour experimental shift in a room-temperature environment. The FM-1 had substantial flexibility for exploring the range of architectural effects referred to in Fig. 2 and Table 1—and it incidentally pioneered the poloidal-field divertor concept.[14]

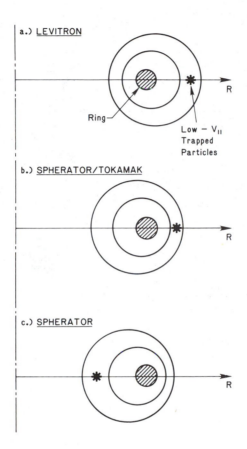

FIGURE 2 Three basic types of axisymmetric toroidal confinement geometry: (a) the levitron, where B_p is generated by a ring current and B_t by external coils; (b) the spherator, where a weak external B_v is added, giving a tokamak-like configuration on the outer plasma surface away from the ring; (c) a spherator with $B_v \sim B_p$ and $B_t \lesssim B_p$, where the mirror trapping region moves to the small-R side of the plasma.

   As illustrated in Fig. 4, the FM-1 studied relatively cold, low-density plasmas and achieved particle confinement times exceeding one second. Particle diffusion was found to decrease with rising T_e, up to roughly the point where the trapped-electron bounce frequency exceeded the collision frequency. Thereafter, diffusion increased again, following a Bohm-like scaling, but with geometrically determined proportionality constants that were as much as 300 times smaller than predicted by Eq. (1).

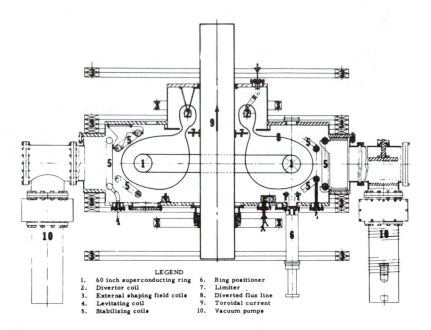

LEGEND

1.	60 inch superconducting ring	6.	Ring positioner
2.	Divertor coil	7.	Limiter
3.	External shaping field coils	8.	Diverted flux line
4.	Levitating coil	9.	Toroidal current
5.	Stabilizing coils	10.	Vacuum pumps

FIGURE 3 The FM-1 spherator. [From Ref. 13]

For a "tokamak-like" electron-cyclotron-wave-heated FM-1 plasma like that of Fig. 2b (with $B_t \sim B_p$), the particle and heat diffusion coefficients are shown in Fig. 5 as functions of T_e. (It should be noted that, in both the spherator studies and the earlier stellarator work, transport coefficients were determined *globally*, on the basis of τ^{-1}, not *locally* as a function of plasma radius.) In the example of Fig. 5, the particle transport D was found to be about 100 times less then D_{Bohm} and the heat transport χ (called $K_\perp$ in the figure) was about 10 times less than D_{Bohm}, or 30 times less than χ_{Bohm} [cf. Eq. (2)].

Turning to the impact of various architectural effects, the FM-1 experiments documented a very marked deterioration of confinement in the limit $B_t \ll B_p$ (Fig. 6) where the shear becomes weak and there is unfavorable magnetic curvature everywhere on the outer plasma surface. (FM-1 confinement was typically very much better on the inner surface of the plasma, facing the ring, which is in a true minimum-B situation.)

The particle confinement times associated with a variety of FM-1 configurations[15] are shown in Fig. 7. Optimum confinement, corresponding to Fig. 4 and to case (c) in Fig. 2, was obtained for ratios of ring current I_p to vertical-field current I_E around unity, and for moderately high ratios of toroidal-field current I_T to

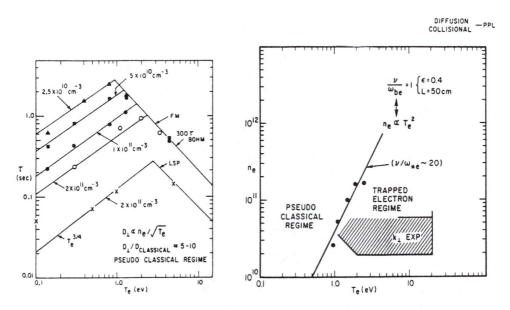

FIGURE 4 Particle confinement in FM-1 (and its smaller prototype, LSP). [From J. Sinnis, M. Okabayashi, J. Schmidt, and S. Yoshikawa, *Physical Review Letters* 29, 1214 (1972)]

ring current $(I_T/I_p \sim 3B_t/B_p \sim 1)$. At lower values of toroidal field, the low-shear regime of Fig. 6 is encountered. At higher values of B_t/B_p, where the configuration becomes increasingly tokamak-like (Fig. 5), confinement is also found to deteriorate. The lower curve in Fig. 7 shows the unfavorable effect of raising I_p/I_E, which produces an outward shift of the poloidal flux surfaces. (The extreme limit $I_p \gg I_E$, shown as case (a) in Fig. 2, corresponds to the earliest levitron experiments,[9] which had no external vertical magnetic field.)

Extensive fluctuation studies were carried out in various spherator experiments,[13–16] mostly by means of Langmuir probes. The magnitude of the fluctuation level $\delta n/n$ was generally correlated with the magnitude of the transport coefficients. For example, fluctuations were found to be smallest in the well-confined case (c) of Fig. 2, and on the inner side (ring side) of the plasma. They were particularly large in case (a) and for weak shear $(B_t \ll B_p)$. An important discovery was that "convective cells," i.e., fluctuations in space with $\omega = 0$, could play a dominant role in accounting for anomalous transport in well-confined and seemingly quiescent spherator plasmas.[17]

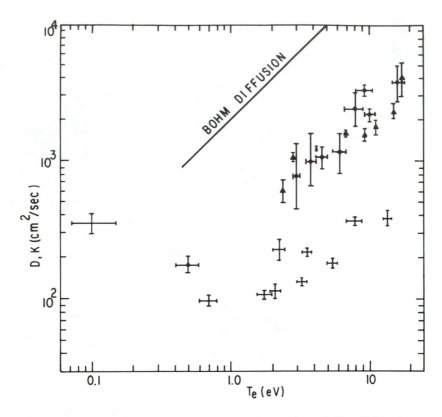

FIGURE 5 Average global transport coefficients in a tokamak-like FM-1 plasma. (The higher set of points is the thermal conductivity K or χ.) [From Ref. 18]

The overall results found for energy confinement in the FM-1 experiments suggest the following simple model (which is based, with some added conservatism, on Ref. 18):

$$\chi = \left(\frac{1}{30}\right) G\,\chi_{\text{Bohm}} = 600\,GT_e(\text{eV})/B(k\Gamma)\,\text{cm}^2/\text{sec}, \tag{3}$$

where G is a number that depends on the magnetic field geometry. There are three architectural cases of principal interest:

1. *Optimal Geometry.* The flux surfaces are like those of case (c) in Fig. 2, with $B_t \lesssim B_p \sim B$, giving high shear and trapping on the small-R side, but only

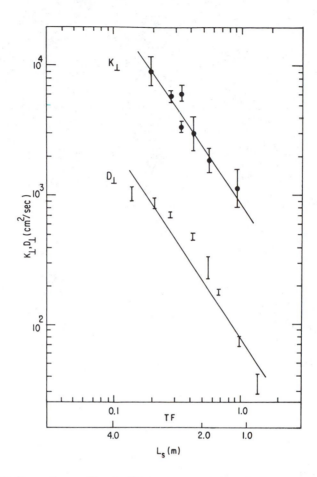

FIGURE 6 Shear-dependence of transport coefficients in low-B_t FM-1 regimes. (The quantity "TF" refers to the ratio of total toroidal field (TF) coil current to ring current I_T/I_p.) [From Ref. 18]

marginally favorable (or even slightly unfavorable) average magnetic well. In this case, G reaches its minimum value of

$$G_{\min} \sim \frac{1}{3}. \tag{4}$$

2. *Approaching Pure Poloidal Field.* The flux surfaces are shaped like those of case (b) or (c) in Fig. 2, but with $B_t \ll B_p \approx B$, so that the magnetic well is

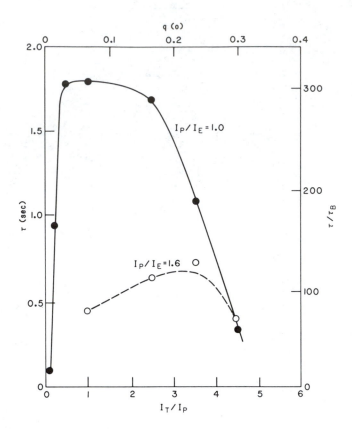

FIGURE 7 Survey of geometric dependences of FM-1 confinement. The parameter I_T/I_p corresponds to roughly three times the "average B_t/B_p." The quantity I_p/I_E measures the ratio of the ring current to the current producing the external vertical field.

strongly unfavorable and the shear length increases as $L_s \propto B_p/B_t$. In this limit we have

$$G \to \frac{B_p}{B_t}. \tag{5}$$

3. *Approaching Tokamak Geometry.* The flux surfaces are shaped like those of case (b) or (c), but with $B_p \ll B_t \approx B$, so that the magnetic well is favorable, but the particle trapping is on the large-R side, and the shear length increases as $L_s \propto B_t/B_p$. In this limit we have

$$G \to \frac{B_t}{B_p}. \tag{6}$$

Let us first consider how the spherator confinement results can be related to the results of the early stellarators, which studied plasmas with somewhat similar parameters. On the basis of the spherator data, we should use Eqs. (3) and (6) to define a "neo-Bohm" formula

$$\chi_B^* \left(B_t / 30 B_p \right) \chi_{\text{Bohm}} = 600 \, T_e(\text{eV}) / B_p(k\Gamma) \, \text{cm}^2/\text{sec}, \tag{7}$$

characterizing low-B_p/B_t devices such as stellarators and tokamaks. For typical effective values of $\langle B_p \rangle / B_t < 10^{-2}$ achieved in early stellarator experiments,[4] the prediction of Eq. (7) is seen to be *more pessimistic* than the simple Bohm prediction. Since the Bohm formula can be identified theoretically as a kind of thermodynamic upper limit for anomalous cross-field transport in MHD-stable configurations, a plausible conclusion is that Eq. (7) can be valid only for $\chi_B^* \leq \chi_{\text{Bohm}}$, and should be replaced by the Bohm formula itself when B_t/B_p is larger than 30.

In seeking to relate Eq. (7) to tokamaks, one notes, first of all, that the plasma parameters are greatly different. Since tokamak plasmas must carry their own "poloidal-field current," there is a minimum condition of the form $\sim 300 < v_{th}/v_s \propto n T_e^{1/2}/J$, which calls for typical plasma parameters that are far from those of Fig. 4. (This point is particularly well documented in Ref. 19, which finds that the addition of as little as 100 A of toroidal plasma current to a spherator plasma with 100 kA of ring current introduces a drastic $(1/n)$-dependent deterioration of confinement.) The requirement that tokamak plasmas must have certain minimum levels of density and temperature entails a second fundamental difference relative to the FM-1 plasmas: in tokamaks, the neutral density is typically "burned out" to a high degree within the plasma core, whereas the plasmas of Fig. 4 are weakly ionized in the low-T_e regime and incompletely burned out even in the highest-T_e experiments.

If we proceed, nonetheless, to make predictions for the tokamak on the basis of Eq. (7), we find, for representative TFTR parameters[20] ($B_p \sim 5 \, k\Gamma$, $B_t \sim 50 \, k\Gamma$, $T_e \sim 3 \, \text{keV}$), that the global energy confinement time is substantially underestimated. In modern tokamaks, where diagnostics are sufficiently good so that transport rates can be determined as a function of plasma radius, one also concludes that χ_e must increase strongly towards the plasma edge—just the opposite of the trend that would be indicated by Eq. (7). In the cold edge region of TFTR ($T_e \sim 100$–$300 \, \text{eV}$), Eq. (7) gives roughly the right answer, but the hot plasma core seems to be governed by different and far more favorable rules.

In summary, we see that spherator data did demonstrate strong architectural effects on confinement, more or less according to theoretical prejudice. That the early stellarator confinement results should have been very poor, is consistent with the geometric dependences inferred from the spherator experiments—indeed one sees that the Bohm formula must represent a kind of benign upper limit on transport that kept Model-C confinement from being even worse than it was. The tokamak results, on the other hand, represent a clear-cut challenge to the notion that architectural optimization is important. Both "architectural theory" and the spherator studies themselves clearly show that the tokamak-type magnetic field configuration is geometrically inferior—whereas actual tokamak confinement is found to be dramatically better than the predictions of Bohm or neo-Bohm scaling. Clearly some *non*-architectural feature of the tokamak plasma regime must be exerting a favorable effect that far exceeds the influence of the purely geometric aspects of toroidal confinement.

TOKAMAK EXPERIMENTS

The success of the T-3 tokamak served to advance $T_i n \tau_E$ to about 5×10^{11} keV cm^{-3} sec in 1969, using a facility of no greater magnitude than had been available in earlier fusion experiments, which only reached the $10^9 - 10^{10}$ keV cm^{-3} sec range. Subsequent years brought continuing progress in the quality of confinement. For example, during the late 1970's and early 1980's, Alcator A and C, which again were facilities comparable in scale to T-3, achieved $T_i n \tau_E$ values[21,22] of about 3×10^{13} and 10^{14} keV cm^{-3} sec, respectively, using ohmic heating at relatively high toroidal magnetic fields and densities.

These experiments pointed to the "neo-Alcator" scaling law

$$\tau_E \propto nqR^2 a, \tag{8}$$

where q is the "MHD safety factor" $2\pi/\iota \approx B_t a / B_p R$, and R and a are the major and minor plasma radii. Equation (8) successfully predicted the confinement scaling observed in the much larger TFTR device at low-to-moderate plasma densities (Fig. 8). Fairly high central densities, up to 4×10^{14} cm^{-3}, have proved to be achievable in TFTR by means of pellet injection,[23] but the favorable n-dependence of Eq. (8) appears to saturate at high densities (cf. Fig. 8).

One interpretation of these ohmic-heating results is that the low-n regime corresponds to the appearance of some extraneous energy-loss channel, similar to the adverse high-v_s/v_{th} dependence reported for spherator plasmas in Ref. 19—though unlikely to be attributable to the same physical mechanism. The high-n portion of the ohmic data in Fig. 8 would then constitute the "normal" confinement regime of the tokamak and would be expected to conform with the general scaling observed for plasmas heated and fueled by a variety of techniques.

Goldston[24] has pointed out that this "normal" tokamak scaling can be approximated by

$$\tau_G \propto I L^{3/2} P_H^{-1/2} \tag{9}$$

or

$$\chi_G \propto P_H^{1/2} L^{-1/2} B_p^{-1}, \tag{10}$$

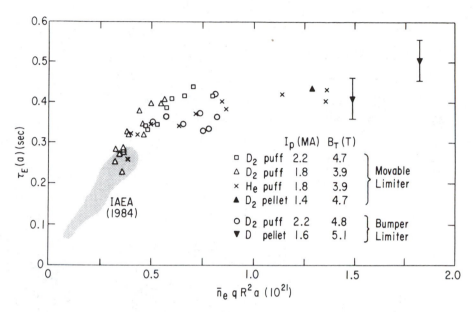

FIGURE 8 Dependence of energy confinement on the neo-Alcator scaling parameter in TFTR ohmic-heating regimes. [From Ref. 23]

where L represents linear size (R, a, L_s, etc.) and P_H is the total plasma-heating power. The tokamak *ohmic*-heating regime has the special advantage that P_H can be raised only by raising I. Pure ohmic heating generally gives the minimal level of P_H for any given I, so that ohmic τ_E-values must be optimal. As described by Eq. (9), they actually tend to "improve with rising P_H" (and I), unlike the typical auxiliary-heating results of Fig. 9, where τ_E deteriorates with rising P_H (for fixed I).

Unhappily, the tokamak reactor regime calls for α-heating powers that greatly exceed the ohmic power level. For economically attractive reactor parameters, one typically requires $V_H \equiv P_H/I \sim (750\,\mathrm{MW})/(15\,\mathrm{MA}) \sim 50\,\mathrm{V}$—in marked contrast with the ohmic regime, where $V_H \lesssim 1\,\mathrm{V}$. The development of practical tokamak reactors depends on the optimization of the high-V_H plasma regime, so as to achieve somewhat more favorable confinement than predicted by the basic Goldston "L-mode" scaling of Ref. 24.

A major step of this kind was the discovery of the "H-mode" in the ASDEX tokamak.[25] Neutral-beam heating in ordinary L-mode operation typically depresses τ_E relative to the pure ohmic-heating case by some factor of order $(P_H/P_{OH})^{1/2}$. When a poloidal divertor separatrix is present, the thermal equilibrium of the tokamak discharge is found to become bi-stable at sufficiently high P_H, with a more favorable upper branch (the H-mode). Recent results of this type, obtained in the

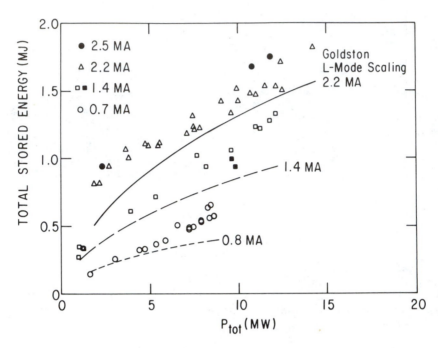

FIGURE 9 Dependence of total stored plasma energy W_{tot} on total input power P_{tot} in TFTR neutral-beam-heating experiments. The ratio $\tau_E = W_{tot}/P_{tot}$ is seen to decrease with rising P_{tot}, consistent with the Goldston L-mode prediction. [Adapted from Ref. 23]

DIII-D tokamak,[26] are shown in Fig. 10. The available range of heating powers in DIII-D has been insufficient, thus far, to determine whether entry into the H-mode improves Eq. (9) by a constant multiplicative factor of 2–3, or whether the adverse $P_H^{-1/2}$ dependence is fundamentally changed. In any case, the H-mode branch is evidently superior to the ohmic-heating regime of DIII-D, since the same τ_E is maintained at much higher V_H.

The happy discovery of the H-mode has served, incidentally, to rekindle interest in the architectural approach to confinement optimization. The presence of a poloidal-field separatrix creates a thin region of enhanced global shear, which also tends to be a region of improved magnetic well, provided that the null points of the separatrix are not located on the large-R side of the plasma. The essential improvements associated with the H-mode are, in fact, found to occur at the separatrix: both D and χ undergo pronounced local decreases, thus permitting relatively high values of n_{edge} and T_{edge}, with resultant increments of n_{core} and T_{core} relative to the L-mode results obtained at the same P_H.

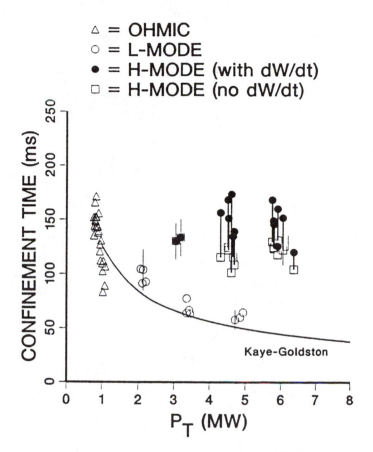

FIGURE 10 Scaling of energy confinement time with P_{tot} in DIII-D neutral-beam-heating experiments. The L-mode plasmas follow a Goldston-type scaling. For $P_{tot} \gtrsim 3\,\text{MW}$, an H-mode branch appears.

While the strong influence of edge-localized phenomena on the *global* confinement time may be surprising at first sight, it is consistent with a wide range of other "tokamak anomalies." Particularly relevant is the finding that the characteristic $T_e(r)$-profile shape in a given L- or H-mode regime with fixed values of $q(0)$ and $q(a)$ cannot be changed readily even by drastic changes in the profile of heating-power deposition.[27,28] This observation implies[29,30] that

$$\tau_E \propto \left(\frac{a^2}{\chi_{\text{edge}}}\right) \frac{n_{\text{core}}}{n_{\text{edge}}}, \tag{11}$$

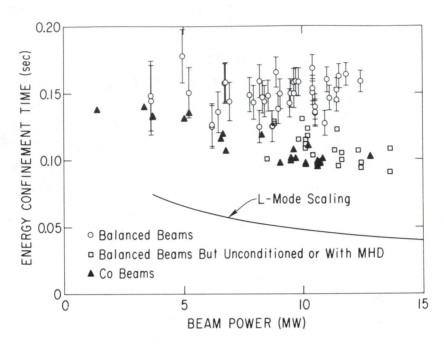

FIGURE 11 Scaling of energy confinement time with neutral-beam power in the TFTR supershot regime. Beams are injected tangentially in the "co" and "counter" directions relative to the plasma current, which is in the range 0.9–1.0 MA. [From Ref. 20]

where the "plasma edge," defined as the maximum radius within which the power deposition profile can be varied without producing much effect on τ_E, turns out to be fairly close to a.

A different kind of "enhanced confinement regime" has been obtained with intense neutral-beam heating in the absence of a divertor separatrix in TFTR[20] (Fig. 11). Again, there is an improvement of τ_E by factors of 2–3 relative to L-mode prediction. The "best data points," corresponding to balanced injection of the tangential neutral beams (so as to avoid driving a rapid toroidal plasma rotation), show little evidence of any unfavorable P_H-dependence. As in the case of the H-mode, the key to this "supershot" regime is an edge effect—the suppression of recycling at the TFTR limiter by using specially conditioned hydrogen-absorbing graphite tiles. The central particle-fueling associated with neutral-beam injection is then able to produce $n(r)$ profiles with large values of $n_{\text{core}}/n_{\text{edge}}$, along with an improvement in global confinement, as predicted by Eq. (11). The success of pellet-injection experiments in Alcator C,[22] TFTR,[23] and ASDEX[31] seems to be due to the same phenomenon.

The TFTR supershot and pellet-injection results, along with initial H-mode results in JET,[32] have pushed the frontier of fusion research to the line marked $T_i n \tau_E \sim 2 \times 10^{14}\,\mathrm{keV\ cm^{-3}\,sec}$ in Fig. 1. There is every reason to believe that significant further advances will occur during the coming year. Breakeven plasmas seem likely to be attainable in TFTR, JET, JT-60,[33] and other large tokamaks— particularly since the relevant breakeven condition in energetic-ion-heated plasmas is somewhat relaxed by comparison with the Lawson curve.[34] If an "enhanced confinement" factor of at least 1.5, relative to the L-mode, can be obtained in the proposed next-generation CIT experiment,[35] the empirical extrapolation to ignition and equilibrium-burn conditions would seem to be favorable.

In this context of programmatic success, it is sobering to reflect that identification of the specific physical phenomena responsible for anomalous transport in toroidal configurations has not progressed decisively since 1956. The physical understanding of Bohm-like transport has never reached the stage of clear, detailed correspondence between experiment and theory. The tokamak confronts us with the additional mystery of a low-level transport mechanism within the plasma core that serves to maintain the $J(r) - T_e(r)$ profile shape. There is some satisfaction in noting that the $J(r)$-constraints prescribed by resistive MHD theory[36] seem to be respected by the experimental plasma—but the transport mechanism whereby the tokamak selects a stable profile has yet to be identified.

COMPARISON OF SPHERATOR AND TOKAMAK RESULTS

During the past two decades, a great deal of detailed tokamak experimental data has accumulated. The translation of data into physical understanding is hampered, however, by the narrowly limited range of architectural and plasma parameters accessible within the inherent tokamak constraints. The ratio B_p/B_t is limited by kink stability, while B_v/B_p is predetermined by the plasma equilibrium in R; the magnitude of J follows from B_p and, in turn, imposes limits on n and T_e.

By way of contrast, in the spherator device discussed earlier, the parameters B_t, B_p, B_v, J, n, and T_e can be varied essentially independently. An appropriately designed spherator could produce an edge plasma region that would be effectively identical with the tokamak edge plasma—and the spherator could then carry out significant parameter variations. Diversion of about one percent of the world tokamak effort to such a study might prove both relevant and enlightening to tokamak research. In the absence of a contemporary spherator project, we may still hope to draw useful inferences from the "fossil" data deposits of the 1970's.

Why did the tokamak emerge as a markedly superior confinement device in the late 1960's? The answer seems to be that, although transport in the tokamak *edge* plasma is actually rather worse than in an architecturally optimal spherator configuration, the tokamak has a relatively well-confined plasma *core*, whereas spherator (and early stellarator) plasmas in some sense were "all edge." In this view, the only architectural merit of the T-3 tokamak was its *simplicity*, which allowed plasmas of relatively large minor radius and high field strength (high plasma density) to be reached with a device of moderate cost. This conjecture is strengthened by the success of modern stellarators in achieving tokamak-like quality of confinement: larger, denser stellarator plasmas exhibit weaker anomalous transport—which is found to resemble tokamak transport in peaking at the plasma edge.[37] (The objection that reversed-field pinches have long had "large dense plasmas," without encountering notably good confinement, can be set aside on the basis that RFP plasmas suffer from an additional energy loss channel—they are not truly MHD-quiescent.)

In this spirit, the neo-Alcator scaling law of Eq. (8) can be seen as representing an increase in the effectiveness of "tokamak plasma-core shielding" as a function of density and size. If we adopt a simple model in which χ_{edge} is given by Eq. (7) and the width of the edge region $\Delta r_{edge} = a - r_{edge}$ is inversely proportional to n_{edge}, the result is

$$\tau_E \propto \left(n_{core} L^3\right) \left(\frac{B_p}{T_{core}}\right). \tag{12}$$

Since B_p/T_e is roughly constant in the data base for ohmic-heated tokamaks, Eq. (12) conforms reasonably well with the neo-Alcator scaling. The *ad hoc* model $\Delta r_{edge} \propto n_{edge}^{-1}$ used above corresponds, for example, to the conjecture that the plasma core should be protected against penetration by neutral atoms, which might help to generate surface noise by irregular deposition.[38] More generally, one would expect the singularity of $d \log T/dr$ and/or $d \log n/dr$ at the plasma edge to drive a variety of noisy phenomena. A specific physical argument in terms of the surface-noise concept has been made by Kadomtsev[39] (who draws the pleasing analogy between anomalous tokamak transport and the turbulence excited by fluid flow through a rusty pipe).

Tokamak confinement data such as that in Fig. 8 indicate that, for sufficiently high plasma density and size, there is no further benefit in raising the ratio $L/\Delta r_{edge}$, presumably because a more fundamental trouble is encountered. In this "normal" tokamak regime, we have seen [cf. Eq. (11)] that maximizing the ratio n_{core}/n_{edge} becomes advantageous for increasing the global τ_E.

In summary, the identification of the spherator plasmas with the *edge* plasmas of modern tokamaks fits in rather neatly with prejudices about tokamak confinement that have evolved in recent years.[24,29,38] If the confinement of these low-Ln plasmas can indeed be characterized in terms of some pervasive form of anomalous transport, we should now inquire as to its nature and parameter dependence.

The simplified account given above has failed to distinguish between the edge heat outflows due to convection ($TD\nabla T$) and conduction ($n\chi\nabla T$), or between the electron and ion channels of heat conduction. Equilibration is sufficiently rapid in the edge region of large tokamaks so that the relative importance of the ion and electron channels is not easy to resolve. In tokamaks, as in other toroidal confinement systems, the particle transport coefficient D is generally smaller than

χ, but since convection is enhanced by recycling at the plasma edge, the convective term cannot normally be neglected in the local heat flow balance.

While keeping these complexities in mind, we can think in terms of an "effective χ_{edge}" and ask how it scales in various cases. On the basis of the spherator results, χ should follow the "neo-Bohm" scaling of Eq. (7) in tokamak-like geometry. Since Eq. (7) refers to the average global $\chi \propto L^2 T_E^{-1}$, we can eliminate the explicit dependence on T by means of $T = P_H \chi^{-1} n^{-1} L^{-1}$, obtaining

$$\chi_B^* \propto P_H^{1/2} L^{-1/2} B_p^{-1/2} n^{-1/2}. \tag{13}$$

(This equivalent *global* expression for neo-Bohm scaling has the incidental merit of offering a much more plausible model for the *local* $\chi(r)$.) There does seem to be a marked similarity between the Goldston scaling of Eq. (10) and the neo-Bohm scaling of Eq. (13):

$$\frac{\chi_G}{\chi_B^*} \propto B_p^{-1/2} n^{1/2}. \tag{14}$$

Since the tokamak data base shows a strong correlation between high-B_p and high-n points, the apparent discrepancy factor in Eq. (14) is never very large.

SUMMARY AND CONCLUSIONS

What explanation can we offer for the major historical event in toroidal confinement research—viz., the advance from Bohm-like transport in early stellarators and spherators to superior confinement in the T-3 tokamak? An explanation in terms of better magnetic architecture is irreconcilable with the spherator data. The most plausible remaining hypothesis is that Bohm-like transport is driven by edge effects (cf. Kadomtsev's "flow through a rusty pipe") and that the real innovation of the T-3 was to provide a thick, dense, MHD-stable plasma. The tokamak *edge* plasma actually bears a strong resemblance to the smaller, lower-density toroidal plasmas of earlier times and seems to exhibit similar anomalous transport. The favorable nL-dependence of neo-Alcator scaling can be seen as resulting from "better shielding," i.e., reduced relative size of the edge-plasma region $(\Delta r_{\text{edge}}/L)$.

The normal high-density tokamak regime is characterized by:

1. A thick plasma-core region, with a rigid $T_e(r)$-profile shape that is maintained by somewhat mysterious but essentially benign transport processes; and
2. A thin Bohm-like outer region that calibrates dT_e/dr in terms of the local χ_{edge} and the global heat throughput P_H, controlling the magnitude of the global energy confinement time according to some model of the form

$$\tau_E = \left(\frac{L^2}{\chi_{\text{edge}}}\right) \frac{n_{\text{core}}}{n_{\text{edge}}} \tag{15}$$

$$\chi_{\text{edge}} = G_{\text{edge}} P_H^{1/2} L^{-1/2} B_p^{-1/2} n^{-1/2}.$$

The quantity G_{edge} refers to geometric dependences: if the tokamak edge plasma resembles the spherator plasma, then one would expect reductions in G_{edge} to result from higher edge shear, improved location of particle trapping, and weaker adverse magnetic curvature.

A more general lesson can also be drawn from the historical experience of toroidal confinement research. During the first part of the 1960's, Bohm diffusion was seen as imposing a prohibitive *upper* limit on confinement, but as of the late 1960's, the prevalence of Bohm diffusion has seemed more likely to reflect a thermodynamic *lower* limit for confinement in MHD-stable plasmas. In the same way, the emergence of Goldston L-mode scaling during the early 1980's was taken to impose a somewhat painful constraint on the prospects of tokamak confinement. Once again, the more likely significance of Goldston scaling is now seen to be that it sets a *lower* limit for global energy confinement in MHD-stable high-density tokamak plasmas. Improvements in the architecture of the tokamak plasma edge have already served to decrease G_{edge} substantially in H-mode operation. Increases in the factor n_{core}/n_{edge} have yielded similar benefits for TFTR supershots.

Figure 1 illustrates that toroidal confinement research, and particularly tokamak research, has made substantial progress towards its goal—even without the guiding light of a true physical understanding. The potential rewards for a rigorously scientific optimization of tokamak confinement, however, remain very large. Achieving ignition at a minimal plasma-current level will be the key to cost-effective steady-state operation (i.e., with noninductive current-drive); moderate reactor unit size and cost is a precondition for timely progress through the developmental phase of fusion power. One good way to achieve the desired quality of insight into the tokamak will be to pursue the search for a consistent overall physics of plasma transport in toroidal geometry.

ACKNOWLEDGEMENTS

I should like to thank Drs. Michio Okabayashi, Derek Robinson, John Schmidt, and Shoichi Yoshikawa for their help in reviewing the results of spherator research. This work was supported by the U. S. Department of Energy Contract No. DE-AC02-76CH03073.

REFERENCES

1. M. N. Rosenbluth, LANL Report LA-2030 (1956); *Second U.N. International Conference on the Peaceful Uses of Atomic Energy* (United Nations, Geneva, 1958), Vol. 31, p. 85.
2. H. P. Furth, J. Killeen, and M. N. Rosenbluth, *Physics of Fluids* 6, 459 (1963).
3. M. N. Rosenbluth, N. A. Krall, and N. Rostoker, *Nuclear Fusion Supplement Part 1*, 143 (1962).
4. E. Hinnov and A. J. Bishop, *Physics of Fluids* 9, 195 (1966).
5. Y. B. Gott, M. C. Ioffe, and V. G. Telkovsky, *Nuclear Fusion Supplement Part 3*, 1042 (1962).
6. H. P. Furth, in *Advances in Plasma Physics*, edited by A. Simon and W. B. Thompson (Interscience, New York, 1968), Vol. 1, p. 67.
7. T. Ohkawa *et al.*, in *Plasma Physics and Controlled Nuclear Fusion Research 1965* (IAEA, Vienna, 1966), Vol. II, p. 531.
8. D. W. Kerst *et al.*, *Physical Review Letters* 15, 396 (1965).
9. D. H. Birdsall *et al.*, in *Plasma Physics and Controlled Nuclear Fusion Research 1965* (IAEA, Vienna, 1966), Vol. II, p. 291.
10. S. Yoshikawa and U. R. Christensen, *Physics of Fluids* 9, 2295 (1966).
11. H. P. Furth, in *Plasma Physics* (IAEA, Vienna, 1965), p. 391.
12. L. A. Artsimovich *et al.*, in *Plasma Physics and Controlled Nuclear Fusion Research 1968* (IAEA, Vienna, 1969), Vol. I, p. 157.
13. S. Yoshikawa, *Nuclear Fusion* 13, 433 (1973).
14. K. Ando *et al.*, in *Plasma Physics and Controlled Nuclear Fusion Research 1974* (IAEA, Vienna, 1975), Vol. II, p. 103.
15. K. Chen, D. Meade, M. Okabayashi, J. A. Schmidt, and S. Yoshikawa, in *Proceedings of the Third International Symposium on Toroidal Plasma Confinement* (Munich, 1973), Paper C8.
16. A. C. Riviere *et al.*, in *Plasma Physics and Controlled Nuclear Fusion Research 1980* (IAEA, Vienna, 1981), Vol. I, p. 855.
17. S. L. Davis, R. J. Hawryluk, and J. A. Schmidt, *Physics of Fluids* 19, 1805 (1976).
18. S. Ejima and M. Okabayashi, *Physics of Fluids* 18, 904 (1975).
19. M. W. Alcock *et al.*, in *Plasma Physics and Controlled Nuclear Fusion Research 1976* (IAEA, Vienna, 1977), Vol. II, p. 305.
20. R. J. Hawryluk *et al.*, in *Plasma Physics and Controlled Nuclear Fusion Research 1986* (IAEA, Vienna, 1987), Vol. I, p. 51.
21. A. Ghondhalekar *et al.*, in *Plasma Physics and Controlled Nuclear Fusion Research 1978* (IAEA, Vienna, 1979), Vol. I, p. 199.
22. M. Greenwald *et al.*, *Physical Review Letters* 53, 352 (1984).
23. M. G. Bell *et al.*, *Plasma Physics and Controlled Fusion* 28, 1329 (1986).
24. R. J. Goldston, *Plasma Physics and Controlled Fusion* 26, 37 (1984).
25. F. Wagner *et al.*, in *Plasma Physics and Controlled Nuclear Fusion Research 1982* (IAEA, Vienna, 1983), Vol. I, p. 43.

26. K. H. Burrell et al., GA Technologies, Inc. Report GA-A18781 (1987).
27. V. V. Alikaev et al., in Plasma Physics and Controlled Nuclear Fusion Research 1984 (IAEA, Vienna, 1985), Vol. I, p. 419.
28. M. Murakami et al., Plasma Physics and Controlled Fusion 28, 17 (1986).
29. H. P. Furth, Plasma Physics and Controlled Fusion 28, 1305 (1986).
30. N. Ohyabu, J. K. Lee, and J. S. deGrassie, GA Technologies, Inc. Report GA-A17890 (1985).
31. M. Kaufman, Plasma Physics and Controlled Fusion 28, 1341 (1986).
32. A. Tanga et al., in Plasma Physics and Controlled Nuclear Fusion Research 1986 (IAEA, Vienna, 1987), Vol. I, p. 65.
33. M. Yoshikawa et al., in Plasma Physics and Controlled Nuclear Fusion Research 1986 (IAEA, Vienna, 1987), Vol. I, p. 11.
34. J. M. Dawson, H. P. Furth, and F. H. Tenney, Physical Review Letters 26, 1156 (1971).
35. J. Schmidt et al., in Plasma Physics and Controlled Nuclear Fusion Research 1986 (IAEA, Vienna, 1987), Vol. III, p. 259.
36. C. Z. Cheng, H. P. Furth, and A. H. Boozer, Plasma Physics and Controlled Fusion 29, 351 (1987).
37. V. Erckmann et al., Plasma Physics and Controlled Fusion 28, 1277 (1986).
38. H. P. Furth, IAEA (INTOR-Related) Specialist Meeting on Confinement in Tokamaks with Intense Heating, Kyoto, Japan, 351 (November 21–22, 1986).
39. B. B. Kadomtsev, Plasma Physics and Controlled Fusion 28, 125 (1986).

Paul H. Rutherford
Plasma Physics Laboratory
Princeton University
Princeton, New Jersey 08544

Resistive Instabilities in Toroidal Confinement

Low-m tearing modes constitute the dominant instability problem in present-day tokamaks. In this paper, the stability criteria for representative current profiles with $q(0)$ values in the vicinity of unity are reviewed; "sawtooth" reconnection to $q(0)$ values just at, or slightly exceeding, unity is generally destabilizing to the $m = 2$, $n = 2$ and $m = 3$, $n = 2$ modes and limits the range of stable profile shapes. Major disruptions can be produced by the simultaneous growth of $m = 2$, $n = 1$ and $m = 3$, $n = 2$ magnetic islands, leading to destabilization of higher-order modes and to the overlapping of several island chains. Internal disruptions—or "sawteeth"—arise in a variety of forms other than that produced by the classically reconnecting $m = 1$ mode. In some cases, the $q(r)$ value is apparently close to unity over a large central part of the plasma; in other cases, the $q(0)$ value remains substantially below unity throughout a sawtooth cycle. Toroidal effects are sufficient to stabilize the resistive $m = 1$ mode in this latter case. Feedback stabilization of $m \geq 2$ modes by rf heating or current drive, applied locally at the magnetic islands, appears feasible; feedback by island current drive is much more efficient, in terms of the radio-frequency power required, than feedback by island heating. Feedback stabilization of the $m = 1$ resistive mode—although yielding particularly beneficial effects for resistive-tearing and high-beta stability by allowing $q(0)$ values substantially below unity—is more problematical, unless the $m = 1$ ideal MHD mode can be given sufficient positive stability. This appears possible, however, either by strong triangular shaping of the central flux surfaces or by appropriate tailoring of the current profile in the vicinity of the $q = 1$ surface.

INTRODUCTION

A small but finite amount of plasma resistivity has a remarkably strong destabilizing effect on the MHD stability of toroidally confined plasmas.[1] In the tokamak case in particular, it can given rise to various modes, especially resistive kink modes, that grow on a time scale that lengthens only gradually with decreasing plasma resistivity.[1] In present-day tokamak experiments, these resistive kinks (or "tearing modes") constitute, by far, the most dominant instability problem. Indeed, a tokamak with a highly non-optimal $q(r)$ profile may encounter gross instability and "disruptive" termination of the discharge.

While major disruptions can be avoided in normal tokamak operation, a low level of resistive-MHD activity generally remains present and often serves to restore the $q(r)$ profile periodically to a preferred form by reconnecting magnetic flux surfaces. For example, the "sawtooth" oscillations shown in Fig. 1 on the left (taken from PLT) serve to maintain $q(0)$ slightly below unity, on average, by reconnecting the magnetic flux arising from $m = 1$ resisitive-kink perturbations, thereby flattening the $j(r)$ profile (and T_e profile) periodically within the region where $q(r) \leq 1$. Such discharges are generally free of significant $m = 2$ activity and exhibit relatively favorable confinement throughout the region $q(r) > 1$. On the other hand, discharges without sawteeth [i.e., with $q(0) > 1$], such as that shown in Fig. 1 on the right, are generally characterized by a high level of $m = 2$ activity, often leading to a major disruption. A "sawtoothing discharge" may undergo a transition to an "$m = 2$ discharge" just after the sawtooth reconnection phase, when $q(0)$ attains its highest value ($\simeq 1$). These observations can be understood in terms of the stability of $m \geq 2$ tearing modes in tokamaks with $q(0)$ values in the general vicinity of unity.

TEARING MODE STABILITY

The stability of a tokamak to $m \geq 2$ tearing modes depends on the radial profile of the "safety-factor" $q(r)$, the m value of the mode, and the position of the singular

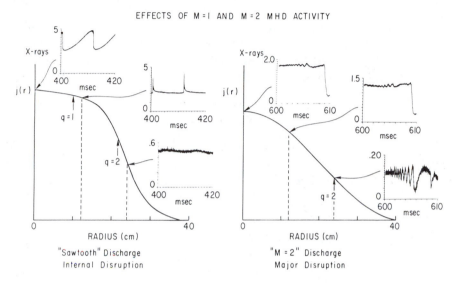

EFFECTS OF M=1 AND M=2 MHD ACTIVITY

FIGURE 1 Illustration of "sawtooth" and "$m = 2$" tokamak discharges (from PLT); the "sawtooth" discharge has $q(0)$ just below unity, leading to $m = 1$ activity at the plasma center, but it is quiescent near the plasma edge; if $q(0)$ rises above unity, strong "$m = 2$" activity near the edge leads to a major disruption. [From H.P. Furth, "The Tokamak," in *Fusion*, edited by E. Teller (Academic Press, New York, 1981), Vol. 1, Pt. A, Chap. 3]

surface r_s with respect to the $q(r)$ profile. Stability criteria can be obtained by varying the perturbed magnetic energy

$$W = (\pi/m^2) \int \left\{ r^3 \left(\frac{\partial \psi}{\partial r} \right)^2 + \left[(m^2 - 1)r + \frac{1}{F} \frac{d}{dr} \left(r^3 \frac{dF}{dr} \right) \right] \psi^2 \right\} dr \quad (1)$$

and solving the resulting Euler-Lagrange equation

$$\frac{\partial}{\partial r} \left(r^3 \frac{\partial \psi}{\partial r} \right) - \left[(m^2 - 1)r + \frac{1}{F} \frac{d}{dr} \left(r^3 \frac{dF}{dr} \right) \right] \psi = 0 \quad (2)$$

on either side of the singular surface r_s. Here, $\psi(r)$ is the radial component B_r of the perturbed magnetic field and

$$F(r) = \iota(r) - n/m, \quad (3)$$

where $\iota(r)$ is the rotational transform (divided by 2π), i.e., $\iota(r) = q^{-1}(r)$. If the Euler-Lagrange equations are satisfied, the perturbed magnetic energy becomes

$$W = -(\pi/m^2)r_s^3\psi_s^2\Delta_s',$$ (4)

$$\Delta_s' = \frac{[\partial\psi/\partial r]_s}{\psi_s},$$ (5)

where $[\ ]_s$ denotes the jump across the singular surface r_s. Modes for which the condition $\Delta_s' > 0$ is satisfied are unstable.

For high values of m, an analytic solution of the Euler-Lagrange equation is possible, giving the result

$$\Delta_s'r_s = -\left(\frac{rq\ dj_z/dr}{B_\theta\ dq/dr}\right)_{r_s} - m$$ (6)

which usually provides a good approximation for $m \geq 3$.

In an early paper on tearing modes in tokamaks,[2] the Euler-Lagrange equations were solved numerically for three representative tokamak current profiles, namely,

$$j_z(r) = j_z(0)/[1 + (r/r_0)^{2p}]^{1+1/p}$$ (7)

corresponding to

$$q(r) = q(0)\left[1 + (r/r_0)^{2p}\right]^{1/p}$$ (8)

for $p = 1$ ("peaked profile"), $p = 2$ ("rounded profile"), and $p = 4$ ("flattened profile").

In the steady-state phase of the tokamak, the current profile invariably contracts until the central q value is in the general neighborhood of unity. (The only exceptions to this pattern of behavior seem to be where high-Z impurity radiation from the central part of the discharge is unusually strong, depressing the central electron temperature and current density, or when noninductive current drive, either by neutral beam or rf techniques, is dominant in the central part of the plasma.) This contraction of the current profile is, no doubt, partly due to the cooling of the outer part of the discharge, but it may also represent a tendency for the discharge to evolve toward a configuration with relatively favorable stability properties against $m \geq 2$ tearing modes. Thus, it is of particular interest to consider the stability of profiles with $q(0) \sim 1$.

Noting that the quantity $r_s\Delta_s'$ provides a measure of the magnetic energy available to a tearing mode, we plot in Fig. 2 the calculated values of $r_s\Delta_s'$ as a function of $q(0)$ for the $m = 2$, $n = 1$ and $m = 3$, $n = 2$ modes for the three representative current profiles already discussed (assuming a conducting wall at $r_w/r_0 = 2.0$). It is evident from Fig. 2(a) that all three profiles are unstable to the $m = 2$ mode if $q(0) > 1$; the "flattened" and "rounded" profiles seem to be relatively unfavorable from an energetic viewpoint, and the effect of the $m = 2$ mode is presumably most severe in these cases. On the other hand, if $q(0)$ values below about 0.9 can be tolerated, the stability of the $m = 2$ mode is much improved, especially in the case

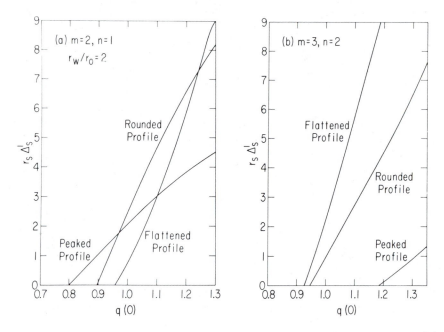

FIGURE 2   Values of $r_s \Delta_s'$ for representative current profiles plotted against $q(0)$, for (a) the $m = 2$ mode and (b) the $m = 3$, $n = 2$ mode; a conducting wall is at $r_w/r_0 = 2$. [Adapted from the results of Ref. 2]

of the "flattened" and "rounded" profiles. For $q(0)$ values above about 0.95, the "flattened" and "rounded" profiles are also strongly unstable to the $m = 3$, $n = 2$ modes, as shown in Fig. 2(b); the *simultaneous* destabilization of the $m = 2$, $n = 1$ and $m = 3$, $n = 2$ modes and the resulting breakup of magnetic surfaces provides a persuasive explanation for the major disruption.

MAJOR DISRUPTIONS

By careful tailoring of the current profile, the $m = 2$ mode can be stabilized even in cases with $q(0)$ above unity and $q(a)$ values in the range $2 < q(a) < 3$. Figure 3 shows the results of an early calculation with $q(0) = 1.05$ and $q(a) = 2.6$, where the stabilization has been brought about by flattening the $j(r)$ profile just inside the $q = 2$ surface.[3] Although all relevant modes — not only the $m = 2$, $n = 1$ mode — are seen to be stable (negative Δ' values), the Δ' values of the $m = 3$, $n = 2$ and

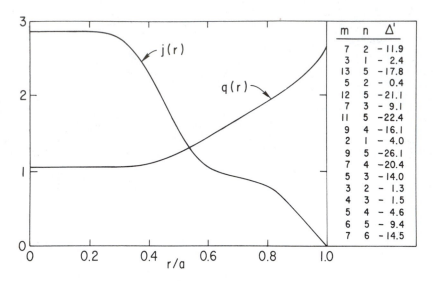

m	n	Δ'
7	2	− 11.9
3	1	− 2.4
13	5	−17.8
5	2	− 0.4
12	5	−21.1
7	3	− 9.1
11	5	−22.4
9	4	− 16.1
2	1	− 4.0
9	5	−26.1
7	4	−20.4
5	3	− 14.0
3	2	− 1.3
4	3	− 1.5
5	4	− 4.6
6	5	− 9.4
7	6	− 14.5

FIGURE 3   Example of a stable profile with $q(a) \simeq 2.6$, without any conducting shell; the values for Δ' are computed for all relevant pairs of (m, n) values. [From Ref. 3]

$m = 4$, $n = 3$ modes are seen to be quite small in magnitude, presumably because of the steepening of the current profile in the region within the $q = 2$ surface.

In the nonlinear regime, flattening of the current profile in the vicinity of the resonant surface may be expected to occur spontaneously, due to the effect of finite-amplitude magnetic islands on the evolution of the current profile. When the width of the magnetic island produced by an unstable $m \geq 2$ tearing mode exceeds the width of the "resistive layer" (typically only a few millimeters in high-temperature tokamaks), the rate of growth of the mode slows dramatically,[4] because of the action of (third-order) nonlinear forces opposing the flow of plasma across the X-point to the interior of the island, as illustrated in Fig. 4. The slow-growing magnetic islands affect the evolution of the current profile, essentially by flattening the profile over a region of order the island width. The growth of the island width w can then be described by an equation of the form

$$\frac{dw}{dt} = \eta \Delta'(w) \tag{9}$$

where the Δ' value has become a (decreasing) function of the island width.[5]

The formulation of "reduced MHD equations" describing the (two-dimensional) evolution of a finite-amplitude perturbation with a single helicity (i.e., involving only harmonics with the same m/n ratio) led to the development of a number

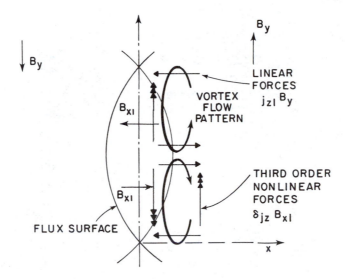

FIGURE 4 Flow of plasma into a growing magnetic island produced by tearing mode instability. The vortex flow is driven by "linear" forces $j_{z_1} B_y$ and impeded by third-order "nonlinear" forces $\delta j_z B_{x_1}$, where δj_z is the second-order induced current, $\delta j_z = -v_{y_1} B_{x_1}/\eta$. [From Ref. 4]

of numerical codes for treating nonlinear tearing modes.[6] Typically, the results of the simple analytic theory were confirmed: as the island width grows, the effective Δ' value decreases, until complete saturation ($\Delta' = 0$) occurs at some finite island size. The "saturated" width of $m = 2$ islands is sensitive to the shape of the initial current profile: the "flattened" current profile of Fig. 2 with a $q(0)$ value just above unity can lead to islands with widths as much as 30% of the minor radius.[7]

The development and application of (three-dimensional) "multi-helicity" numerical codes[8] showed that large $m = 2$ islands will, in general, destabilize other tearing modes with higher-order helicities. Figure 5 shows the result of one such calculation, beginning with the "flattened" profile of Fig. 2 with $q(0) \simeq 1.05$. In this case, the $m = 2$, $n = 1$ and $m = 3$, $n = 2$ modes are both unstable initially; after a period of slow growth lasting for about a thousandth of the global skin time, the $m = 5$, $n = 3$ mode is also destabilized, leading to the overlap of the three sets of islands. The overlap of islands with disparate helicities produces highly stochastic fields, thereby providing a plausible explanation of the rapid "thermal quench" in major disruptions. The destabilization of the $m = 5$, $n = 3$ mode in Fig. 5 is presumably due to the steepening of the current profile at the $q = 1.67$ surface,

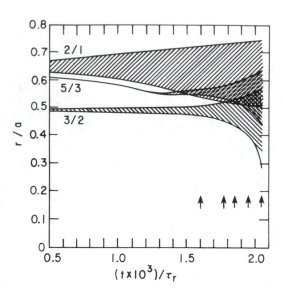

FIGURE 5 Multi-helicity simulations of nonlinear island growth, for the "flattened" profile case of Fig. 2 with $q(0) \simeq 1.05$. Both $m = 2$, $n = 1$ and $m = 3$, $n = 2$ modes are unstable initially. [From Ref. 8]

which is an inevitable consequence of the flattening of the profile at the $q = 1.5$ and $q = 2.0$ surfaces.

In general, the onset of major disruptions in large tokamaks such as TFTR can be correlated closely with the transgression of the tearing mode stability boundary.[9] However, there are also regimes in which the $m = 2$, $n = 1$ and/or $m = 3$, $n = 2$ modes are excited strongly but do not lead to disruptions. Figure 6 shows observations of $m = 2$, $n = 1$ and $m = 3$, $n = 2$ activity in high-$q(a)$, "supershot" discharges in TFTR.[10] The onset of the MHD-like activity is seen to produce a substantial deterioration in confinement, evidenced by a dropoff in the neutron emission. The "supershot" regime in TFTR is characterized by low collisionality, high β_p value, and a substantial component of beam-driven current — and also "bootstrap" current — in the central part of the plasma.

Figure 6(b) shows a case in which a strong $m = 3$, $n = 2$ mode arises by itself, i.e., without any $m = 2$, $n = 1$ activity. This is unusual and difficult to explain theoretically, because the Δ' criteria predict instability of the $m = 3$, $n = 2$ mode typically only in cases already unstable to the $m = 2$, $n = 1$ mode (cf. Fig. 2). In this context, an interesting effect has been identified recently,[11,12] in which the reduction in bootstrap current within a magnetic island (because of flattening of the

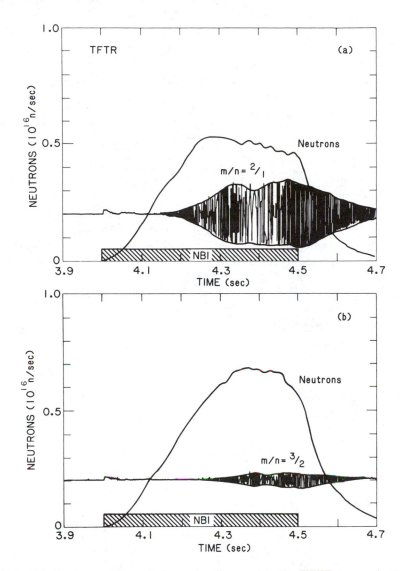

FIGURE 6   Effect of magnetic activity on confinement in the TFTR "supershot" regime. Cases (a) and (b) show the onset of $m = 2$, $n = 1$ and $m = 3, n = 2$ modes, respectively, and their effect on the neutron emissivity arising from beam-plasma and thermonuclear reactions.

pressure profile) leads to destabilization of tearing modes, even those with $\Delta' < 0$.

With the bootstrap current included, the evolution of finite amplitude magnetic islands in low-collisionality plasmas is found to be described by an equation of the form

$$\frac{dw}{dt} = \eta \left[\Delta' - \sqrt{\epsilon} \left(\frac{rp'}{B_\theta^2} \right) \left(\frac{q}{rq'} \right) \frac{1}{w} \right] \tag{10}$$

where the second term on the right arises from the bootstrap current and is always destabilizing for normal profiles. This effect might well account for the prevalence of $m = 3$, $n = 2$ modes in TFTR's high-β_p, low-collisionality "supershot" regime.[13]

SAWTEETH OSCILLATIONS

Tokamak discharges with $q(0) < 1$ exhibit "sawtooth" behavior, in which a strongly-growing $m = 1$ resistive kink reconnects the magnetic surfaces in the central region of the plasma such that $q(0)$ is periodically restored to unity. The process of reconnection can be described quantitatively[14] in terms of the ($m = 1$, $n = 1$) helical flux function

$$\chi(r) = \int^r B_\theta \, dr - (r^2 B_z / 2R) \tag{11}$$

which undergoes the transformation illustrated in Fig. 7. Initial flux elements $d\chi$ at r_1 and r_2 combine into the final flux element at r in such a way that the toroidal flux (area) is conserved:

$$r \, dr = r_1 \, dr_1 + r_2 \, dr_2. \tag{12}$$

The final flux function $\chi_f(r)$ can be obtained from the initial flux function $\chi_i(r)$ by means of the relation

$$r \frac{dr}{d\chi_f} = r_2 \frac{dr_2}{d\chi_i} \bigg|_{r_2(\chi)} - r_1 \frac{dr_1}{d\chi_i} \bigg|_{r_1(\chi)}. \tag{13}$$

The reconnected $q(r)$ profile has $q(0) = 1.0$; the reconnected $j(r)$ profile has a reversed surface current at the outermost radius of reconnection, but this will survive only transiently.

Numerical simulations of the nonlinear evolution of the $m = 1$ tearing mode confirm the validity of the Kadomtsev-Monticello model: the $m = 1$ tearing mode grows rapidly — even after the width of the magnetic island far exceeds the width of the resistive layer — and the final result is complete reconnection of the region within the $q = 1$ surface.[15] Such "complete" internal reconnection can clearly explain the enhanced thermal transport and rapid drop in central temperature associated with the sawtooth "crash."

A "simple" sawtooth of this type is illustrated in Fig. 8(a), which shows X-ray traces from the central region of the TFTR tokamak. A pronounced $m = 1$ "precursor" oscillation is evident, which grows in amplitude until "complete relaxation" occurs, causing a sharp drop in the temperature within the central part of

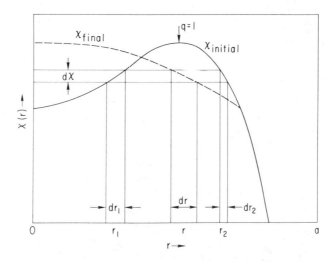

FIGURE 7 Illustration of sawtooth reconnection according to the Kadomtsev model. Initial flux elements at r_1 and r_2 combine into the final flux element at r in such a way that the toroidal flux (area) of an element $d\chi$ of helical flux is conserved.

the plasma and a corresponding rise in the temperature outside the $q = 1$ surface. However, other types of sawteeth are also observed on TFTR, including "small sawteeth" [Fig. 8(b)], which are generally similar to simple sawteeth except for the absence of $m = 1$ precursor oscillations, and "compound sawteeth" [Fig. 8(c)], which take the form of a periodic sequence of 'subordinate" and "main" relaxations. The subordinate relaxation of compound sawteeth is characterized by only a very small drop in the central temperature; although precursor oscillations are generally absent, strong $m = 1$ "successor" oscillations appear immediately after the sawtooth crash. The main relaxation is characterized by a much larger drop in the central temperature; again, precursors are normally absent, but successors sometimes appear.

Several theoretical models have been advanced to explain these novel observations.[16,17] The essential difference between the high-temperature plasmas in large tokamaks such as TFTR and the more resistive plasmas in earlier tokamaks is probably the very long skin time in TFTR, which prevents any significant change in the $q(r)$ profile during a sawtooth period. Accordingly, most theories of compound sawteeth have been based on the idea that the $q(r)$ profile can remain very flat — with a value close to unity — in the central part of the plasma, throughout the entire sawtooth cycle. In contrast to normal sawtooth relaxation [Fig. 9(a)], it

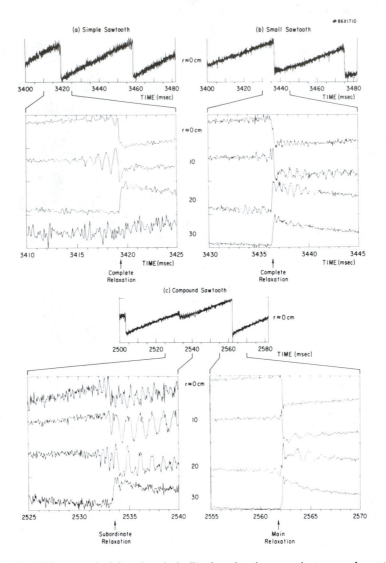

FIGURE 8 Soft X-ray emissivity signals indicating the three main types of sawteeth observed in TFTR: (a) simple or normal sawteeth; (b) small sawteeth, without precursors; and (c) compound sawteeth. [From Ref. 16]

is hypothesized that the relaxations in compound sawteeth leave the current profile slightly "hollow." In the subordinate relaxation [Fig. 9(b)], the q value drops first below unity off axis, and reconnection of the helical flux is limited to a region between the two resonant $q = 1$ surfaces, not including the magnetic axis. In the main relaxation [Fig. 9(c)], the hollowness of the current profile is more pronounced and, although the q value again drops first below unity off axis, the

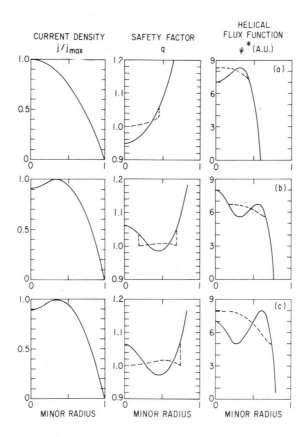

FIGURE 9   Illustration of three different types of sawtooth reconnection; in each case, the solid and broken curves represent profiles before and after reconnection, respectively. Case (a) is the normal reconnection of monotonic $j(r)$ and $q(r)$ profiles. Case (b) has a slightly hollow $j(r)$ profile and a $q(r)$ profile with an off-axis minimum; reconnection of the helical flux by "double" $m = 1$ tearing modes does not extend to the magnetic axis, consistent with the subordinate relaxation of compound sawteeth. Case (c) has a more deeply hollow $j(r)$, for which reconnection can extend to the axis, as in the main relaxation of compound sawteeth.

consequent reconnection of the helical flux extends to include the axial region. At least one numerical treatment has been successful in following an entire cycle of a compound sawtooth — requiring only a suitable parametric dependence of the anomalous crossfield thermal diffusivity in the central part of the plasma.[19]

THE TOKAMAK WITH $q(0) < 1$

An even more intriguing recent experimental observation has resulted from the first successful application of the Faraday rotation technique to the direct measurement of the $q(r)$ profile in ohmically heated discharges in the TEXTOR tokamak.[20] In the low-$q(a)$ case shown in Fig. 10, the measured central current density corresponds to a $q(0)$ value of 0.63 [see Fig. 10(c)]; the possible experimental errors in the measurement are significant, but correspond only to a range 0.54-0.73 in possible $q(0)$ values. Small sawteeth are apparent on the interferometer density traces [Fig. 10(b)]; the sawtooth inversion radius is in agreement with the location of the $q = 1$ resonance inferred from the measured $j(r)$ profile. The measured $q(0)$ value does not change appreciably ($10 \pm 5\%$) during a sawtooth cycle. Soft X-ray measurements of the $T_e(r)$ profiles [Fig. 10(e)] indicate that the measured $j(r)$ profile is consistent with neoclassical resistivity with a uniform Z_{eff} of about 2. At higher $q(a)$ values (~ 4), the measured $j(r)$ profile is much narrower, but the $q(0)$ value is again substantially below unity. The tearing modes whose resonances are indicated in Fig. 10(d) are all stable (note the negative Δ' values). However, despite the observed flattening of the $j(r)$ profile in the vicinity of the $q = 1$ surface, the $m = 1$, $n = 1$ mode remains unstable — at least in the usual cylindrical approximation.

The introduction of toroidal coupling can have an important stabilizing effect on the $m = 1$, $n = 1$ tearing mode. At sufficiently low aspect ratio, the flattening (or slight inversion) of the current profile in the vicinity of the $q = 1$ surface can yield complete stabilization of the mode.[21] Figure 11 shows a stable TEXTOR-like case with an aspect ratio of 3.25 and with $q(0)$ and $q(a)$ values of 0.66 and 1.96, respectively. The harmonic components of the radial field perturbation B_r are shown in Fig. 11(b), indicating the important role of the $m = 2$ harmonic. The $q(r)$ profile is flattened considerably in the vicinity of the $q = 1$ surface, but not so much as to reduce the shear to zero. When the aspect ratio is increased, keeping the profile shapes constant, the mode is destabilized, although the Δ' values remain modest in magnitude (see Fig. 12).

If the dominant $m = 1$ resistive mode can be suppressed as $q(0)$ is progressively lowered below unity, higher-order resistive modes will be encountered with resonant surfaces falling in the region where $q(r) < 1$; the most relevant of these will be the modes with $m/n = 4/5$, $3/4$, and $2/3$. Figure 13 (solid curves) shows an "optimized" case with $q(0) \simeq 0.5$ and $q(a) \simeq 2.0$, in which all of these higher-order modes (including modes such as $m/n = 5/4$, $4/3$, and $3/2$, whose resonant surfaces fall in the region outside the $q = 1$ surface) are stable.[9] (Note the slight flattening of the profiles in the vicinity of the $q = 2/3$ and $3/4$ resonances.) The solid-curve profiles in Fig. 13 remain, of course, strongly unstable to the $m = 1$, $n = 1$ mode. However, when a TEXTOR-like "plateau" is formed on the current profile in the vicinity of the $q = 1$ surface, as illustrated, for example, by the broken line in Fig. 13, the higher-order modes tend to be destabilized. The particular case shown in Fig. 13 (broken curves) results from increasing the width of the plateau until the modes $m/n = 2/3$, $4/5$, and $5/4$ are all unstable.[22] This suggests an

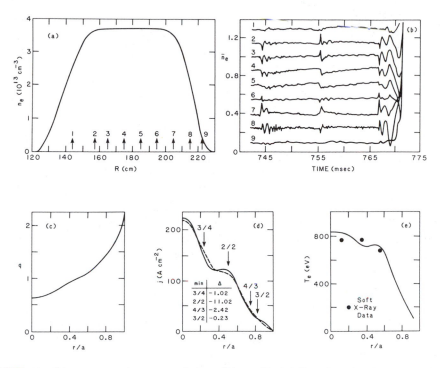

FIGURE 10   Direct measurements of the $q(r)$ profile by Faraday rotation in the TEX-TOR tokamak. (a) The density profile showing the location of nine channels of Faraday rotation; (b) small sawteeth observed on the interferometer density traces; (c) the measured $q(r)$ profile; (d) the $j(r)$ profile constructed from the measured $q(r)$ profile, and the Δ' values for all relevant modes other than the $m/n = 1/1$ mode; and (e) the electron temperature profile constructed from the $j(r)$ profile with the assumption of neoclassical resistivity with constant Z_{eff}, compared with soft X-ray data. [From Ref. 20]

unorthodox, but intriguing, alternative interpretation of sawtooth relaxation: the sawtooth crash results from enhanced thermal transport due to overlapping higher-order (e.g., $m/n = 2/3$, $3/4$, $4/5$) magnetic islands, which are destabilized by the steepening of the current profile in the $q(r) < 1$ region, brought about by the widening plateau at the $q = 1$ surface. Such an interpretation could be consistent with the observation on TEXTOR that $q(0)$ remains well below unity throughout

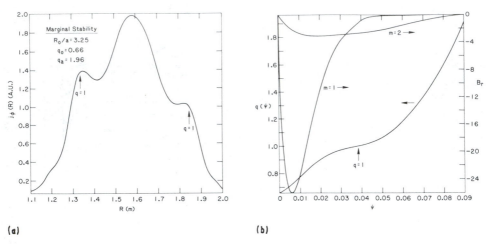

(a)   **(b)**

FIGURE 11   (a) TEXTOR-like $j(r)$ profile, modified to provide stability against the $m = 1$, $n = 1$ resistive mode; the difference between the theoretical profile and the measured profile lies well within the experimental error bars. (b) Fourier decomposition of the poloidal harmonics of the $n = 1$ resistive mode, superimposed on the $q(r)$ profile. [From Ref. 21]

the sawtooth cycle; it has many features in common with the widely accepted interpretation of the major disruption in terms of overlapping $m/n = 2/1$, $3/2$, and $5/3$ islands (cf. Fig. 5).

With the assumption that the TEXTOR-like low-$q(0)$ regime can be attained in higher-temperature and higher-beta plasmas, the successful suppression of the $m = 1$ mode and the associated strong sawteeth could have substantial benefits for tokamak performance: (i) it could provide a significant improvement in the limiting beta value for ballooning instabilities by allowing reduced $q(0)$ and $q(a)$ values; (ii) it could provide indirect stabilization of $m = 2$ external kinks by allowing more centrally-peaked $j(r)$ profiles than would otherwise be possible at low $q(a)$ values; and (iii) it could provide an improvement in confinement by allowing increased plasma current. To explore the possibilities of this sort, the ideal-MHD stability of a conventional D-shaped tokamak ($\kappa = 1.6$, $\delta = 0.3$, $R/a = 3.2$) with $q_\psi(0) = 0.6$ and $q_\psi(a) = 1.8$ has been examined with a code that adjusts local pressure gradients to provide marginal stability against Mercier and ballooning modes (see Fig. 14).[21] While the Mercier criterion imposes the more demanding requirement over much of the region where $q(\psi) < 1$, shear stabilization is strong enough to allow a substantial pressure gradient to be supported in this central region (Fig. 14). The overall (stable) $\langle \beta \rangle$ value is 10.0%, which exceeds the Troyon limit $[\langle \beta \rangle_{\text{Troyon}} (\%) = 3.5\ I(\text{MA})/a(\text{m})B(\text{T})]$ by a factor of 1.35. The $n = 1$ external kink, which has a

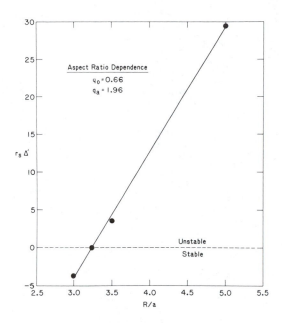

FIGURE 12 Effect of varying the aspect ratio on the stability of the $m = 1$ resistive mode for the TEXTOR-like $j(r)$ profile shown in Fig. 11. Even when the mode is unstable, the Δ' value is finite, as indicated in the figure. [Adapted from the results of Ref. 21]

dominant $m = 2$ component, is stabilized by a conducting wall placed at $r_w/a = 1.15$, but would otherwise be strongly unstable; triangularity is sufficient to stabilize the $n = 1$ internal kink.

FEEDBACK STABILIZATION OF TEARING MODES

Feedback stabilization of $m \geq 2$ modes by rf heating and/or current drive has been proposed.[23,24] To produce a stabilizing effect, the feedback technique must *increase* the plasma current density at the O-point of the magnetic islands associated with the tearing mode and *decrease* the current density at the X-point (separatrix) of the island.

There are two principal options for producing the desired current perturbations by rf feedback techniques:

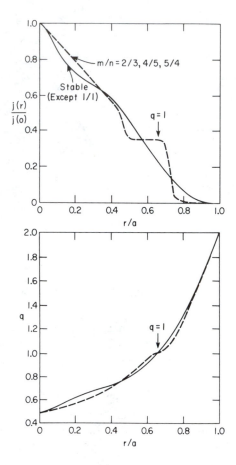

FIGURE 13 Effect of a "plateau" in the $j(r)$ profile at the $q = 1$ surface on the stability of higher-order resistive modes. The solid curves are profiles that are stable to all resistive modes other than $m/n = 1/1$. Note the slight flattening of the profiles in the vicinity of the $q = 2/3$ and $q = 3/4$ surfaces. These profiles are then modified to produce a plateau at the $q = 1$ surface. When the plateau is as large as that shown by the broken curves, the higher-order modes $m/n = 2/3$, $4/5$, and $5/4$ have all been destabilized.

1. Heat the magnetic islands (O-points) by localized rf heating, thereby lowering the local resistivity.
2. Drive additional noninductive currents within the magnetic islands (O-points).

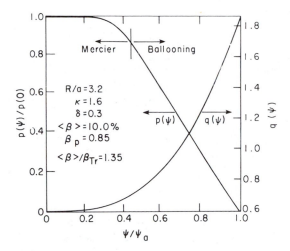

FIGURE 14 Profiles of $p(\psi)$ and $q(\psi)$ versus poloidal flux ψ for a D-shaped tokamak that is everywhere marginally stable to localized pressure-driven modes, with $q(0) = 0.6$ and $q(a) = 1.8$. The radius of the $q = 1$ surface is $\sqrt{\langle r^2 \rangle}/a = 0.72$. The $\langle \beta \rangle$ value is 10.0% and exceeds the "Troyon limit" β_{Troyon} by a factor 1.35. [From Ref. 21]

In both cases, the rf power must be phase modulated to match a perturbation signal from some suitable detector (for example, the electron temperature measured by electron cyclotron emission). Feedback techniques based on lower hybrid waves (for current drive) or electron cyclotron waves (for heating) are theoretically capable of providing the required localization of the rf power,[25] and the application of these techniques to the control of the $m = 1$ mode is illustrated in Fig. 15.

The theory of feedback stabilization of tearing modes by island heating is based on the standard treatment of $m \geq 2$ modes in their slow-growing (nonlinear) phase, except that the resistivity on flux surfaces interior to magnetic islands is allowed to be perturbed relative to that on exterior flux surfaces. The rf power density is modulated in phase with the rotating island,

$$P_{rf} = \tilde{P}_{rf} \cos(m\theta - n\phi - \omega t), \tag{14}$$

and the radial profile of power deposition is assumed to be quite narrow, but not as narrow as the island itself. The width w of the magnetic island is found to grow according to a relation of the form:

$$\frac{dw}{dt} = \eta \left(\Delta' - C_h \frac{\tilde{P}_{rf} w}{K_{\perp e} T_e} \right), \tag{15}$$

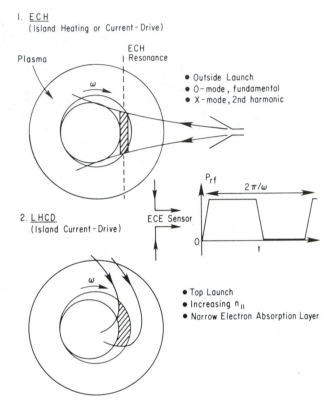

1. ECH
   (Island Heating or Current-Drive)

ECH
Resonance

Plasma

ω

- Outside Launch
- O-mode, fundamental
- X-mode, 2nd harmonic

P_{rf}

$2\pi/\omega$

ECE Sensor

0   t

2. LHCD
   (Island Current-Drive)

ω

- Top Launch
- Increasing $n_{\parallel}$
- Narrow Electron Absorption Layer

FIGURE 15   Schematic illustration of feedback stabilization of $m = 1$ rotating magnetic islands by rf heating or current drive using electron cyclotron heating (ECH) or lower hybrid current drive (LHCD) techniques. [From Ref. 25]

where $K_{\perp e}$ is the electron thermal diffusivity and C_h is a numerical constant given by

$$C_h = 0.75(r j_z / B_\theta)(q / r q'). \tag{16}$$

Unfortunately, the power requirements for this type of rf feedback turn out to be prohibitively large, especially if the cross-field electron thermal conductivity within the magnetic island is as large as the observed global thermal conductivity.

The theory of feedback stabilization by island current drive proceeds along similar lines. The rf-driven current density is modulated in phase with the rotating island,

$$j_{rf} = \tilde{j}_{rf} \cos(m\theta - n\phi - \omega t), \tag{17}$$

and the radial deposition profile is assumed to be quite narrow, but not as narrow as the island itself. The island width w grows according to

$$\frac{dw}{dt} = \eta \left(\Delta' - C_d \frac{\tilde{j}_{rf}}{j_{z_0}} \frac{1}{w} \right), \tag{18}$$

where j_{z0} is the unperturbed local current density and C_d is a numerical constant given by

$$C_d = 8(rj_z/B_\theta)(q/rq'). \tag{19}$$

Unlike feedback by island heating, where only islands above a certain size can be stabilized, feedback by current drive will suppress all islands smaller than some critical size.

The successful feedback stabilization of the $m = 1$ mode and the associated "sawteeth" would have more substantial benefits than the suppression of the $m = 2$ mode. However, in a "cylindrical" tokamak with circular cross-section, the ideal-MHD $m = 1$ mode ("internal kink") is marginally stable, and the resistive mode becomes strongly unstable and does not enter a slow-growing nonlinear phase. The rapid growth of the $m = 1$ mode makes rf feedback stabilization somewhat problematical.[26]

However, the ideal-MHD mode can become positively stable in a tokamak with a strongly shaped (triangular) plasma cross-section.[27] (Toroidicity will also stabilize the internal kink at low β_p values, but this effect does not seem, of itself, strong enough for rf feedback to be feasible unless it is accompanied by a TEXTOR-like flattening of the current profile in the vicinity of the $q = 1$ surface.) If the ideal-MHD mode is positively stable, an effective Δ' value can be calculated, and the rf feedback theory can be applied.[24] For present purposes, we neglect toroidal effects and consider a D-shaped plasma boundary of the form

$$r/a = 1 - \xi_{2a} \cos 2\theta + \xi_{3a} \cos 3\theta. \tag{20}$$

(The quantities ξ_{2a} and ξ_{3a} are related to the more familiar elongation κ and triangularity δ by $\kappa \simeq 2\xi_{2a}$ and $\delta \simeq 3\xi_{3a}$.) For this case, the effective Δ' value is given by

$$r_s \Delta'_s = \frac{(r_s q'_s)^2 (r_s/a)^2}{(\delta W_3 \xi_{3a}^2 - \delta W_2 \xi_{2a}^2)}, \tag{21}$$

where the numerical quantities δW_2 and δW_3 (both positive) have been calculated previously for representative current profiles,[27] as well as similar quantities describing toroidal effects at finite β_p. (The effect of reduced shaping of flux surfaces near

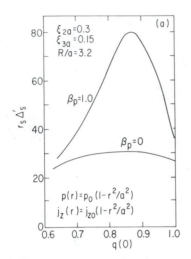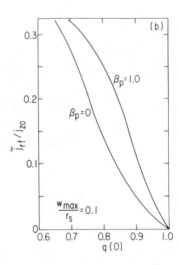

FIGURE 16   (a) Values of the effective $\Delta'_s r_s$ for the $m = 1$ mode in the case of a shaped cross section, with triangular deformation $\xi_{3a} = 0.15$ ($\delta = 0.45$) and elliptic deformation $\xi_{2a} = 0.3$ ($\kappa = 1.6$); the dependence on β_p arises from toroidal effects. (b) Feedback current density $\tilde{\jmath}_{rf}$ required to stabilize the $m = 1$ mode for island widths up to a maximum $w_{\max}/r_s = 0.1$; $\tilde{\jmath}_{rf}$ is expressed as a fraction of the initial local current density j_{z0} and is plotted as a function of the $q(0)$ value. [From Ref. 24]

the magnetic axis, relative to that of the plasma boundary, is included in the numerical values of δW_2 and δW_3, and it is calculated from an equilibrium that is consistent with the assumed current profile.)

To provide a concrete example,[24] we consider a D-shaped cross-section with $\xi_{2a} = 0.3$ and $\xi_{3a} = 0.15$ (corresponding to $\kappa \simeq 1.6$ and $\delta \simeq 0.45$). Figure 16(a) shows the values of $r_s \Delta'_s$ for this case, with the assumption of a parabolic current profile, and Fig. 16(b) shows the feedback current density $\tilde{\jmath}_{rf}$ (expressed as a fraction of the initial local current density j_{z0}) required to stabilize the $m = 1$ mode as a function of the $q(0)$ value, for a maximum permitted island width of $w_{\max}/r_s = 0.1$. We see that suppression of $m = 1$ islands with widths up to $w/r_s \sim 0.1$ requires values of $\tilde{\jmath}_{rf}/j_{z0}$ in the range 0.1–0.3—a demanding requirement, but not entirely impossible. Thus, feedback stabilization of the $m = 1$ mode by means of rf current drive will require a shaped plasma cross section and substantial feedback power, but it may be a feasible option for realizing and sustaining the $q(0) < 1$ regime in a tokamak.

Moreover, the TEXTOR results provide encouragement that the $m = 1$ tearing mode can enter a slow-growing or saturated phase, even when stabilizing mechanisms such as triangular shaping are absent. Indeed, for the TEXTOR-like current profile shown in Fig. 11, the values of $r_s \Delta'_s$ are modest ($\sim 10 - 30$), even if the aspect ratio is not small enough to produce complete stabilization (see Fig. 12). For such values of $r_s \Delta'_s$, the feedback-modulated fractions of the current density, i.e., the values of $\tilde{j}_z / j_{z_0}$, needed to stabilize islands with widths up to $w_{\max}/r_s \sim 0.1$ are in the range 0.05 - 0.1, a modest requirement. If the presence of a finite-amplitude $m = 1$ island is sufficient to produce the required local flattening of the $j(r)$ profile, then feedback control of this mode by rf techniques should be feasible, leading to a low-$q(0)$, sawtooth-suppressed regime of the standard tokamak configuration.

ACKNOWLEDGEMENTS

The author is grateful for useful discussions of many aspects of the subject matter of this paper with H.P. Furth, C.-Z. Cheng, J. Manickam, A.M.M. Todd, and R.B. White and for discussions of TFTR sawtooth data with K.M. McGuire and of TEXTOR data with W. Stodiek.

This work was supported by the United States Department of Energy Contract No. DE-AC02-76-CH0-3073.

REFERENCES

1. H.P. Furth, J. Killeen, and M.N. Rosenbluth, *Physics of Fluids* **6**, 459 (1963).
2. H.P. Furth, P. Rutherford, and H. Selberg, *Physics of Fluids* **16**, 1054 (1973).
3. A.H. Glasser, H.P. Furth, and P.H. Rutherford, *Physical Review Letters* **38**, 234 (1977).
4. P.H. Rutherford, *Physics of Fluids* **16**, 1903 (1973).
5. R.B. White, D.A. Monticello, M.N. Rosenbluth, and B.V. Waddell, *Physics of Fluids* **20**, 800 (1977).
6. M.N. Rosenbluth, D.A. Monticello, H.R. Strauss, and R.B. White, *Physics of Fluids* **19**, 1987 (1976).
7. R.B. White, D.A. Monticello, and M.N. Rosenbluth, *Physical Review Letters* **39**, 1618 (1977).
8. B. Carreras, H.R. Hicks, J.A. Holmes, and B.V. Waddell, *Physics of Fluids* **23**, 1811 (1980).
9. C.-Z. Cheng, H.P. Furth, and A. H. Boozer, *Plasma Physics and Controlled Fusion* **29**, 351 (1987).

10. A.W. Morris, E.D. Fredrickson, K.M. McGuire *et al.*, in *Proceedings of the Fourteenth European Conference on Controlled Fusion and Plasma Physics* (Madrid, 1987).
11. J.D. Callen *et al.*, in *Plasma Physics and Controlled Nuclear Fusion Research 1986* (IAEA, Vienna, 1987), Vol. II, p. 157.
12. R. Carrera, R.D. Hazeltine, and M. Kotschenreuther, *Physics of Fluids* **29**, 899 (1986).
13. M. Kotschenreuther, private communication.
14. B.B. Kadomtsev, *Soviet Journal of Plasma Physics* **1**, 389 (1975); also D.A. Monticello, unpublished (1975).
15. B.V. Waddell, M.N. Rosenbluth, D.A. Monticello, and R.B. White, *Nuclear Fusion* **16**, 528 (1987).
16. K.M. McGuire *et al.*, in *Plasma Physics and Controlled Nuclear Fusion Research 1986* (IAEA, Vienna, 1987), Vol. I, p. 421.
17. V.V. Parail and G.V. Pereverzev, *Soviet Journal of Plasma Physics* **6**, 14 (1980).
18. J.A. Wesson, P. Kirby, and M.F. Nave, in *Plasma Physics and Controlled Nuclear Fusion Research 1986* (IAEA Vienna, 1987), Vol. II, p. 3.
19. J.E. Drake *et al.*, in *Plasma Physics and Controlled Nuclear Fusion Research 1986* (IAEA, Vienna, 1987), Vol. I, p. 387.
20. H. Soltwisch, W. Stodiek, J. Manickam, and J. Schlueter, in *Plasma Physics and Controlled Nuclear Fusion Research 1986* (IAEA, Vienna, 1987), Vol. I, p. 263.
21. J. Manickam, C.-Z. Cheng, P.H. Rutherford, W. Stodiek, and A.M.M. Todd, in *Proceedings of the Fourteenth European Conference on Controlled Fusion and Plasma Physics* (Madrid, 1987).
22. C.-Z. Cheng, private communication.
23. Y. Yoshioka, S. Konoshita, and T. Kobayashi, *Nuclear Fusion* **24**, 565 (1984).
24. P.H. Rutherford, in *Course and Workshop on Basic Physical Processes of Toroidal Fusion Plasmas*, Varenna, Italy, 1985 (CEC EUR 10418 EN, Brussels, 1986), p. 531.
25. D.W. Ignat, P.H. Rutherford, and H. Hsuan, in *Course and Workshop on Application of rf Waves to Tokamak Devices*, Varenna, Italy, 1985 (CEC EUR 10333 EN, Brussels, 1986), p. 525.
26. R.B. White, in *Workshop on Magnetic Reconnection and Turbulence* (Cargese, France, July 7-13, 1985).
27. D. Edery, G. Laval, R. Pellat, and J.L. Soule, *Physics of Fluids* **19**, 260 (1975); also J.W. Conner, and R.J. Hastie, unpublished (1977).

Ravindra N. Sudan
Laboratory of Plasma Studies and
 Cornell Theory Center
Cornell University
Ithaca, New York 14853

Strong Plasma Turbulence

A systematic procedure is described for deriving a spectral equation from the renormalized theory of strong turbulence based on the direct interaction approximation of Kraichnan. This method is applied to drift wave turbulence in plasmas.

INTRODUCTION

From the very early beginnings of plasma physics as a separate, well-identified discipline, it was recognized that a plasma is rarely in a state of thermodynamic equilibrium. It is capable of releasing its free energy through the excitation of unstable eigenmodes or waves. The nonlinear interaction of these modes leads eventually to plasma turbulence. When the amplitudes of the dispersive eigenmodes are small, the state of the plasma may be treated as a linear superposition of these modes with amplitudes which are slowly varying functions of time governed by the weak nonlinear interaction. If, in addition, the random phase approximation is made, one obtains a kinetic equation for mode amplitudes which describes the state of "weak turbulence." Such an equation based on quantum mechanical perturbation theory was derived for the phonon gas in a solid by Peierls,[1] and has received widespread application in plasma physics.[2]

When the modes are neither dispersive nor are the amplitudes small, i.e., the conditions for weak turbulence do not obtain, the situation is inherently closer to fluid turbulence and therefore vastly more intractable. The fully developed turbulence in this limit is now considered "strong" and its mathematical treatment proceeds as follows.

In general terms we seek the average properties of a set of random variables $C(\mathbf{x}, t)$ whose time dependence is governed by

$$\frac{\partial}{\partial t} C - LC = V(C, C) + f, \tag{1}$$

or, in its Fourier transform representation,

$$(\omega - \omega_{\mathbf{k}}) C(\mathbf{k}, \omega) = \int d\mathbf{k}' d\omega' V(\mathbf{k}, \omega; \mathbf{k}', \omega') C(\mathbf{k}', \omega') C(\mathbf{k} - \mathbf{k}', \omega - \omega') + f(\mathbf{k}, \omega),$$

$$C(\mathbf{x}, t) = \int d\mathbf{k}\, d\omega\, C(\mathbf{k}, \omega) \exp\left[i(\mathbf{k} \cdot \mathbf{x} - \omega t)\right], \tag{2}$$

where L is a linear integro-differential operator, V is a quadratic nonlinear operator, and f is an externally applied stochastic force with known properties; C, f, k, and x may be vectors in which case L and V are tensors of appropriate rank. In what follows we shall restrict C to be a scalar unless otherwise noted. The solution of Eq. (1) or (2) involves generating relations for the averaged quantities $\langle C \rangle$, $\langle CC \rangle$, $\langle CCC \rangle$, etc., and we rapidly run into the usual problem of obtaining an adequate method for closure.

The quasi-normal hypothesis[3] demands that the fourth-order cumulant vanishes, i.e.,

$$\langle\!\langle CCCC \rangle\!\rangle \equiv \langle CCCC \rangle - \Sigma \langle CC \rangle \langle CC \rangle = 0. \tag{3}$$

Equation (3) provides one technique for closure, but has some serious defects because it can violate the nonnegativity of the spectrum.[3]

The method that has found greatest popularity in plasma physics is Kraichnan's Direct Interaction Approximation[4] (DIA), which deals with the correlation function $Q \equiv \langle C(\mathbf{x}, t) C(\mathbf{x}', t') \rangle$ and the response function defined through the functional derivative $R \equiv \langle \delta C(\mathbf{x}, t) / \delta f(\mathbf{x}', t') \rangle_{\delta f \to 0}$. The response function contains the system dynamics, and it is in the retention of the dynamical element that DIA differs from purely statistical closure techniques. This method of closure delivers a pair of coupled integral equations for Q and R. The proper justification for DIA is, however, neither apparent nor easily demonstrated. A systematic method for establishing DIA and the next-order equations is that due to Martin, Siggia, and Rose[5] (MSR) which relies heavily on Schwinger's formalism for quantum electrodynamics[6] and the concepts of mass and charge renormalization. Subsequently, Jensen[7] has developed a formal analogue to the method of MSR through the use of Feynman's phase integral approach. The reader anxious to learn more about these formal approaches is referred to the excellent article by Krommes.[8]

DIRECT INTERACTION APPROXIMATION

The basic postulates of DIA are (i) maximal randomness and (ii) weak dependence. By the former is meant that the stochastic nature of the problem results from nonlinear interactions between modes; indeed, the distribution for C is as close to Gaussian as possible without violating the governing equation. Weak dependence implies that the dynamic coupling among a small finite number of modes vanishes as the number of modes N becomes very large. More precisely, if the energy density E is invariant, i.e.,

$$\sum_{\mathbf{k}}^{N} \left\langle |C_{\mathbf{k}}|^2 \right\rangle = E, \tag{3}$$

then

$$\sum_{\mathbf{k}}^{N} \sum_{\mathbf{k}'}^{N} \sum_{\mathbf{k}''}^{N} \langle C_{\mathbf{k}} C_{\mathbf{k}'} C_{\mathbf{k}''} \rangle \, \delta(\mathbf{k} + \mathbf{k}' + \mathbf{k}'') = \text{const.} \tag{4}$$

and

$$\frac{\langle C_{\mathbf{k}} C_{\mathbf{k}'} C_{\mathbf{k}''} \rangle}{\left[\left\langle |C_{\mathbf{k}}|^2 \right\rangle \left\langle |C_{\mathbf{k}'}|^2 \right\rangle \left\langle |C_{\mathbf{k}''}|^2 \right\rangle \right]^{1/2}} \to N^{-1/2}, \tag{5}$$

i.e., the triple correlation vanishes as $N^{-1/2}$ for $N \to \infty$.

Kadomtsev[9] formulated the DIA in Fourier $(\mathbf{k}, \omega)$ space in which the correlation function is related to the spectral function $I(\mathbf{k}, \omega)$

$$\langle C(\mathbf{k}, \omega) C^*(\mathbf{k}', \omega') \rangle = I(\mathbf{k}, \omega) \delta(\mathbf{k} - \mathbf{k}') \delta(\omega - \omega'), \tag{6}$$

and the response function R is related to the nonlinear frequency shift $\widetilde{\Gamma}(\mathbf{k}, \omega)$ [so-called "mass renormalization" in MSR terminology], through

$$2\pi R(\mathbf{k}, \omega) = i \left[\omega - \omega_{\mathbf{k}} + \widetilde{\Gamma}(\mathbf{k}, \omega) \right]^{-1}. \tag{7}$$

The DIA equations for $\widetilde{\Gamma}$ and I are given by

$$\widetilde{\Gamma}(\mathbf{k}, \omega) = - \int d^3 k' d\omega' \frac{W_{\mathbf{k},\omega;\mathbf{k}'',\omega''} W_{\mathbf{k}'',\omega'';\mathbf{k},\omega} I(\mathbf{k}', \omega')}{\omega'' - \omega_{\mathbf{k}''} + \widetilde{\Gamma}(\mathbf{k}'', \omega'')}, \tag{8}$$

$$\left| \omega - \omega_{\mathbf{k}} + \widetilde{\Gamma}(\mathbf{k}, \omega) \right|^2 I(\mathbf{k}, \omega) = \frac{1}{2} \int d^3 k' d\omega' \left| W_{\mathbf{k},\omega;\mathbf{k}',\omega'} \right|^2 I(\mathbf{k}', \omega') I(\mathbf{k}'', \omega''), \tag{9}$$

with $\mathbf{k}'' = \mathbf{k} - \mathbf{k}'$, $\omega'' = \omega - \omega'$ and $W_{\mathbf{k},\omega;\mathbf{k}',\omega'} = V(\mathbf{k}, \omega; \mathbf{k}', \omega') + V(\mathbf{k}, \omega; \mathbf{k}'', \omega'')$. If $I(\mathbf{k}, \omega)$ is a bell-shaped function of ω with width Γ centered on $\omega = \omega_{\mathbf{k}}$, then from Eq. (8) it is easy to see that $\operatorname{Im} \widetilde{\Gamma}$ and $\operatorname{Re} \widetilde{\Gamma}$ will have the form shown in Fig. 1. $\operatorname{Im} \widetilde{\Gamma}$ is approximately constant over most of the spectrum, and the width Γ of the spectrum is in fact

$$\Gamma(\mathbf{k}) \approx \operatorname{Im} \widetilde{\Gamma}(\mathbf{k}, \omega_{\mathbf{k}}). \tag{10}$$

As originally presented by Kraichnan,[4] the equation for the response function R, or alternatively the equation for $\widetilde{\Gamma}$, is singular. The singularity has been removed by Kraichnan in a modified version of the DIA which adopts a Lagrangian framework.[10-12] This singularity has been attributed by Kadomtsev[9] to the inclusion of adiabatic interactions in the integral over $\mathbf{k}'$ space. In the interaction of $C(\mathbf{k})$, $C(\mathbf{k}')$, and $C(\mathbf{k}'')$ satisfying the triangle equality $\mathbf{k} = \mathbf{k}' + \mathbf{k}''$, those interactions (see Fig. 2) in which $|\mathbf{k}'| \ll |\mathbf{k}|$, $|\mathbf{k}''|$ are to be excluded because they represent the interaction of a wave packet $\mathbf{k}$ with a very long wavelength disturbance $\mathbf{k}'$ that results only in a local distortion of the wave packet $\mathbf{k}$, rather than a resonant transfer of energy which takes place when $\mathbf{k}$, $\mathbf{k}'$, and $\mathbf{k}''$ are of comparable magnitude. Kadomtsev's prescription for eliminating the singularity and restoring the correct physics is to cut off the $\mathbf{k}'$ integration from below at αk (see Fig. 3). With this correction and writing

$$I(\mathbf{k}, \omega) = I(\mathbf{k}) \, g\left[(\omega - \omega_\mathbf{k})/\Gamma(k)\right], \tag{11}$$

where $g(\omega)$ is a resonant function, e.g., a Lorentzian or a Gaussian, Sudan and Keskinen[13] have solved Eq. (8) to obtain

$$\Gamma(k_\perp) = D k_\perp^2 I^{1/2}(k_\perp) \tag{12}$$

for two-dimensional $\mathbf{E} \times \mathbf{B}$ ionospheric turbulence; D depends on α and parameters peculiar to the physical situation. Similar expressions are obtained for other cases, but it is important to note that an inspection of Eq. (8) will quickly yield $\Gamma(k) \propto I^{1/2}$ provided

$$\Gamma(k) > |\gamma_\mathbf{k}|, \Delta\omega_\mathbf{k}, \tag{13}$$

where $|\gamma_\mathbf{k}|$ is the linear damping or growth rate and $\Delta\omega_\mathbf{k} = \omega_{\mathbf{k}'} - \omega_{\mathbf{k}-\mathbf{k}'}$ is the mode dispersion. In the opposite limit of $\Gamma(k) \ll |\gamma_\mathbf{k}|$, $\Delta\omega_\mathbf{k}$, it can be shown that weak turbulence is recovered.[14] The quantity $\Gamma(k)$ in Eq. (12) can also be identified with the inverse eddy turnover time[15] $k_\perp \Delta V_{k_\perp}$ where $\Delta V_{k_\perp}$ is the root-mean-square $\mathbf{E} \times \mathbf{B}$ velocity of eddies of scale size $k_\perp^{-1}$. Since $\left\langle |\Delta V_{k_\perp}|^2 \right\rangle \propto I(k_\perp)$, it is now apparent why Γ is proportional to $I^{1/2}$.

SPECTRAL EQUATIONS

Having solved for Γ from Eq. (8), we may obtain $I(k)$ from Eq. (9), employing the relation of Eq. (11). This straightforward procedure is, however, not very useful to obtain a workable spectral equation. A better approach is to start with an equation for a system quadratic invariant such as the energy density. Take, as an example,

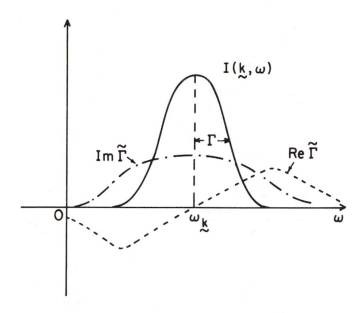

FIGURE 1 Schematic plot of $I(\mathbf{k},\omega)$, $\mathrm{Re}\,\widetilde{\Gamma}(\mathbf{k},\omega)$, and $\mathrm{Im}\,\widetilde{\Gamma}(\mathbf{k},\omega)$ as functions of ω.

isotropic turbulence governed by the incompressible Navier-Stokes equations. We have

$$\left(\frac{\partial}{\partial t} + 2\nu k^2\right) E(k,t) = \frac{\partial \Pi}{\partial k}, \tag{14}$$

where

$$E(k) = 4\pi k^2 I(k),$$

$$\langle u_i(\mathbf{k},t)u_i(-\mathbf{k},t')\rangle = \left(\delta_{ij} - \frac{k_i k_j}{k^2}\right) I(k, t - t'),$$

$$u_i(\mathbf{x},t) = \int d^3k\, u_i(\mathbf{k},t)\exp(i\mathbf{k}\cdot\mathbf{x}).$$

Also, $u_i(\mathbf{x},t)$ is the random velocity, ν is the kinematic viscosity, and $\Pi(k)$, the energy flux in k space, is given by[4]

$$\Pi = \Pi^+ - \Pi^- = \frac{1}{2}\int_k^\infty dk' \int^k \int^\Delta dp\,dr\,R(k'\mid p,r)$$

$$- \frac{1}{2}\int_0^k dk' \int_k^\infty dp \int_k^\Delta dr\,R(k'\mid p,r), \tag{15}$$

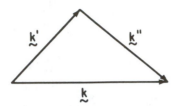

Resonant Interaction

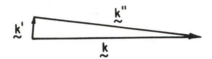

Non-resonant , adiabatic Interaction

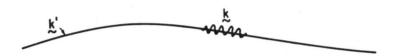

FIGURE 2 Typical triangle representation of $k = k' + k''$ for resonant and nonresonant interaction of modes.

with

$$R(k \mid p, r) + R(p \mid r, k) + R(r \mid p, k) = 0.$$

The $\triangle$ symbol in Eq. (15) denotes that the triangle equality (cf. Fig. 3) must be obeyed, i.e.,

$$|k' - r| \leq p \leq k' + r. \tag{16}$$

We must also observe Kadomtsev's cutoff, and this is displayed in Fig. 3, where the hatched area is defined by Eq. (16) and the double hatched regions are cut off by noting that

$$\alpha k' < p \qquad \text{and} \qquad \alpha k' < r, \tag{16'}$$

with α an arbitrary factor less than unity. Taking account of Kadomtsev's cutoff, Sudan and Pfirsch[14] have shown that

$$\Pi^+ \gg \Pi^-, \tag{17}$$

and

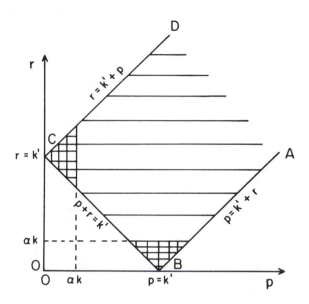

FIGURE 3 The region of integration denoted by $\triangle$ in Eq. (15) is the interior of A B C D described by $|k' - r| \leq p \leq k' + r$; the hatched regions $\propto k' < p$ and $\propto k' < r$ represent Kadomtsev cutoff; $\triangle_\alpha$ denotes A B C D without "cutoff" regions. [Adapted from Ref. 14]

$$\Pi \approx \Pi^+(k) = \int_{\alpha k}^{k} dk' \Gamma(k') E(k'), \tag{18}$$

with

$$\Gamma(k) = D k^{3/2} E^{1/2}(k) \propto k \left[k^2 \left\langle |u_k|^2 \right\rangle \right]^{1/2} \propto k \triangle u_k. \tag{19}$$

In the stationary state from Eqs. (18), (19), and (14), we obtain a power balance equation,[14]

$$\int_{\alpha k}^{k} dk' \Gamma(k') E(k') + 2\nu \int_{0}^{k} dk' k'^2 E(k') = \varepsilon, \tag{20}$$

where ε is the energy input per unit mass per unit time at $k \to 0$ by an external agent. The first term in Eq. (20) represents the power transferred to modes of length scale shorter than k^{-1} by the interaction of scales greater than k^{-1}, and the second term is the viscous dissipation of modes between 0 and k. Equation (20) may also be written, with $Y(k) = k\Gamma E$, as

$$\frac{d\Pi}{d \ln k} = \frac{d}{d \ln k} \int_{\alpha k}^{k} (d \ln k') Y(k') = -2\nu k^2 E(k). \tag{21}$$

In the limit $\nu = 0$, i.e., in the inertial range, we obtain the self-similar relation

$$Y(k) = \alpha Y(\alpha k), \tag{22}$$

whose solution is $Y(k) = k^{-1}$ and where $E(k) \propto k^{-5/3}$ [from Eq. (19)] is the Kolmogorov-Obukhov Law. Beyond the inertial range, Eq. (21) yields[15]

$$\frac{d}{dk}(k\Gamma E) = -\frac{2\nu k^2 E}{\ln(1/\alpha)}, \tag{23}$$

which is easily solved to give

$$E(k) = [\nu/2D\ln(1/\alpha)]^2 \, k^{-5/3} \left(k_m^{4/3} - k^{4/3}\right)^2. \tag{24}$$

Here, $k_m = [D\ln(1/\alpha)]^{1/2} \left(8\varepsilon/\nu^3\right)^{1/4}$ is the Kolmogorov cutoff. A detailed solution of Eq. (20) furnishes the following:[14]

1. The solution of Eq. (24) is valid for $k < 0.38 k_m$; the exact solution does not give $E(k) = 0$ at $k = k_m$ but indicates that it falls off very rapidly.
2. For $k \ll k_m$, we have $E(k) = K_0 \varepsilon^{2/3} k^{-5/3}$. The Kolmogorov constant K_0 is related to α through

$$\ln(1/\alpha) = 8 \left[K_0^3 h(\alpha)\right]^{-1}$$

and

$$D\ln(1/\alpha) = K_0^{-3/2},$$

where $h(\alpha)$ is a known function. For $K_0 = 1.44$, we have $\alpha = (3.5)^{-1}$.

With the employment of a similar logical procedure, a spectral equation in two or three dimensions for low-frequency isotropic turbulence in a plasma can be derived if a quadratic invariant like the energy density exists in the limit of zero damping/growth. Such an invariant, $J(k) = \int d\omega \, d^2\Omega \, k^n I(\mathbf{k}, \omega)$, where $d^3k = k^2 dk \, d^2\Omega$, exists if the nonlinear matrix elements of Eq. (2) obey the following relation:[14]

$$\int d^2\Omega' d^2\Omega'' \left[k^n V_{\mathbf{k'},\mathbf{k''}} + k'^n V_{\mathbf{k''},\mathbf{k}} + k''^n V_{\mathbf{k},\mathbf{k'}}\right] V^*_{\mathbf{k'},\mathbf{k''}} \delta(\mathbf{k} + \mathbf{k'} + \mathbf{k''}) = 0. \tag{25}$$

For a plasma we may replace the externally injected energy ε by the free energy released through a linear instability with growth rate $\gamma(\mathbf{k})$. Thus, we have[14]

$$\frac{d}{dk}(k\Gamma J) = \frac{2\gamma J}{\ln(1/\alpha)}. \tag{26}$$

EXAMPLES

The turbulence driven by crossed $\mathbf{E} \times \mathbf{B}$ drifts in a collisional, weakly ionized plasma[16,17] and drift waves in a collisionless, magnetized plasma are two examples where Eq. (26) may be applied in a two-dimensional physical situation.

1. $\mathbf{E} \times \mathbf{B}$ TURBULENCE The two-fluid collisional equations with the neglect of both electron and ion inertia yield[13]

$$\frac{\partial n}{\partial t} + \frac{\mathbf{B} \times \nabla \phi}{B^2} \cdot \nabla n = D_1 \nabla^2 n, \qquad (27)$$

$$\nabla \cdot \left[n \left(\nabla \phi + \frac{\nu_i}{\Omega_i} \frac{\mathbf{B} \times \nabla \phi}{B^2} \right) \right] = D_2 \nabla^2 n. \qquad (28)$$

Here, $n_e = n_i = n$, the plasma β is negligible, ϕ is the electrostatic potential, ν_j is the collision frequency of the jth specie with neutrals, Ω_j is the cyclotron frequency, and D_1 and D_2 are appropriate diffusion coefficients; also $\nu_e \ll \Omega_e$ and $\nu_i > \Omega_i$. Equation (28) follows from taking the divergence of the current density to vanish. From Eqs. (27) and (28) we may derive[13]

$$\left[\frac{\partial}{\partial t} + i \left(\omega_{\mathbf{k}} + i \gamma_{\mathbf{k}} \right) \right] n(\mathbf{k}, t) = \frac{i}{2} \frac{\nu_i}{\Omega_i} \sum_{\mathbf{k}=\mathbf{k}'+\mathbf{k}''} \left(\frac{\mathbf{k}' \cdot \mathbf{v}_d}{k'^2} - \frac{\mathbf{k}'' \cdot \mathbf{v}_d}{k''^2} \right)$$

$$\widehat{B} \cdot \mathbf{k}' \times \mathbf{k}'' n(\mathbf{k}') n(\mathbf{k} - \mathbf{k}'), \qquad (29)$$

where

$\mathbf{k} \cdot \mathbf{B} = 0,$

$\omega_{\mathbf{k}} = \mathbf{k} \cdot \mathbf{v}_d / (1 + \psi),$

$\gamma_k = \gamma_0 - k^2 D_0; \qquad \gamma_0 \propto L^{-1},$

$\mathbf{v}_d = \dfrac{\widehat{B} \times \nabla \phi_0}{B}, \qquad \psi = \nu_e \nu_i / \Omega_e \Omega_i, \qquad L^{-1} = \dfrac{1}{n_0} \dfrac{dn_0}{dz}.$

The gradients $\nabla \phi_0$ and ∇n_0 are maintained by external influence (see Fig. 4). Notice that the modes are nondispersive, and strong turbulence will therefore prevail. Equation (29) is in the canonical form; $(\delta n)^2 = (n - n_0)^2$ is a conserved quantity and Eq. (26) applies. Define $I(\mathbf{k}, \omega)$ from

$$\langle \delta n(\mathbf{k}, \omega) \delta n^*(\mathbf{k}', \omega') \rangle = I(\mathbf{k}, \omega) \delta(\mathbf{k} - \mathbf{k}') \delta(\omega - \omega'),$$

with $I(\mathbf{k}) = \int_{-\infty}^{\infty} d\omega\, I(\mathbf{k}, \omega)$. The solution of Eq. (26) is given by[14]

$$\widehat{E}(x) = x^{-5/3} \left\{ 1 - x^{-2/3} - \frac{1}{2S} \left(x^{4/3} - 1 \right) \right\}^2, \qquad (30)$$

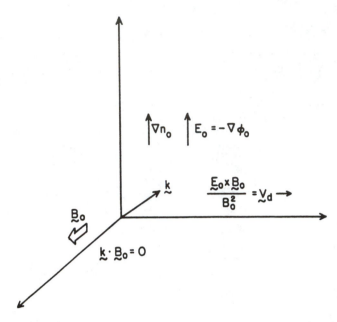

FIGURE 4 Geometry of the $\mathbf{E} \times \mathbf{B}$ instability which occurs for $\nabla n_0 \cdot \nabla (e\phi_0) > 0$; the ion drift vanishes because they are unmagnetized.

with $\widehat{E} = [D \ln(1/\alpha)]^2 (k_0^2/\gamma_0^2) E(k)$, $E(k) = kI(k)$, $x = k/k_0$, the equivalent Reynolds number is $S = \gamma_0/k_0^2 D_0$, and the boundary condition is $E(k_0) = 0$. Figure 5 plots $\widehat{E}(x)$. Note that $\gamma(x) < 0$ for $x > x_d = S^{1/2}$. Also, $\widehat{E}(x)$ drops catastrophically at $x = x_m = (2S)^{3/4}$. The spectrum between $x_d = S^{1/2}$ and $x_m = (2S)^{3/4}$ is populated by the nonlinear interaction of unstable waves below $x = x_d$. The solution of Eq. (30) is valid so long as

$$\Gamma(x) > |\gamma(x)| . \tag{31}$$

We have assumed in the above treatment that the turbulence is two-dimensional and isotropic with $\mathbf{k} \cdot \widehat{B} \equiv k_\| = 0$. While this is a plausible assumption, it does not hold rigorously. Departures from isotropy have been observed in numerical calculations[18,19] of Eq. (29). The anisotropy results from the term $\left(\mathbf{k}' \cdot \mathbf{v}_d/k'^2 - \mathbf{k}'' \cdot \mathbf{v}_d/k''^2\right)$ in the matrix element of Eq. (29).

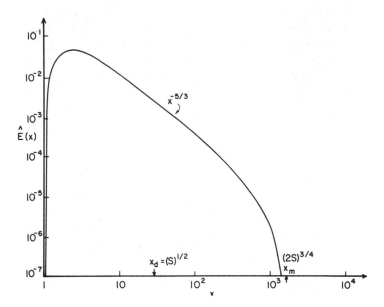

FIGURE 5 Plot of $\widehat{E}(x)$ as a function of x from Eq. (30) for $S = 10^4$. [From Ref. 14]

2. DRIFT WAVE TURBULENCE In the limit of $T_i \rightarrow 0$ and $\beta < m_e/m_i$, the Hasegawa-Mima equation describing the nonlinear interaction of drift waves (see Horton[20] for details) is applicable when the magnetic field has no shear and the electron dynamics are linearized:

$$\left[\frac{\partial}{\partial t} + i\left(\omega_{\mathbf{k}} + i\gamma_{\mathbf{k}}\right) \right] \phi(\mathbf{k}, t) = \frac{ic}{B} \sum_{\mathbf{k}=\mathbf{k}'+\mathbf{k}''} \widehat{B} \cdot \mathbf{k}' \times \mathbf{k}'' \left(|k'\rho|^2 - |k''\rho|^2 \right) \phi(\mathbf{k}')\phi(\mathbf{k}-\mathbf{k}'),$$

(32)

where

$$\omega_{\mathbf{k}} = \mathbf{k} \cdot \mathbf{v}_d(1 - k^2\rho^2), \qquad \mathbf{v}_d = (cT_e/eB)\left(\nabla n_0 \times \widehat{B} \right)/n_0,$$

$$\rho^2 = m_i c^2 T_e/e^2 B^2,$$

and the growth rate $\gamma_{\mathbf{k}}$ depends upon the details of the case. Drift modes are also non-dispersive for $k^2\rho^2 \ll 1$, and Eq. (32) again can be put in the canonical form of Eq. (26). Unlike the $\mathbf{E} \times \mathbf{B}$ case discussed above, Eq. (32) supports not one but two quadratic invariants, the energy density $E = \int d^2k(1 + k^2\rho^2)\left\langle |\phi(\mathbf{k})|^2 \right\rangle$

and the enstrophy density $\varepsilon = \int d^2k\, k^2(1 + k^2\rho^2)\langle |\phi(\mathbf{k})|^2 \rangle$. To proceed further we make the following conjecture inspired by the exact results of two-dimensional hydrodynamics:[21]

In the stationary state the power spectrum of the fluctuations is exclusively determined by only the cascade of one invariant in any particular region of **k** *space. If the system supports more than one invariant, then each invariant determines a particular region of* **k** *space.*

In the context of the drift wave problem, the above conjecture leads to the following spectral equations in the limit $k^2\rho^2 \ll 1$:

$$\frac{d}{dk}(k\Gamma\varepsilon) = 2\gamma(k)\varepsilon/\ln(1/\alpha), \qquad \text{for} \qquad k > k_*, \tag{33}$$

$$\frac{d}{dk}(k\Gamma E) = -2\gamma(k)E/\ln(1/\alpha), \qquad \text{for} \qquad k < k_*, \tag{34}$$

where k_* is located in the region where $\gamma(k) > 0$ as shown in Fig. 6; also, $E(k) = 2\pi k I(k)$, $\varepsilon(k) = 2\pi k^3 I(k)$, $I(k) = \langle |\phi(k)|^2 \rangle$, and

$$\Gamma(k) = D\frac{c}{B}\,(k^2\rho^2)\,k^3 I^{1/2}(k) \equiv Ck^5 I^{1/2}. \tag{35}$$

Equations (33) and (34) are valid only when the conditions for strong turbulence, Eq. (31) and

$$\Gamma(k) > |\Delta\omega_{\mathbf{k}}| \sim k^2\rho^2\,|\mathbf{k}\cdot\mathbf{v}_d|, \tag{36}$$

are satisfied. Recalling that the wave amplitude $\tilde{\varphi}_{\mathbf{k}} \sim k I^{1/2}(k)$, we find that Eq. (36) furnishes the approximate saturation amplitude:

$$\frac{e\tilde{\phi}}{T_e} \gtrsim (kL)^{-1}. \tag{37}$$

Since $\widetilde{\delta n}/n_0 = e\tilde{\phi}/T_e$ for drift waves, Eq. (37) implies

$$k\widetilde{\delta n} \gtrsim \frac{n_0}{L}, \tag{38}$$

which is a well-known result that the gradient of the fluctuating density just exceeds the gradient of the mean density. Neglecting the right-hand sides of Eqs. (33) and (34), we obtain the result of Fyfe and Montgomery,[22] viz.,

$$I(k) \propto k^{-6} \qquad \text{for} \qquad k > k_*,$$

and

$$I(k) \propto k^{-14/3} \qquad \text{for} \qquad k < k_*. \tag{39}$$

If $\gamma(\mathbf{k}) > 0$ [with $\gamma_m = \max \gamma(\mathbf{k})$] for $k_1 < k < k_2$ and $\gamma(\mathbf{k}) < 0$ everywhere else, the solution of Eqs. (33) and (34) is

$$I(k) = \left(\gamma_m/k_1^5 C\right)^2 g(x), \qquad x > 1,$$

$$= \left(\gamma_m/k_1^5 C\right)^2 f(x), \qquad x < \frac{k_2}{k_1}, \qquad (40)$$

where $x = k/k_1$,

$$g(x) = \frac{4}{9} x^{-6} \left[\int_1^x dx \, x^{-3} \gamma/\gamma_m \right]^2,$$

and

$$f(x) = \frac{4}{9} x^{-14/3} \left[\int_1^{k_2/k_1} dx \, x^{-11/3} \gamma/\gamma_m \right]^2.$$

To make this result self-consistent, Eqs. (31) and (36) must be satisfied. Thus we require

$$\gamma_m > \frac{3}{2} (k_1 \rho)^2 \omega_* \left[\int_1^x dx \, x^{-3} \gamma/\gamma_m \right]^{-1}, \qquad \text{for} \qquad \frac{k_d}{k_1} > x > 1,$$

$$\gamma_m > \frac{3}{2} (k_1 \rho)^2 \omega_* \left[\int_x^{k_2/k_1} dx \, x^{-11/3} \gamma/\gamma_m \right]^{-1}, \qquad \text{for} \qquad \frac{k_D}{k_1} < x < 1, \qquad (41)$$

where k_d and k_D are wave numbers at which the spectrum drops off sharply.

ALMOST TWO-DIMENSIONAL TURBULENCE

Frequently in strongly magnetized plasmas the phase of low-frequency fluctuations varies slowly along the magnetic field but rapidly across the field, i.e., $k_\| \ll k_\perp$. The turbulence would be strictly two-dimensional only in the limit $k_\| = 0$. In this case the correlation length is infinite along the magnetic field. However, this case is basically unstable and tends to evolve to the other limit where the parallel correlation length is zero. This occurs because the modes scatter off each other during one eddy turnover time $\tau \sim \Gamma^{-1}$, leading to fluctuations with finite $k_\|$ spread. A quantitative measure of this effect is obtained by expanding $I(\mathbf{k}, \omega)$ as follows:

$$I\left(\mathbf{k}_\perp'', k_\| - k_\|'\right) = I\left(\mathbf{k}_\perp'', k_\|\right) - k_\|' \frac{\partial I}{\partial k_\|} + \frac{1}{2} \left(k_\|'\right)^2 \frac{\partial^2 I}{\partial k_\|^2} + \cdots$$

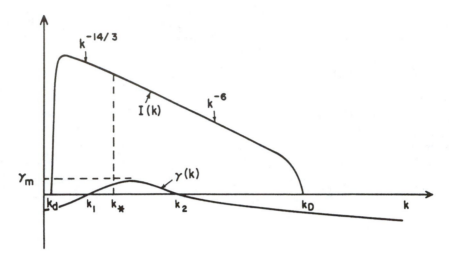

FIGURE 6 Schematic plot of $\gamma(k)$ and $I(k)$ for long-wavelength drift waves.

and substituting this expression into Eq. (9) to obtain, after considerable algebra, the result of Rosenbluth and Sudan:[23]

$$\frac{\partial I}{\partial t} + \frac{1}{k_\perp} \frac{\partial (k_\perp \Pi)}{\partial k_\perp} = 2\gamma(\mathbf{k})I + \frac{\partial}{\partial k_\parallel} \mathcal{D} \frac{\partial}{\partial k_\parallel} I, \tag{42}$$

where

$$\mathcal{D} = \frac{1}{2} \left\langle \frac{(\Delta k_\parallel)^2}{\tau} \right\rangle \approx \Gamma(k_\perp) \int dk_\parallel \frac{1}{2} k_\parallel^2 I(k_\perp, k_\parallel) \bigg/ \int dk_\parallel I(k_\perp, k_\parallel) .$$

It follows from the nonlinear diffusion term in Eq. (42) that the width $\Delta k_\parallel$ of the $k_\parallel$ spectrum will increase monotonically until linear damping due to finite $k_\parallel$ sets in. This opens up a new energy loss channel at large parallel wavelengths, distinct from the energy sink at large $k_\perp$ (see Fig. 7).

NUMERICAL SIMULATION

In recent years the advent of supercomputers has opened up the study of turbulence through numerical simulation of the basic fluid equations. However, even with the power of the presently available computing machines, the time required for a single run lasting an eddy turnover time may take some 30 hours of computer time.

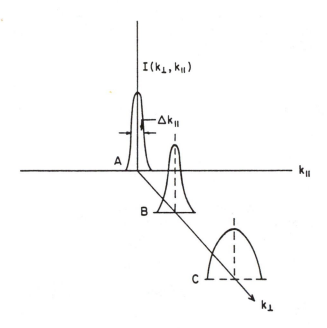

FIGURE 7 Schematic plot of $I\left(k_\perp, k_\parallel\right)$ from Eq. (42) in the stationary state. Note that $\triangle k_\parallel$ increases with $k_\perp$ until strong $k_\parallel$-dependent damping sets in at C. [From Ref. 23]

A three-dimensional spectral code for Navier-Stokes turbulence simulating a cube of size L, with N^3 grid points, requires a storage of $8N^3$ words and CPU time $\propto N^4 \log_2 N$. For $N = 128$, the CRAY XMP time/step is $\sim 20\,\mathrm{sec}$ and a run of 5×10^3 steps takes approximately 30 hours. Even so, the Reynolds number achieved is only $\sim 10^3$. The simulation of large Reynolds number flows which exist in space plasma physics and perhaps in the large fusion machines requires:

1. Big jumps in processor power and the development of massively parallel architecture;
2. Improvement in numerical algorithms; and
3. Better models to describe the physical situation to be simulated.

We discuss two examples that employ DIA-based models in plasma turbulence:

1. A DIA-based model for sub-grid modelling in a spectral code through a renormalized damping which takes into account the coupling of modes actually computed (see Fig. 8), with short-wavelength modes outside the domain included in the computation.[24]
2. A model developed by Albert, Similon, and Sudan[25] for computing two-dimensional drift wave turbulence with a renormalized growth rate that represents

the coupling of modes in the k_x, k_y plane at one value of $k_\|$ to the modes in the k_x, k_y plane at some other value of $k_\|$. Thus, a two-dimensional computation includes an important three-dimensional effect discussed in the preceding section.

SUB-GRID MODELLING

In the equation

$$\left[\frac{\partial}{\partial t} + i\left(\omega_{\mathbf{k}} + i\gamma_{\mathbf{k}}\right)\right]\phi_{\mathbf{k}} = \sum_{\mathbf{k}=\mathbf{k'}+\mathbf{k''}} V_{\mathbf{k},\mathbf{k'},\mathbf{k''}}\phi_{\mathbf{k'}}\phi_{\mathbf{k''}},\tag{43}$$

the summation in $\mathbf{k} = (k_x, k_y)$ space extends over a large domain, but in the numerical computation we retain only $N \times N$ modes, extending to $\mathbf{k}_N$. The interaction of the $N \times N$ modes with $|\mathbf{k}| > |\mathbf{k}_N|$ modes is described by a damping term $-\widetilde{\widetilde{\Gamma}}(k)\phi_{\mathbf{k}}$ included in the left-hand side of Eq. (43), where $\widetilde{\widetilde{\Gamma}}$ is given by

$$\widetilde{\widetilde{\Gamma}}(k) = \Gamma\left(\mathbf{k},\omega_{\mathbf{k}}\right) = -\frac{1}{2}\sum_{\mathbf{k'}}^{\Delta_\alpha} \frac{W_{\mathbf{kk'}}W_{\mathbf{k'k}}I_{\mathbf{k-k'}}\operatorname{sgn}\Gamma_{\mathbf{k'}} + \{\mathbf{k'} \leftrightarrow \mathbf{k}-\mathbf{k'}\}}{|\Gamma_{\mathbf{k'}}| + |\Gamma_{\mathbf{k-k'}}|}.\tag{44}$$

In the summation over $\mathbf{k'}$, only terms with either $\mathbf{k'}$ or $|\mathbf{k} - \mathbf{k'}| > k_m$ are included, as shown in Fig. 8. The expression for $I(k)$ for $k > k_m$ is obtained by extrapolation of the computed $I(k)$ for $k < k_m$. Several iterations are needed before $\widetilde{\widetilde{\Gamma}}$ and the spectra settle down to the stationary state shown in Figs. 9 and 10. Notice that the index of the isotropic spectrum is -3.47, which is close to Kolmogorov index of -3 for the inertial range. In the absence of the $\widetilde{\widetilde{\Gamma}}$ term, one would obtain an equipartition of energy between the modes,[21,26] and the spectrum is considerably flatter. If sufficient viscosity is introduced to prevent equipartition, then the inertial range is not even recovered because the spectrum drops rapidly. Thus the scheme outlined above appears to yield the correct eddy damping for large Reynolds number turbulence.[24]

WEAK COUPLING TO $k_\|$ MODES

Here the coupling of $\mathbf{k}_\perp$ to $k_\|$ modes is again obtained through the DIA of Eq. (9). Without going through the details, we quote the result obtained by Albert, Similon, and Sudan.[25] This coupling is taken into account through the term $\gamma_{\text{eff}}\phi_{\mathbf{k}}$ introduced in the left-hand side of Eq. (32) where

$$\gamma_{\text{eff}} = \int dk_\| \, \gamma\left(k_\perp, k_\|\right) g\left(k_\perp, k_\|\right),$$

$$g\left(k_\perp, \zeta\right) = \int dk_\| \, g\left(k_\perp, k_\|\right)\exp\left(ik_\|\zeta\right),\tag{45}$$

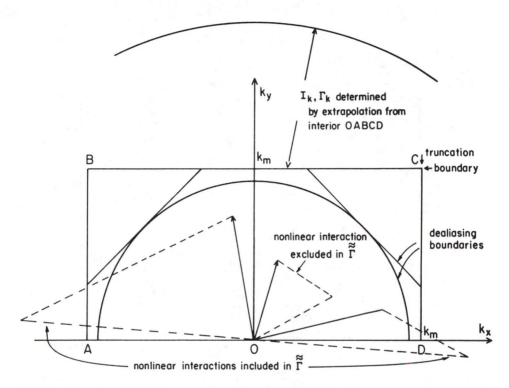

FIGURE 8 Region A B C D in the $k_x - k_y$ plane in which Eq. (43) in numerically computed. The interaction of modes within A B C D with modes outside A B C D (sub-grid modes) is accounted for by calculating $\tilde{\tilde{\Gamma}}$ according to Eq. (44).

and

$$bg\left(k_{\perp},\zeta\right)+\int ds\,\gamma\left(k_{\perp},\zeta-s\right)g\left(k_{\perp},s\right)$$

$$=\left\{b+\left[\int ds\,\gamma\left(k_{\perp},-s\right)g\left(k_{\perp},s\right)\right]\right\}g^{2}\left(k_{\perp},\zeta\right)$$

$$2b\equiv\left|-i\left(\omega-\omega_{\mathbf{k}}\right)+\Gamma\left(k_{\perp}\right)\right|^{2}. \tag{46}$$

Equation (46) is solved numerically at each time step.

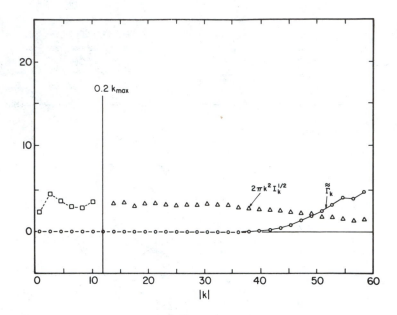

FIGURE 9 Plot of numerically computed $\tilde{\tilde{\Gamma}}$ as a function of $|k|$.

CONCLUDING REMARKS

In addition to numerical simulation, powerful new analytical tools are being brought to bear on the problem of turbulence. Theories of chaos in dynamical systems appear to shed some light on the transition from laminar to turbulent flow.[27] Renormalized group theory, so successfully applied by Wilson[28-29] to phase transitions, has been used by Ma and Mazenko[30] to treat the dynamics of spin systems and by Foster, Nelson, and Stephen[31] to obtain the long-wavelength properties of a fluid stirred randomly at short wavelengths. Very briefly, the renormalized group method as applied to an incompressible Navier-Stokes fluid consists of the following steps:

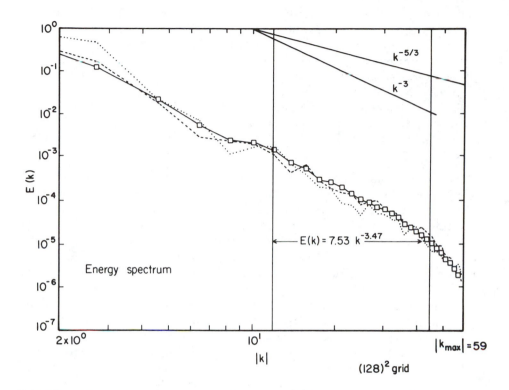

FIGURE 10 Plot of the spectrum as a function of $|k|$ from a spectral code including the eddy damping term $\widetilde{\Gamma}$ to represent sub-grid modelling.

1. In d dimensions, the lth component of the velocity field is given by

$$v_l(\mathbf{k}) = G^0 f_l(\mathbf{k}) - \frac{i}{2} G^0 P_{lmn}(\mathbf{k}) \int \frac{d^d q}{(2\pi)^{d+1}} v_m(\hat{q}) v_n \left(\hat{k} - \hat{q} \right)$$

$$\langle f_i f_j \rangle = (2\pi)^{d+1} 2 D_0 k^{-\gamma} P_{ij}(\mathbf{k}) \delta(\mathbf{k} + \mathbf{k}')$$

$$P_{ij}(\mathbf{k}) = \delta_{ij} - k_i k_j / k^2; \qquad P_{lmn} = k_m P_{ln} + k_n P_{lm}$$

$$G^0 = \left[-i\omega + \nu k^2 \right]^{-1}; \qquad \hat{k} \equiv (\mathbf{k}, \omega).$$

Also, f_l is a solenoidal external random force which cuts off above Λ.

2. Eliminate modes $v_l^>(\mathbf{k})$ with $\Lambda e^{-p} < k < \Lambda$, $p > 0$, from the equation for modes $v_l^<(\mathbf{k})$ with $0 < k < \Lambda e^{-p}$.

3. Average the reduced set over $f_l^>(\hat{k})$ that acts in $\Lambda e^{-p} < k < \Lambda$: this operation redefines coefficients which enter the reduced equations of motion. The fluctuating remainder is added to the noise $f_l^<$; p is a measure of the fraction of degrees of freedom eliminated.

4. Rescale space, time, $\mathbf{v}$, and $\mathbf{f}$ so that the new equations are similar to the primitive Navier-Stokes equations.

5. Obtain recursion relations for the propagator G, correlation functions, $\langle v_l v_m \rangle$, viscosity, etc., and obtain renormalized parameters as $p \to \infty$. The analysis generally reduces to a convergent perturbation series in the expansion parameter $\epsilon = d - d_*$.

This yields renormalized expressions for the viscosity and the external noise or stirring force. The recent effort by Yakhot[32] and Yakhot and Orzag,[33] although yielding some interesting results, especially, for the numerical constants associated with Navier-Stokes turbulence, is fraught with difficulties. Indeed, one serious difficulty is apparent in the algebra, because for $d = 3$, the small parameter used in the expansion turns out to be 4! A deeper understanding of the renormalized group method is necessary before it can be successfully employed for attacking plasma turbulence.

Finally, let us recapitulate the successes and deficiencies of the DIA, which has received wide usage in plasma physics. In the modified form due to Kadomtsev[9] or in the Lagrangian version due to Kraichnan,[10,11] it confirms the Kolmogorov-Obukov law and, indeed, Lagrangian DIA also delivers the Kolmogorov constant. *Thus, it is safe to say that for plasma systems which are homologous to the Navier-Stokes fluid, similar results may be obtained in the inertial range.* In situations where the inertial range does not really exist, the nonlinear frequency shift $\tilde{\Gamma}(\mathbf{k}, \omega)$ or damping $\Gamma(\mathbf{k})$ is an extremely useful quantity with which to determine the saturation level of a weakly driven system,[34] i.e., $\Gamma(\mathbf{k}) = \gamma(\mathbf{k}) > 0$, and the spectrum extends only to the region of $\mathbf{k}$ space where $\gamma(\mathbf{k}) > 0$. Modes outside $\gamma(\mathbf{k}) > 0$ may only be weakly excited. However, for strongly driven systems, spectral equations of the kind in Eq. (26) become applicable. More importantly, DIA can be used to incorporate important physical effects in numerical algorithms or models for simulating turbulence.

On the other hand, DIA is unable to explain the presence of structures in turbulence, e.g., intermittency and vortex structures in a Navier-Stokes fluid[35] and cavitons in high-frequency Langmuir turbulence in a plasma.[36] This weakness is due to the fact that phase correlation of nearby modes is not properly taken into account by DIA, which is the "most Gaussian" approximation that correctly furnishes energy cascade. It is of interest to note that the nonlinearity in the equations leads not only to the stochastic nature of turbulence, but it also determines the close phase correlation of neighbouring modes, as in a soliton. A truly successful theory of turbulence will require both effects to be properly accounted.

ACKNOWLEDGEMENTS

I am indebted to all my collaborators—Michael Keskinen, Russell Kulsrud, Dieter Pfirsch, Marshall Rosenbluth, Göran Schultz, and Philippe Similon—in various aspects of the work described in this paper. In particular, I wish to express my gratitude to our guru, Marshall Rosenbluth, for inspiration over the long period of our association. This work was supported by the National Science Foundation and Office of Naval Research.

REFERENCES

1. R. Peierls, *Quantum Theory of Solids* (Oxford University Press, Oxford, England, 1965).
2. See, for example, A. A. Galeev and R. Z. Sagdeev, "Theory of Weakly Turbulent Plasma" in *Handbook of Plasma Physics, Volume I, Basic Plasma Physics*, edited by A. A. Galeev and R. N. Sudan (North Holland, New York, 1983), p. 679.
3. Y. Ogura, *Journal of Fluid Mechanics* **16**, 33 (1963).
4. R. H. Kraichnan, *Journal of Fluid Mechanics* **5**, 497 (1959).
5. P. C. Martin, E. D. Siggia, and H. A. Rose, *Physical Review A* **8**, 423 (1973).
6. J. Schwinger, *Proceedings of the National Academy of Sciences of the United States of America* **37**, 452 (1951).
7. R. J. Jensen, *Journal of Statistical Physics* **25**, 183 (1981).
8. J. A. Krommes, "Statistical Descriptions and Plasma Physics" in *Handbook of Plasma Physics, Volume II, Basic Plasma Physics*, edited by A. A. Galeev and R. N. Sudan (North Holland, New York, 1984), p. 184.
9. B. B. Kadomtsev, *Plasma Turbulence* (Academic Press, New York, 1965).
10. R. H. Kraichnan, *Physics of Fluids* **8**, 575 (1965).
11. R. H. Kraichnan, *Physics of Fluids* **9**, 1728 (1966).
12. Y. Kaneda, *Physics of Fluids* **29**, 701 (1986).
13. R. N. Sudan and M. J. Keskinen, *Physical Review Letters* **38**, 966 (1977); *Physics of Fluids* **22**, 2305 (1979).
14. R. N. Sudan and D. Pfirsch, *Physics of Fluids* **28**, 1702 (1985).
15. R. M. Kulsrud and R. N. Sudan, *Comments on Plasma Physics and Controlled Fusion* **7**, 47 (1982).
16. A. Simon, *Physics of Fluids* **6**, 382 (1963).
17. F. C. Hoh, *Physics of Fluids* **6**, 1184 (1963).
18. M. J. Keskinen, R. N. Sudan, and R. L. Ferch, *Journal of Geophysical Research* **84**, 1419 (1979).
19. S. Zargham and C. E. Seyler, Laboratory of Plasma Studies Report, Cornell University, Ithaca, New York, Report No. LPS-364 (1986).

20. W. Horton, "Drift Wave Turbulence and Anomalous Transport," in *Handbook of Plasma Physics, Volume II, Basic Plasma Physics*, edited by A. A. Galeev and R. N. Sudan (North Holland, New York, 1984), p. 383.
21. See, for example, R. H. Kraichnan and D. Montgomery, *Reports on Progress in Physics* **43**, 547 (1980).
22. D. Fyfe and D. Montgomery, *Physics of Fluids* **22**, 246 (1979).
23. M. N. Rosenbluth and R. N. Sudan, *Physics of Fluids* **29**, 2347 (1986).
24. G. Schultz and R. N. Sudan, *Bulletin of the American Physical Society* **31**, 1497 (1986).
25. J. Albert, P. Similon, and R. N. Sudan, *Bulletin of the American Physical Society* **31**, 1522 (1986).
26. J. B. Taylor and B. McNamara, *Physics of Fluids* **14**, 1492 (1971).
27. See, for example, the compendium of papers in H. Bai-lin, *Chaos* (World Scientific Publishing Company, Singapore, 1984).
28. K. G. Wilson, *Physical Review B* **4**, 3174 (1971).
29. K. G. Wilson and J. Kogut, *Physical Review C* **12**, 77 (1974).
30. S. K. Ma and G. Mazenko, *Physical Review B* **11**, 4077 (1975).
31. D. Forster, D. Nelson, and M. Stephen, *Physical Review A* **16**, 732 (1977).
32. V. Yakhot, *Physical Review A* **23**, 1486 (1981).
33. V. Yakhot and S. Orszag, *Journal of Sci. Comp.* **1**, 1 (1986).
34. See, for example, T. S. Hahm, P. H. Diamond, P. W. Terry, L. Garcia, and B. A. Carreras, *Physics of Fluids* **30**, 1452 (1987); P. L. Similon and P. H. Diamond, *Physics of Fluids* **27**, 916 (1984).
35. A. K. M. F. Hussain, *Physics of Fluids* **26**, 2816 (1983); *Journal of Fluid Mechanics* **173**, 303 (1986).
36. V. E. Zakharov, *Zh. Eksp. Teor. Fiz.* **62**, 1745 (1972) [*Soviet Physics-JETP* **35**, 908 (1973)].

Nicholas A. Krall
Krall Associates
1070 America Way
Del Mar, California 92014

Not Quite Universal Instabilities

The physics of the drift wave and the universal instability, its historical development, and the extent to which this research continues today are reviewed.

INTRODUCTION

The influence of Marshall Rosenbluth has been felt in many areas of plasma physics, but perhaps nowhere more greatly than in the field of microinstabilities. He virtually invented the drift wave and its natural consequence the "universal instability," which seems to dominate the residual transport in macroscopically stable confinement geometries.

This paper is intended to give a hint of the thought processes that started Rosenbluth and his collaborators in this direction of research and to give a brief survey of drift wave and universal instability physics itself. Finally, I will indicate the extent to which this line of work still thrives in confinement physics research.

In many ways, the choice of a physics problem is as crucial as the solution of it. For myself at least, it was always more difficult to figure out what to work on than to do the work itself. A special genius of Marshall Rosenbluth has always been the insight to pick out the most immediate and far reaching problem. What follows may not show how Marshall decided to work on microinstabilities, but at least it will show how he explained the importance of the problem to those of us who were fortunate enough to share in this work.

IDENTIFICATION OF THE PROBLEM

The basic problem of confinement physics was and is that we are attempting to create and maintain a nonequilibrium state. This is clear when one notes that the Gibbs distribution is not spatially localized:

$$f(v, r) = \exp(-H/kT), \quad \text{with } H = mv^2/2 + q\phi(r) \tag{1}$$

$\phi = (+)$ confines electrons only

$\phi = (-)$ confines ions only

$\phi = 0$ confines nothing

where H is the Hamiltonian of the electron-ion system, m is the particle mass, v is the velocity, q the charge, and ϕ the electrostatic potential. Thus, Rosenbluth pointed out that the question was not one of confinement, but rather one of rate of relaxation.

There are two basic mechanisms for relaxation:

1. **Hydrodynamic processes**
   Characteristically, these are rapid relaxation processes, which are global and controllable (in principle) by a suitable choice of magnetic geometry.
2. **Microscopic processes**
   These are basically local processes, often involving finite Larmor radius effects, kinetic effects, resonant particles, parallel electric fields, and so on; although they may produce only slow relaxation, their dependence on details of the particle distribution make them difficult or impossible to be controlled.

The energy source for these micro-processes, as Rosenbluth explained, might involve such vital features of a confined particle distribution that they could not be eliminated; then the instability would truly be "universal." One particular feature he had in mind was that of expansion free energy. Consider a confined density distribution such as that shown in Fig. 1 by the solid line, and consider a perturbation as indicated by the dashed line. An estimate of the energy released by the interchange is

$$\frac{\Delta E}{E} \approx \left(\frac{\Delta r}{\rho} \frac{\partial \rho}{\partial r} \right)^2, \tag{2}$$

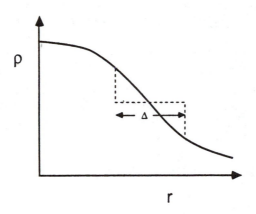

FIGURE 1   Mass density ρ as a function of the plasma radius r: unperturbed profile (solid curve) and perturbed profile (dashed curve).

where ρ is the mass density of the plasma. In order to be unstable, the energy released must be greater than the kinetic energy that is required to move the particles a distance Δr in a time $1/\omega$:

$$\Delta E > \rho V^2, \text{ with } V \sim \omega \Delta r. \tag{3}$$

Finally, comparing these two requirements gives a condition on the perturbation frequency, ω, in order that the expansion energy be adequate to drive instability namely,

$$\omega < V_T \frac{1}{\rho} \frac{\partial \rho}{\partial r} < \Omega_i, \tag{4}$$

where the condition that the confinement region be many gyroradii in size gives the last inequality.

With this discussion, Rosenbluth identified a driver for a universal instability and the frequency range in which to look for it.

The presentation in this section is amplified in the review article by Rosenbluth,[1] which was presented at the 1973 Kiev Conference.

INSIGHT INTO PROBABLE SOLUTIONS OF THE PROBLEM

One of the things that made working with Marshall Rosenbluth so pleasant was that, having identified the problem, he had the knack of being able to guess not only the form of the final solution, but also where in "physics space" to look for this solution. In this instance, he explained the situation to me (note that this statement was made in 1959) as follows:[2]

> "At present there is a lot known about nonuniform plasmas in the MHD limit (long space scale, fluid time scale, zero Larmor radius limit), and a lot known about kinetic effects of a uniform plasma, including waves of many different time scales, space scales, and involving finite Larmor radius effects. The place to look for new effects is clearly in combining the nonuniform features of MHD theory with the kinetic effects studied for uniform plasmas."

Thus the study of drift waves and universal instability began with three essential elements:

- low frequency waves;
- nonuniform plasmas; and
- resonant particles with finite Larmor radius.

The first two elements were to tap expansion free energy, and the third was to include effects not already present in existing MHD theory.

EARLY EFFORTS IN UNIVERSAL INSTABILITY THEORY

The first question Rosenbluth asked was: Will a mild nonuniformity in plasma density, viz.,

$$\epsilon = \frac{R_L + \lambda}{n} \frac{\partial n}{\partial r} \ll 1, \tag{5}$$

destabilize a plasma wave that was stable for a uniform plasma?

A perturbation analysis[3] to fourth order in ϵ indicated that the answer was No. This important result was generalized as a property of all Lagrangian systems by Francis Low,[4] after Rosenbluth described the perturbation result to him during a lunchtime card game.

Undaunted by this negative result, Rosenbluth pointed out that not all systems are Lagrangian. Indeed, a wave characterized by a group velocity, ω/k, similar to that of particle drifts, V_D, will have the property that the number of particles in resonance, f_R, although vanishingly small as the nonuniformity goes to zero,

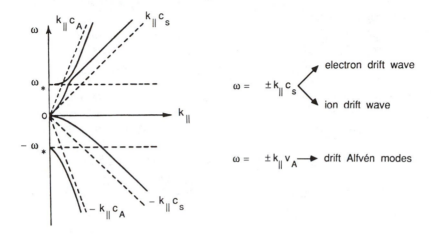

FIGURE 2   Dependence of drift wave frequency on the parallel wave number. [From Ref. 6]

contains an essential singularity, which makes the perturbed system to be non-Lagrangian and allows instabilities without violating "Low's Theorem": i.e.,

$$\frac{\omega}{k} \simeq V_D \propto \epsilon V^2, \tag{6}$$

$$f_R \simeq e^{-V^2/\overline{V}^2} \sim e^{-1/\epsilon}. \tag{7}$$

In fact, both the generation of new modes, termed drift waves because of their common feature, and the destabilization of known modes followed from a kinetic approach that took into account this resonance.[5] The connection between the drift waves produced by a density nonuniformity and the modes in a uniform plasma is nicely shown in Fig. 2, taken from the book by Kadomtsev.[6]

FORMALISM FOR MICROSTABILITY THEORY

Rosenbluth's ideas of (possibly) kinetic instabilities that tap expansion free energy led to the formal development of a technique to explore such effects. In this section, we sketch this technique and list some of the important early results.

Appropriate to exploring a new effect, the early work was simplified by considering a plasma of low pressure ($nT/B^2 \ll 1$), uniform magnetic field ($B = B_z = $ constant), weak nonuniformity ($\epsilon \ll 1$), and slab-like geometry with $n = n(x)$, where n is the plasma density.

The treatment was based on describing the plasma meta-equilibrium (i.e., equilibrium until collisions) by a distribution of the form

$$f(E, P_z, P_y), \quad \text{with} \quad E = \frac{mV^2}{2} + mgx, \quad \text{and} \quad P_y = mV_y + \Omega mx. \tag{8}$$

Note that $P_y = mV_y + eAy/c$, with $A_y(x = 0) = 0$, where m is the mass, V is the particle velocity, g is an effective gravitational force to simulate the effect of magnetic curvature, P is the canonical momentum constant of the motion, and A_y is the vector magnetic potential. The equilibrium distribution was specifically taken to be

$$f_0 = n_0[1 + \epsilon(x + V_y/\Omega)]e^{-(mV^2/2T+mgx/T)} \left(\frac{m}{2\pi T}\right)^{3/2}. \tag{9}$$

The analysis begins by considering a perturbation to this equilibrium,

$$f = f_0 + \delta f \tag{10}$$

$$\mathbf{E} = \delta\mathbf{E} = -\nabla\phi_0(x)e^{i(kx - Kz + \omega t)} \tag{11}$$

and calculating the development of the perturbation through the Vlasov equation,

$$\left[\frac{\partial}{\partial t} + \mathbf{v} \cdot \nabla + \frac{q}{m}\left(\frac{\mathbf{v} \times \mathbf{B}}{c}\right) \cdot \nabla_{\mathbf{v}}\right]\delta f = -\frac{q}{m}\delta\mathbf{E} \cdot \nabla_{\mathbf{v}} f_0. \tag{12}$$

This is solved by the method of characteristics to give the perturbed distribution function,

$$\delta f = \frac{e}{m} \int_{-\infty}^{t} \nabla\phi \cdot \nabla_{\mathbf{v}} f_0 dt = \frac{e}{m} f_0 \left[-\frac{m}{T}\phi + i\left(\frac{\omega m}{T} + \frac{\epsilon k}{\Omega}\right)\int_{-\infty}^{t} \phi(r', t')dt'\right], \tag{13}$$

and the problem closed using Maxwell's equations,

$$\nabla^2\phi = -4\pi \sum_j q_j \, \delta f_j. \tag{14}$$

The key feature in making further progress is to note that $r'(t')$, which is the orbit of a plasma particle in the unperturbed fields, contains the features which can tap the expansion free energy, viz.,

- ∇B drifts;
- $v_{\parallel}$ currents;
- ∇n and ∇T drifts (in ϵ); and
- magnetic curvature effects through g.

It can also include more effects, such as trapped particles or collisions, as required by the geometry or the parameter range. The same formalism can be used to treat higher pressure plasmas by the use of the full set of perturbed fields and the full set of Maxwell's equations.

A full description of this approach can be found in the Rosenbluth papers cited previously or in his Trieste lecture on microinstabilities.[7]

THE UNIVERSAL INSTABILITY

Solving the equations (13) and (14) with the restrictions mentioned earlier gives the original equation for the universal instabilities:

$$\lambda_D^2 k^2 = \sum_j \left\{ 1 - \left[1 + \frac{Tk}{m\Omega\omega} \left(\frac{mg}{kT} + \frac{1}{n}\frac{dn}{dx} \right) \right] \sum_n \frac{\omega e^{-k^2 T/m\Omega^2} I_n(b)}{\omega + n\Omega + kg/\Omega} \right. $$

$$\left. \times \left[1 + W \left(\frac{\omega + n\Omega + kg/\Omega}{K(2T/m)^{1/2}} \right) \right] \right\}. \tag{15}$$

Here Ω is the gyrofrequency, I_n is a Bessel function of imaginary argument $b = k^2 R_L^2$, and W is the dispersion integral defined in any of the Rosenbluth papers as

$$W(y) \equiv -\lim_{\Delta\to 0} \frac{1}{\sqrt{\pi}} \int_{-\infty}^{\infty} \frac{xe^{-x^2}dx}{x + y - i\Delta}. \tag{16}$$

Equation (15) was solved in the following three limits to give results which are now very familiar:

Case I. $K = 0, \; \omega \ll \Omega$

Case II. $\overline{V}_i < \omega/K < \overline{V}_e, \; \omega \ll \Omega$

Case III. $\omega \simeq \Omega_i \sim KV_D$

CASE 1:

This case should have reduced to the MHD limit, but instead it contained an important modification. Because it retains the particle orbits, it keeps the fact that the electron and ion $E \times B$ drifts, when averaged over a particle orbit, are not the same, the ion orbit being averaged over a larger spatial region:

$$\left\langle \frac{E \times B}{B^2} c \right\rangle_{R_{Li}} \simeq \left\langle \frac{E \times B}{B^2} c \right\rangle_{R_{Le}} \left[1 - \frac{k^2 R_{Li}^2}{2} \right]. \tag{17}$$

This gives a change in the charge separation, neglected by MHD, and alters the stability behavior. This gave, almost as an automatic byproduct, the limit now known as finite Larmor radius MHD, which has low frequency and long wavelength, as in MHD, but with a particular ordering:

$$\underset{\text{MHD}}{k^2 R_{Li}^2 \; \ll \; \frac{\omega}{\Omega} < 1} \implies \underset{\text{FLR/MHD}}{\frac{\omega}{\Omega} \sim \; k^2 R_{Li}^2 \ll 1.} \tag{18}$$

A consequence of this is that while MHD predicted instability for any equilibrium with magnetic mirror curvature, the FLR/MHD theory gave stability when the curvature was weak:

$$\text{FLR/MHD}: \qquad R_c/L_n > 2\sqrt{R_c L_n}/R_{Li} \qquad \text{stable} \qquad (19)$$

$$\text{MHD}: \qquad R_c/L_n > 0 \qquad \text{unstable.} \qquad (20)$$

Here R_c is the radius of magnetic curvature, L_n is the density gradient length, and R_{Li} is the ion Larmor radius. Following the publication of this groundbreaking result,[8] many researchers have invoked FLR effects to explain the unexpected stability behavior of experiments with weak curvature or large ion orbits.

CASE 2:

This is the "standard" limit for calculating universal instabilities, with the ions behaving as a fluid and the electrons behaving as particles:

$$\overline{V}_i < \omega/K_\parallel < \overline{V}_e \quad \text{(fluid ions, kinetic electrons).} \qquad (21)$$

The structure of plasma waves in this limit, viz.,

$$\omega \sim k v_D \left\{ 1 - \frac{i k V_D}{K \overline{V}_e} \left[1 - e^{-k^2 R_{Li}^2} I_0 \left(k^2 R_{Li}^2 \right) \right] \right\}, \qquad (22)$$

contains several features:

- The modes are universal, in that any $v_D = (cT/eBL_n)$ is unstable.
- Finite Larmor radius is required for instability, i.e., $I_0 = 1$ is stable.
- Parallel (to **B**) currents can be added to allow instability even in the zero Larmor radius limit.

CASE 3:

The flute limit of this case, $K_\parallel = 0$, also gave a universal instability,

$$K_\parallel = 0, \ \omega \sim \Omega_i$$

$$\omega_i \simeq \Omega_i(m/M), \ k R_{Le} \sim \frac{\overline{V}_e}{V_D} \frac{m}{M} \geq 1, \qquad (23)$$

and was the first discovered. A byproduct of this result[5] was the idea that instabilities, even when universal, might not be catastrophic, but rather that, because of their localized nature, they might produce a "local boiling" effect. This is the first mention, to my knowledge, of the "anomalous resistivity" phenomena, which is now such a basic part of plasma modelling.

THE NOT QUITE UNIVERSAL INSTABILITY

One of the frustrating things about working with Marshall Rosenbluth was that he always knew more than he would tell you. Whenever, in a fit of algebraic frenzy, one had apparently caught up with him, he would quietly turn the page and show that there was more to the problem than meets the eye. So it was no surprise that in the case of the universal instability, by the time we had worked out the instability, he had clear insight into the physical phenomena that would stabilize it.

FINITE LENGTH

A vital feature of drift waves is that they (often) include a perturbed electric field parallel to $\mathbf{B}_0$, which is not Landau damped by the ions because of the long wavelength and rapid phase velocity parallel to $\mathbf{B}_0$:

$$K_{\parallel}\overline{V}_i < \omega \approx kV_D \lesssim cT/eBL_n^2. \tag{24}$$

For this to be possible, the plasma geometry and parameters must be essentially unchanged over a distance $1/K_{\parallel}$, allowing the wave to balloon in the unstable region. A short fat plasma is therefore more likely to be stable to drift waves:

$$L < L_n^2/R_{Li}, \quad \text{stable to low } m \text{ modes.} \tag{25}$$

This was first discussed in a paper by Rosenbluth *et al.*[9] Incidentally, this may also have been the first notice of ballooning instabilities which were subsequently developed by Coppi and Rosenbluth,[10] among others.

SHEAR

If finite length can stabilize, then magnetic shear can also stabilize, since this places a limit on the size of the parallel wave number $K_{\parallel}$. Characterizing the shear distance by a length $1/S$, where

$$\mathbf{B} = B_0\hat{\mathbf{z}} + S \times B_y\hat{\mathbf{y}}, \tag{26}$$

we note that if $K_{\parallel} = 0$ at $x = 0$, then at the plasma surface $K_{\parallel} = SkL_n$, where k is the wave number k_y, which was perpendicular to $\mathbf{B}_0$ at $x = 0$. This gives a crude estimate of the strength of the shear required for stability:

$$S > R_{Li}/L_n^2. \tag{27}$$

This result was also included in Ref. 9.

LOW FREQUENCY EFFECTS

The stability of distributions which are a function of magnetic moment and energy had been discussed in general by J.B. Taylor. Rosenbluth demonstrated this effect explicitly for drift waves.[11]

FINITE PLASMA PRESSURE

The effect of finite plasma pressure can also stabilize drift waves, by introducing an electromagnetic component to the perturbed field. It is beyond the scope of this paper to discuss the many studies of finite beta stabilization.

UNIVERSAL INSTABILITIES ON THE CONTEMPORARY SCENE

One sign of a really good idea is its longevity, and the drift wave has shown great staying power. This is evident both in the volume of research still exploring this branch of plasma physics knowledge and in the repeated application of this effect to the understanding of plasma experiments. In this section, we mention only three examples of recent work based on these early ideas.

TRANSPORT IN TOKAMAKS

A recent review[12] recognized that "drift waves have long been considered the most likely cause of anomalous transport in tokamaks." Table 1 shows the variety of drift wave branches explored in the search for an adequate description of tokamak transport. It is instructive to identify the classes of drift waves in tokamak and to correlate some of the past and present terminology. The general range of frequency and wave number are those of the original universal instability,

$$\omega \sim k V_D, \quad V_i < \omega/K_\parallel < V_e, \quad K_\parallel \ll k_\perp, \tag{28}$$

but the embodiment includes specific features of the tokamak, namely:

- The local approximation is often invalid, $\mathbf{B} = \hat{\imath}(r)B$.
- Flux surface averages are often required.
- Nonlinear effects may be important.
- Trapped particles must be included.
- Collisionality must also be considered.

Three branches of the electron drift wave are prevalent in a tokamak, in various limits of collisionality:

TABLE 1 Schematic classification of drift wave branches and various theories for turbulent transport. [From Ref. 12]

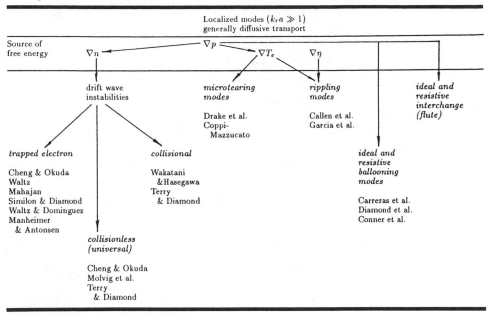

DRIFT DISSIPATIVE WAVES: These are collisional drift waves, characterized by temperature gradients, ∇T, and parallel currents. They reduce to the MHD modes in suitable limits. The collisionality regime is

$$\nu_{ei} > \overline{V}_e/qR, \qquad (29)$$

where q is the safety factor and R is the major radius of the torus.

COLLISIONLESS DRIFT WAVES: This is the "universal instability," characterized by resonant electrons, in a collisionality regime

$$(r/R)^{1.5}\overline{V}_e/Rq < \nu_{ei} < \overline{V}_e/qR. \qquad (30)$$

TRAPPED ELECTRON MODES: With trapped electrons providing a particle species whose drifts are average over orbits that bounce between two turning points, a new way to destabilize drift waves arises in the limit

$$\nu_{ei} < (r/R)^{1.5} \overline{V}_e / qR. \tag{31}$$

In addition to the electron drift waves, at least three other modes have emerged as influencing the behavior of tokamak experiments, including:

- Ion drift waves, in the form of the ion mixing mode discussed by Coppi and his group;
- Trapped ion modes, which emerge in some ranges of collisionality and wavelength; and
- Drift Alfvén modes, which appear as β increases and are related to MHD tearing-ballooning modes.

This work is summarized in the review article by Rosenbluth and Rutherford.[13]

VERY HIGH β DRIFT INSTABILITIES

In another recent work,[14] Rosenbluth and his collaborators investigated the drift instability in a plasma with $\beta \gg 1$. In such a system, the pressure is nearly uniform, $\nabla n/n = -\nabla T/T$, with the magnetic field acting only to reduce the heat flux to the walls. This study centered on waves in the usual range of universal instability, $\omega < K_{\parallel} \overline{V}_e$, and included both electric and magnetic fluctuations, δE and δB, which is a general requirement for a high β calculation. The survey identified stable and unstable parameter ranges for the drift wave, described by the parameter

$$\alpha \equiv \frac{\partial \ln B_0}{\partial \ln T_i}. \tag{32}$$

The results included the following:

- $T_i/T_e < 2.4$
  The FLR mode is stable for $\alpha < 0$ or $\alpha > 4$.
  The "zeroth-order FLR" mode is stable for $\alpha > \frac{1}{2(1+T_i/T_e)}$
- $T_i/T_e > 2.4$
  Both drift waves are stable for $-|\alpha_{c2}| < \alpha < 0$,
  where $\alpha_{c2} \to 1$ as $T_i/T_e \underset{\to}{\sim} 100$,
  and $\alpha_{c2} \to 1/5$ as $T_i/T_e \sim 10$.

The significance of this result is that it indicates the optimum parameter range for magnetic insulation and suggests that the transport scaling will be

$$\tau_L \sim \frac{a_p^2 B}{T} \sim \text{Bohm}. \tag{33}$$

This application is given here to indicate the very wide range of plasma parameters in which the drift wave has an important effect.

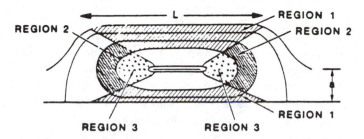

FIGURE 3   The FRC magnetic geometry and stability regions. [From Ref. 15]

DRIFT WAVES IN THE FIELD REVERSED CONFIGURATION (FRC)

A persistent mystery in the FRC (Table 1 shows the magnetic field lines for the top half of that geometry) has been that the mechanism identified to explain particle transport, namely, lower hybrid instability, is predominantly an edge effect and does not explain magnetic flux loss from the interior. A recent exploration of low frequency drift waves in this geometry showed that it is characterized by three regions,

    Region 1: Racetrack, from $\beta \ll 1$ to $\beta > 1$
    Region 2: Mirror, $\beta < 1$
    Region 3: Mirror, $\beta > 1$

and that it is unstable to three drift waves in various regions,

    Collisionless drift instability
    Drift dissipative instability
    Trapped electron instability.

Estimates of the transport from these modes give the following scaling,

$$\tau_\phi \sim \tau_N \sim \frac{a_p^2 B}{T} \tag{34}$$

$$\tau_E \sim \frac{a^2 B}{T}\left(1 + \frac{T^{1/2}}{a_p B}\right)^{-1} > \tau_N, \tau_\phi \text{ for } a_p \text{ small,} \tag{35}$$

which is in substantial agreement with experiment and may explain the observed anomaly.[15,16]

CONCLUSION

A general feature of Marshall Rosenbluth's work is that there is no conclusion; rather, there is an ongoing, ever deepening process of understanding nature. In the case of drift waves and universal instabilities in plasma, he has given us an idea based on clear physical principles, which opened up a new branch of plasma physics. Through a combination of insight and powerful formalism, he developed this branch in much detail, while moving on to initiate many other lines of important research. This branch of plasma theory still dominates a substantial number of the present models for the relaxation of confined plasma toward equilibrium, which, as he pointed out to begin with, is still one of the basic problems of controlled fusion.

REFERENCES

1. M.N. Rosenbluth, in *Advances in Plasma Physics*, ed. A. Simon and W.B. Thompson (John Wiley, New York, 1974), Vol. 5, p. 75.
2. M.N. Rosenbluth, private communication.
3. N.A. Krall and M.N. Rosenbluth, *Physics of Fluids* **4**, 163 (1961).
4. F. Low, *Physics of Fluids* **4**, 842 (1961).
5. N.A. Krall and M.N. Rosenbluth, *Physics of Fluids* **5**, 1435 (1962).
6. B.B. Kadomtsev, *Plasma Turbulence* (Academic Press, New York, 1962), Fig. 18, p. 84.
7. M.N. Rosenbluth, in *Plasma Physics* (IAEA, Vienna, 1965), p. 485.
8. M.N. Rosenbluth, N.A. Krall, and N. Rostoker, *Nuclear Fusion Supplement*, Pt. **1**, 143 (1962).
9. N.A. Krall and M.N. Rosenbluth, *Physics of Fluids* **8**, 1488 (1965).
10. B. Coppi and M.N. Rosenbluth, in *Plasma Physics and Controlled Nuclear Fusion Research 1965* (IAEA, Vienna, 1966), Vol. I, p. 617.
11. N.A. Krall and M.N. Rosenbluth, *Physics of Fluids* **7**, 1094 (1964).
12. P.C. Liewer, *Nuclear Fusion* **25**, 543 (1985).
13. M.N. Rosenbluth and P.H. Rutherford, in *Fusion*, ed. E. Teller (Academic Press, New York, 1981), Vol. 1, Pt. A, p. 32.
14. A.Y. Aydemir, H.L. Berk, V. Mirnov, O.P. Pogutse, and M.N. Rosenbluth, *Physics of Fluids* **30**, 3083 (1987).
15. N.A. Krall, *Physics of Fluids* **30**, 878 (1987).
16. A.L. Hoffman, J.T. Slough, L.C. Steinhauer, N.A. Krall, and S. Hamasaki, in *Plasma Physics and Controlled Nuclear Fusion Research 1986* (IAEA, Vienna, 1987), Vol. II, p. 541.

William B. Thompson
Department of Physics
University of California, San Diego
La Jolla, California 92093

Stimulated Bremsstrahlung—a Free Electron Laser?

It is shown that in a plasma with a sufficiently anisotropic electron distribution, induced bremsstrahlung can lead to super-radiance, hence to a possible "free electron laser." Gain is small, but amplification can be in a broad band, and the process shows some promise.

INTRODUCTION

The observation that the stimulated emission of bremsstrahlung can exceed absorption was made in 1962 by Marcuse.[1] We present here a re-examination of this process, using both classical and quantum arguments, and consider the role of the electron distribution, determining the condition needed for amplification.

We will show that the attenuation due to inverse and induced bremsstrahlung may be written as

$$\frac{1}{\varepsilon}\frac{d\varepsilon}{dt} = -\nu\left(\frac{\omega_p}{\omega}\right)^2 \Lambda = -\left(\frac{4\pi n Z e^4}{m^2 v_\theta^3}\right)\left(\frac{4\pi n e^2}{m\omega^2}\right)\Lambda, \qquad (1)$$

where n is the electron density, v_θ the electron thermal speed, ω the frequency of the radiation, and Λ a numerical factor depending on the nature of the electron distribution, the remaining symbols having their usual meaning. The quantity Λ is positive for an isotropic distribution, but may become negative if the distribution is sufficiently anisotropic.

An inevitable competing process is Compton scattering (in the non-relativistic case, Thomson scattering):

$$\frac{1}{\varepsilon}\frac{d\varepsilon}{dt} = -n\frac{8\pi}{3}\frac{e^4}{m^2c^3}. \tag{2}$$

Amplification necessarily requires

$$Z\left(\frac{\omega_p}{\omega}\right)^2 > \frac{2}{3}\left(\frac{v_\theta}{c}\right)^3$$

or

$$nZ > 4 \times 10^{11}\left(\frac{\hbar\omega}{kT}\right)^2 T^{7/2}, \tag{3}$$

where $\hbar\omega$ is the energy of the photon, and T is measured in electron volts and n in particles/cm^3. Thus, high densities are required for amplification, and for high frequencies, high temperatures as well.

The possible gain can be characterized by a length

$$L = \left(\frac{c}{v_\theta}\right)\lambda_f\left(\frac{\omega}{\omega_p}\right)^2 = 1.18 \times 10^{31}\frac{T^{3/2}}{n^2\lambda^2}, \tag{4}$$

where T is in electron-volts and lengths are in centimeters. For example, if $n = 10^{22}$, $T = 100$, and $\lambda = 0.1\mu$, then $L \simeq 1.18$ cm. If $n = 10^{18}$, $T = 10$, and $\lambda = 10\mu$, then $L = 3.7$ meters. Clearly, high densities are required if the gain length is to be reasonable.

We turn now to the calculation of the gain factor, Λ, and to the conditions on the distribution function needed for amplification.

GAIN FACTOR

A classical calculation is appropriate for the case of radio-frequency waves and should be reasonable if $\hbar\omega \ll kT$. To approach this, we use standard perturbation theory, assuming that in addition to the radiation field $\mathbf{E}(\omega, \mathbf{k})$, there is a fluctuating noise potential $\sum \phi(\Omega, \mathbf{K})$. To second order in both ϕ and E, the energy absorption becomes

$$\frac{\partial \varepsilon}{\partial t} = \mathbf{j}^3 \cdot \mathbf{E} = e \int d^3 v \, \mathbf{E} \cdot \mathbf{v} \, \delta f^{(3)}$$

$$= i \frac{e^4}{m^3} \int d^3 v \, \mathbf{E} \cdot \mathbf{v} \int d^3 K \, d\Omega \frac{\langle \phi \phi^*(\mathbf{K}, \Omega) \rangle}{(\omega + \mathbf{k} \cdot \mathbf{v})} \frac{\mathbf{k}}{[\Omega + \omega + (\mathbf{K} + \mathbf{k}) \cdot \mathbf{v}]} \cdot \frac{\partial}{\partial \mathbf{v}}$$

$$\times \left\{ (\mathbf{E} + \mathbf{v} \times \mathbf{B}) \cdot \frac{\partial}{\partial \mathbf{v}} \frac{\mathbf{K}}{(\Omega + \mathbf{K} \cdot \mathbf{v})} + \mathbf{K} \cdot \frac{\partial}{\partial \mathbf{v}} \frac{(\mathbf{E} + \mathbf{v} \times \mathbf{B})}{(\omega + \mathbf{k} \cdot \mathbf{v})} \right\} \cdot \frac{\partial}{\partial \mathbf{v}} f, \quad (5)$$

where $\mathbf{B}$ is the radiation magnetic field.

To evaluate this we neglect $\mathbf{K} \cdot \mathbf{v}$ and $\mathbf{B}$, since $v \ll c$, and, for the noise fluctuations, use the static screened potential of the ions:

$$\langle \phi \phi \rangle = \frac{2}{\pi} \frac{n_+ Z^2 e^2}{(K^2 + K_0^2)^2} = \frac{2}{\pi} \frac{n_- Z e^2}{(K^2 + K_0^2)^2}. \quad (6)$$

Here, K_0 is the Debye wave number. Keeping the dominant term in the integral over K yields

$$\frac{\partial \varepsilon}{\partial t} = \left(\frac{4\pi n_- Z e^4}{m^2 v_\theta^3} \right) \left(\frac{4\pi n_- e^2}{m \omega^2} \right) \frac{E^2}{8\pi}$$

$$\times \left\{ \frac{v_\theta^3}{\pi} \int \frac{d^3 v}{v^2} \log \left(1 + \frac{K^2 v^2}{\omega^2 + \omega_p^2} \right) \left(\frac{1}{2} e_\perp^2 \frac{\partial f_+}{\partial v} + e_\| e_\perp \frac{\partial f}{\partial v} \right) \right\}, \quad (7)$$

where K_M is a cutoff at large wave numbers, which is either $K_M = m v^2 / 2 e^2$, the classical distance of closest approach, or $m v / \hbar$, the deBroglie wavenumber, while $e_\|$ and $e_\perp$ are projections, parallel and perpendicular to $\mathbf{v}$, of the unit vector $\mathbf{e}$ along $\mathbf{E}$. Thus $e_\| = \mathbf{e} \cdot \hat{\mathbf{v}}$ and $e_\perp = \mathbf{e} - e_\| \hat{\mathbf{v}}$, with $\mathbf{v} = v \hat{\mathbf{v}}$. If f is isotropic the quantity in braces in Eq. (7), which is equal to $-\Lambda$, becomes

$$-\Lambda = \frac{4\pi}{3} (\log A) v_\theta^3 f_0(0), \quad (8)$$

where

$$A = \left[1 + \frac{(2kT)^2}{\hbar^2 (\omega^2 + \omega_p^2)} \right]$$

and the radiation field is attenuated.

Suppose that f is anisotropic but has axial symmetry about an axis $\mathbf{b}$, so that it can be expanded in Legendre polynomials as

$$f = \sum f_n P_n (\mathbf{b} \cdot \hat{\mathbf{v}}).$$

Then, using the combination theorem for spherical harmonics allows one to perform the integrals over the solid angles, leaving

$$\Lambda = -\frac{4\pi}{3}\left\{v_\theta^3 f_0(0) + \frac{3}{5}v_\theta^3\left[1 - 3(\mathbf{e}\cdot\mathbf{b})^2\right]\int\frac{dv}{v}f_2(v)\right\}. \tag{9}$$

Since the quantity $(\mathbf{e}\cdot\mathbf{b})^2$ varies in value between 0 and 1, the sign of the second term in Eq. (9) is determined by the orientation of $\mathbf{e}$ with respect to $\mathbf{b}$. Amplification requires

$$\frac{6}{5}\int\frac{dv}{v}f_2 > f_0(0). \tag{10}$$

This classical calculation applies only for small values of ω. However, for small ω, the approximations used in evaluating $\langle\phi\phi^*\rangle$ are dangerous, since ω lies near the natural resonant frequencies of the plasma—viz., the plasma frequency, the gyro frequency, or the ion acoustic frequency—and the time behavior of the fluctuating field becomes important. Moreover, plasmas with the necessary degree of anisotropy are often unstable, and the large amplitude oscillations arising spontaneously may scatter the radiation.

At optical frequencies, amplification may occur before instabilities develop, and all resonances lie at frequencies much lower than that of the radiation, but for optical frequencies a quantum mechanical calculation is essential. We turn now to this, but for simplicity stay in the extreme nonrelativistic limit, $v \ll c$.

The well-known transition probabilities for bremsstrahlung[2] may be written

$$w = \frac{1}{n_+ n_-}\nu\frac{\omega_p^2}{\omega^2}\left[\frac{kT}{\hbar\omega}\frac{v_\theta}{q^4}(\mathbf{e}\cdot\Delta\mathbf{p})^2\delta(\varepsilon_f - \varepsilon_0 - \hbar\omega)\right]d^3p_f\,d^3p_0, \tag{11}$$

where $\mathbf{p}_0$ and $\mathbf{p}_f$ are the initial and final momenta, $\Delta\mathbf{p} = \mathbf{p}_f - \mathbf{p}_0$, and $\mathbf{q} = \Delta\mathbf{p} - \hbar\mathbf{k} \simeq \Delta\mathbf{p}$. If screening is significant, then q^4 in the denominator in Eq. (11) is replaced by $(q^2 + q_d^2)^2$ where $q_d = \hbar K_\theta$. This transition probability gives either the probability of a photon being absorbed by an electron whose initial and final momenta are $\mathbf{p}_0$ and $\mathbf{p}_f$, or the probability of it being emitted by one with initial and final momenta interchanged. The net rate of energy absorption is then

$$\frac{1}{\varepsilon}\frac{d\varepsilon}{dt} = \nu\frac{\omega_p^3}{\omega^2}\left\{\frac{kT}{\hbar\omega}v_\theta\int d^3p_f\,d^3p_0\frac{\delta(\varepsilon_f - \varepsilon_0 - \hbar\omega)}{q^4}(\mathbf{e}\cdot\Delta\mathbf{p})^2\left[f(p_f) - f(p_0)\right]\right\}. \tag{12}$$

We evaluate this using the following devices. In the integral involving $f(p_0)$, we integrate over the magnitude of p_f, then replace the solid angle integration by an integration over q, using $\mathbf{p}_0$ to define the angles, while in that involving $f(p_f)$ we interchange the role of $\mathbf{p}_f$ and $\mathbf{p}_0$. Again, for an anisotropic distribution we introduce the Legendre expansion and, after some reasonably straightforward manipulation, reduce the expression for Λ to

$$-\Lambda = \frac{kT}{\hbar\omega}\frac{8\pi^2}{3}\left\{\int d\varepsilon \log \tilde{A}\left[f_0(\varepsilon + \hbar\omega) - f_0(\varepsilon)\right]\right\}$$

$$+ \frac{3}{5}\left[1 - 3(\mathbf{e}\cdot\mathbf{b})^2\right]\left[\left(\frac{1}{3} + \frac{\hbar\omega}{\varepsilon + \hbar\omega}\right)f_2(\varepsilon + \hbar\omega) - \left(\frac{1}{3} - \frac{\hbar\omega}{\varepsilon}\right)f_2(\varepsilon)\right].$$

(13)

Here, we have

$$\tilde{A}^2 = \frac{\left(p_f + p_0\right)^2 + q_d^2}{\left(p_f - p_0\right)^2 + q_d^2} = \frac{\left(\sqrt{1 + \dfrac{\hbar\omega}{\varepsilon}} + 1\right)^2 + \dfrac{\hbar^2\omega_p^2}{2\varepsilon kT}}{\left(\dfrac{\hbar\omega}{k\varepsilon}\right)^2\left[\sqrt{1 + \dfrac{\hbar\omega}{\varepsilon}} + 1\right]^{-2} + \left(\dfrac{\hbar^2\omega_p^2}{2\varepsilon kT}\right)}$$

or, replacing ε by kT, we find

$$\tilde{A}^2 = \frac{\left(\sqrt{1 + \dfrac{\hbar\omega}{kT}} + 1\right)^2 + \dfrac{\hbar^2\omega_p^2}{2(kT)^2}}{\left(\dfrac{\hbar\omega}{kT}\right)^2\left[\sqrt{1 + \dfrac{\hbar\omega}{kT}} + 1\right]^{-2} + \left(\dfrac{\hbar^2\omega_p^2}{2(kT)^2}\right)}.$$

(14)

The requirement for amplification, in this case, is

$$\frac{6}{5}\int d\varepsilon\left[\left(\frac{\hbar\omega}{\varepsilon + \hbar\omega} + \frac{1}{3}\right)f_2(\varepsilon + \hbar\omega) + \left(\frac{\hbar\omega}{\varepsilon} - \frac{1}{3}\right)f_2(\varepsilon)\right]$$

$$> \int d\varepsilon\left[f_0(\varepsilon + \hbar\omega) - f_0(\varepsilon)\right].$$

(15)

SOME EXAMPLES

We consider now a couple of examples of amplifying distributions. For simplicity, the frequency-independent classical condition of Eq. (10) will be used.

1. A Gaussian Electron Beam

Consider a beam of electrons with a Gaussian profile moving through stationary ions. Then we may write

$$f(v) = \frac{1}{\pi^{3/2}} \frac{1}{v_\theta^3} \exp\left[-\frac{(\mathbf{v} - \mathbf{v}_D)^2}{v_\theta^2}\right]$$

$$v_\theta^3 f_0(0) = \frac{1}{\pi^{3/2}} \exp\left(-\frac{v_D^2}{v_\theta^2}\right).$$

The quantity $\int (f_2/v)\, dv$ can be evaluated, after the integrals over the solid angle have been carried out, with the use of the saddle point method for large v_D/v_θ, which yields

$$\frac{\sqrt{\pi}}{\pi^{3/2}} \left(\frac{v_D}{v_\theta}\right)^3.$$

Hence the criterion of Eq. (10) becomes

$$\frac{6}{5}\sqrt{\pi} \left(\frac{v_D}{v_\theta}\right)^3 > e^{-(v_D/v_\theta)^2} \tag{16}$$

which holds even for values of $v_D/v_\theta < 1$, where the saddle point approximation is questionable.

2. A Non-Symmetric Temperature: $T_\perp \gg T_\parallel$

Here we have

$$f = \frac{\alpha\sqrt{\beta}}{\pi^{3/2}} \exp\left[-\left(\alpha \sin^2\theta + \beta \cos^2\theta\right) v^2\right]$$

$$f_0 = \frac{\alpha\sqrt{\beta}}{\pi^{3/2}}.$$

After an almost straightforward calculation, we obtain

$$\int \frac{f_2(v)}{v}\, dv = \frac{2\alpha\sqrt{\beta}}{\pi^{3/2}}\left[\frac{2}{3} + \frac{\alpha}{\beta - \alpha} - \frac{\beta}{\beta - \alpha}\sqrt{\frac{\alpha}{\beta - \alpha}}\,\tan^{-1}\sqrt{\frac{\beta - \alpha}{\alpha}}\right],$$

which varies between 0 and $4\alpha\sqrt{\beta}/3\pi^{3/2}$ as the difference $\beta - \alpha$ varies from 0 to ∞. Hence the criterion of Eq. (10) becomes

$$\frac{12}{5}\left[\frac{2}{3} + \frac{\alpha}{\beta - \alpha} - \frac{\beta}{\alpha - \alpha}\sqrt{\frac{\alpha}{\beta - \alpha}}\,\tan^{-1}\sqrt{\frac{\beta - \alpha}{\alpha}}\right] > 1. \tag{17}$$

The maximum value of the bracket is $2/3$; thus, since $24/15 > 1$, gain is possible for some values of $T_\perp/T_\parallel$.

CONCLUSIONS

We have shown that for two examples, it is possible for amplification to occur, at least in the classical limit. Gain is low except for very dense plasmas, but if the distributions can be produced, it may be significant. Plasma conditions, however, may not be easy to obtain, especially since the gain scales with the collision frequency, which also determines the rate at which the plasma relaxes to equilibrium and isotropy.

Gain is most easily reached for low frequency, but here collective plasma phenomena can get in the way, although it may be possible to analyze and avoid such effects. On the other hand, many methods are being developed for the production of extremely dense plasmas, at high temperatures, so that experiments in the optical region may be possible. More analysis is also needed, particularly an extension toward the relativistic case and a more complete examination of the quantum result.

REFERENCES

1. D. Marcuse, *Bell System Technical Journal* **41**, 1557 (1961).
2. R. Feynman, *Quantum Electrodynamics* (W. A. Benjamin, New York, 1961), p. 110.

Herbert L. Berk
Institute for Fusion Studies
The University of Texas at Austin
Austin, Texas 78712

Effect of Energetic Particles on Plasma Stability

The theory of hot particle stabilization is reviewed. The usual MHD theory may not be applicable to systems containing a very energetic minority particle component whose energy density is greater than that of the background plasma. When the curvature drift frequency of the hot component is greater than the predicted MHD growth rate, a stabilizing decoupling of the hot species from the plasma can occur, leading to stable configurations that would otherwise be predicted to be MHD unstable. The regimes of stable confinement are somewhat restricted, however, and mechanisms for instability still exist. The stabilizing and the destabilizing effects of hot particles have been observed in bumpy torus, mirror, and tokamak experiments.

INTRODUCTION

The principal approach for achieving configurations with stable plasma confinement is to seek magnetohydrodynamic (MHD) stable systems in accordance with the conventional MHD energy principle that was first fully elucidated by Bernstein, Frieman, Kruskal and Kulsrud.[1] A principal result of the MHD analysis is that a stable plasma needs to be confined in a system with average concave magnetic field line curvature. This result leads to restrictions on plasma confinement that affect the simplicity that would be desirable in reactor engineering applications.[2] For example, for open systems (mirror machines), a simple axisymmetric mirror must have convex curvature at its mid-plane,[2] leading to the predictions[3] and observations[4,5] of MHD instability. Convex curvature (i.e., minimum-B) in vacuum fields is only achieved by introducing more complicated three-dimensional magnetic fields. In tokamaks, where average minimum-B can be achieved, the plasma is normally restricted to relatively low beta values, $\beta \leq 10\%$, with the plasma at higher beta values being susceptible to ballooning[6] and kink[7] instabilities, which are primarily driven by the unfavorable convex curvature regions of the magnetic field.

Though the MHD principle is quite general, it is still based on idealizations such as: the plasma is described by a single set of fluid equations; the magnetic fields do not reconnect inside the plasma; the Larmor radius is infinitesimal; etc. Deviations from the ideal MHD model may lead either to additional instability effects or to possible stabilization effects. Thus, by examining systems that deviate from ideal MHD theory, one might find alternative approaches to establishing stable plasma systems.

One promising alternative is the speculation that hot particles can be used to build stable systems, since their rapid motion through the plasma will not allow them to respond to slow MHD motion. Hence, the dynamics of the hot particles might decouple from that of the plasma, and their main contribution would be to establish an embedded current that forms a mininum-B configuration for the background plasma. This approach was first promoted by Christofilis[8] in the formulation of the Astron concept and was continued by Fleischmann[9] who first demonstrated field reversal in the Astron configuration. The Elmo mirror machine, first investigated by Dandl,[10] which was extended to a bumpy torus configuration,[11,12] is an approach with a principle similar to that of Astron: i.e., a hot particle component can establish embedded passive currents that form stabilizing minimum-B fields for the background plasma. In addition, it has been suggested that hot particles can be effective in field-reversed theta pinches[13,14] and in tokamaks[15] for establishing more stable conditions than would be possible from what is inferred from MHD theory.

It is straightforward to show that MHD stable systems are established if the hot particle currents are taken to be passive.[16,17] More complicated analysis is involved if the hot particle dynamics are self-consistently included. One finds that there are regimes of parameter space where such a simple view is correct, but there are other regimes where the hot particle dynamics introduce their own type of instability. In this review we summarize the current understanding of the effects of hot particles on MHD stability. In the development of the understanding of this field, the contributions of Marshall Rosenbluth and his collaborators have played an important role.

HOT PARTICLE DECOUPLING CONDITION

After the formulation of the MHD energy principle, it was understood that the kinetic behavior of plasmas must be taken into account in order to obtain a more accurate description of a plasma with long-mean-free-path. The first comprehensive theories of describing this behavior were developed by Kruskal and Oberman[18] and by Rosenbluth and Rostoker.[19] These investigations indicated that stability for a long-mean-free-path plasma was quite similar to that (though slightly easier to achieve than) in the short-mean-free-path case, in which conventional MHD theory would be applicable. This kinetic theory still indicated the need to have favorable concave curvature to achieve stability. Both theories indicated that stability at finite beta would not be substantially improved if a diamagnetic well, which has a minimum-B configuration, were created. Thus the first comprehensive kinetic theory did not indicate any favorable stability properties for a diamagnetic minimum-B system if the field line curvature remains convex.

After the development of ideal MHD theory, considerable theoretical effort was expended to understand nonideal effects, which was principally directed in understanding finite resistivity effects that allow for field line reconnection[20] and finite Larmor radius effects.[21] These pioneering contributions by M.N. Rosenbluth and his collaborators are still considered paradigm models for these problems.

The original finite Larmor radius calculation contained the mechanism whereby one could understand hot particle stabilization. In that calculation it was shown that a curvature-driven mode would be stabilized if the growth rate, as predicted by MHD theory, could be reduced to be less than half the curvature drift frequency of the thermal species. Normally, this requires extremely low density, such that the ion Debye length is considerably greater than the ion Larmor radius.[21] However, Krall[22] extended the model of the original finite Larmor radius theory to where there was a hot species present that carried nearly all the pressure, along with a cold background species with low pressure but with most of the density that governs the plasma's inertial response. Krall used this model to explain the stability of hot electron plasmas as observed in the Elmo experiment by Dandl.[10] Krall noted that the growth rate predicted by MHD theory for the normal interchange mode of a plasma consisting of a set of species j, with density n_j and temperature T_j, is

$$\gamma_{\mathrm{MHD}} = \left[\frac{\sum_j n_j T_j}{n_i M_i R_c r_p} \right]^{1/2} , \tag{1}$$

where R_c is the field line curvature radius, r_p the plasma radius, n_i the background ion density, and M_i the ion mass. The curvature drift frequency ω_{dj} of species j is given by

$$\omega_{dj} = \frac{\ell T_j}{M_j R_c r_p \omega_{cj}} ,$$

with ω_{c_j} the cyclotron frequency and ℓ the mode number. If we suppose that one species (denoted by subscript h) is extremely hot, so that it carries most of the plasma pressure (in Dandl's experiment the hot electrons had an energy of several hundred keV) but only a fraction of the density, then the stability condition is given by

$$
\gamma_{\mathrm{MHD}} \equiv \left[\frac{n_h}{n_i} \frac{R_c r_p}{\ell^2 a_h^2} \frac{M_h}{M_i} \right]^{1/2} < \frac{1}{2} , \tag{2}
$$

where $a_h = (T_h/M_h)^{1/2}$ is the hot particle Larmor radius. The stability condition of Eq. (2) can be rewritten as

$$
\frac{n_h}{n_i} < \frac{\ell^2}{4} \frac{a_H^2}{R_c r_p} \frac{M_i}{M_h} , \tag{3}
$$

with the longest wavelength, viz., $\ell = 1$, being the most difficult to stabilize.

This stability theory was subsequently modified by Berk,[23] who noted that the hot electron curvature drift frequency in these experiments was frequently comparable to or greater than the ion cyclotron frequency. Hence, the character of the inertial response of the ions changes, giving rise to an altered stability condition, which was found to be given by

$$
\frac{n_h}{n_i} < 2 \left(\frac{\omega_{ci}}{\omega_{dh}} \right) \left[1 + \frac{1}{2} \left| \frac{\omega_{dh}}{\omega_{ci}} \right| - \left(1 + \left| \frac{\omega_{dh}}{\omega_{ci}} \right| \right)^{1/2} \right] . \tag{4}
$$

This formula reproduces Eq. (3) if $\omega_{dh}/\omega_{ci} \ll 1$. A plot of the stability criterion is given in Fig. 1.

This curve indicates that if n_h/n_i is small enough, stability can be achieved, and that the more energetic the species, the larger the stability window.

Thus, with a small enough ratio of hot to background particles, relatively simple theory indicates that the hot particles decouple from the background plasma. Once this decoupling condition can be achieved, it was originally assumed that the hot particle component could be treated as a passive current source, which can then form a minimum-B system that can stabilize the background plasma.[16,17] In fact, the original Elmo experiments[10] claimed that a sufficiently high-β plasma was achieved ($\beta \simeq 50\%$) and that a diamagnetic minimum-B well was formed in a simple mirror configuration. At low background density, this experiment, as well as other hot electron experiments by Lichtenberg and Lieberman[24] and by Hiroe et al.,[25] demonstrated instability if n_h/n_i is too large (the so-called T-M transition). The observed transition between stable and unstable behavior was roughly in accord with the Krall-Berk theories, as well as later improved theories.[26–28] Thus, several experiments have established that plasmas with hot particles can store considerable kinetic energy, even though the system is predicted to be MHD unstable, and that the transition to an unstable system is observed if n_h/n_i is too large.

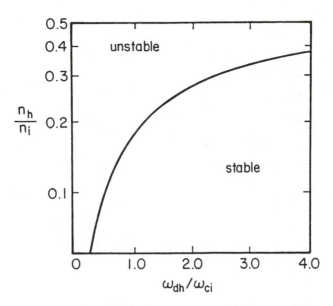

FIGURE 1   Stability diagram for the fraction of hot electron density versus the hot particle drift frequency needed to stabilize a hot electron plasma with respect to a hot electron interchange mode.

HOT PARTICLE INSTABILITY SOURCES

Unfortunately, for the purposes of plasma containment, this stability picture is too simple, since the hot particles still have means by which to interact with the plasma. For example, it was noted by Greene and Coppi[29] that the FLR stabilization mechanism of Ref. 21 produces positive and negative energy waves, with the negative energy waves being susceptible to instabilities that are driven by positive energy dissipation sources. Furth[30] argued that such a mechanism should be applicable in Astron. Indeed, experiments indicated rapid losses unless the negative energy waves were eliminated.[31] The nature of this negative energy wave is important for any hot particle scheme. To help understand this mechanism, we shall now discuss the similarities and differences of the Astron and Elmo experiments, as well as the features of a possibly related mirror experiment.

   In Fig. 2 we show schematic diagrams of the Astron, Elmo, and mirror (with a disc-shaped plasma) experiments.

   In these experiments, hot particles are injected into a simple mirror machine whose magnetic field lines have convex curvature. In the Astron experiment, depicted in Fig. 2(a), the hot particle orbits basically consist of one Larmor radius that is concentric about the midplane. As a result, a current forms that is peaked about a radius $r = r_L$, where r_L is the hot particle Larmor radius. A magnetic field

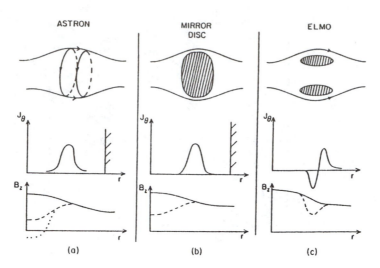

FIGURE 2   Schematic diagrams of hot particle configurations in a mirror machine with: a) single Larmor radius orbits, as in Astron; b) small Larmor radii orbits, in a disc-shaped mirror plasma; c) small Larmor radii orbits in a hot electron annulus, as in Elmo.

is formed that depresses the vacuum magnetic field and produces a minimum-B geometry. Indeed, the objective of Astron was to produce a reversed field, which has been accomplished by Fleischmann.[9] One may also observe that if a metallic conductor is placed outside the current layer, reversed currents form in the conductor that induce stabilizing magnetic fields that are proportional to the strength of the internal current and inversely proportional to the square of the axial length of the internal current layer.

A similar current profile develops in a mirror plasma with a disc-shaped plasma, even with small Larmor radius. The shaded area in Fig. 2(b) denotes the region occupied by plasma, which in this diagram is assumed to be flat in the center and to drop sharply at the edge at a radius r_0. The current induced is proportional to the local pressure gradient, so that the current is then peaked around the radius r_0. This current, which is similar in shape to the Astron current, likewise causes a magnetic field depression in the center and induces a stabilizing magnetic field on external conductors.

In the Elmo experiment, the situation is somewhat different in that the hot particle plasma is confined to an annulus localized around the radius r_0. The current then has the dipole shape shown in Figure 2(c), which causes the magnetic field first to decrease more rapidly in the radial direction and then to increase. The region where the magnetic field radially increases is intended to supply the stabilizing magnetic field for the background plasma. However, since the current is dipole in nature, the effect of this current on external conductors, which in turn could feed back on the plasma, is considerably less than in the other two cases.

Returning to the problem of negative energy waves, we note that Furth[30] pointed out that an Astron configuration would excite a negative energy precessional mode, which is destabilized when dissipation is present. The frequency of this mode is the curvature drift frequency of the external magnetic field. Christofilis[32] observed that this mode can be converted to a positive energy wave if conducting walls are present and if the hot particle currents are sufficiently large so as to stabilize the system with passive feedback. Indeed, stability was demonstrated quite dramatically.[31] An Astron E-layer surrounded by conductors was shown to have a quiescent decay, until the E-layer's current became too weak to induce the stabilizing effects to counter the vacuum field's unfavorable curvature. At this point, rapid loss of the E-layer was observed.

Recently, it has been shown that a small Larmor radius plasma with hot particles, which has the same current profile as the Astron configuration, will exhibit the identical precessional mode with the identical stability condition if all quantities are expressed in terms of the curvature drift and the equilibrium hot particle currents.[33] Hence, conducting walls in disc-shaped mirror plasmas can also be used to convert the negative energy waves to positive energy waves, as in the Astron experiment.

However, in the Elmo and EBT experiments, the conducting walls cannot be effectively used for stabilization of the precessional mode because the dipole nature of the current layer reduces the magnetic interaction with external conductors. Furthermore, finite Larmor radius stabilization of higher mode-number precessional modes,[33] which is an extremely strong mechanism in Astron, is not predicted to be as important for moderate mode-number Elmo and EBT experiments.[34] Thus, the precessional mode might be expected to be a limitation on the hot particle containment in such experiments.

Other theories also indicated stability problems with a hot particle annulus. Van Dam and Lee[35] and also Nelson[36] observed that the decoupling condition fails if the beta of the background plasma becomes too large, i.e.

$$\frac{4\pi}{B^2} \frac{\partial p_c}{\partial r} > \frac{1}{R_c} \, , \tag{5}$$

where p_c is the background plasma pressure and R_c is the unfavorable radius of curvature. When such core pressure gradients are exceeded, the plasma responds in an unstable manner quite similar to conventional MHD predictions; in other words, the hot particles are no longer decoupled from the background plasma.

In a general study, Van Dam, Rosenbluth, and Lee[37] developed a very low frequency energy principle. The theory is for the small Larmor radius limit. It is assumed that the plasma response frequency, ω, is much less than the diamagnetic drift frequency and the gradient-B drift frequency. This model generalized the previous kinetic energy principles,[19,20] which assumed ω to be greater than the diamagnetic drift frequency. The exhibition of negative energy perturbations on plasma modes described by this new energy principle was shown to be necessary and sufficient for instability by Antonsen and Lee.[38] Van Dam, Rosenbluth, and Lee[37] also showed that the low frequency energy principle predicted more instability than the Kruskal-Oberman model. This result is somewhat paradoxical, since the new theory developed to explain hot particle stability actually predicts more restrictive stability conditions.

The stability difficulties with the low frequency energy principle are, in fact, related to the presence of the negative energy precessional mode when the hot particles are decoupled, i.e., when Eq. (5) is not fulfilled. The dissipative damping mechanisms from the background plasma then cause destabilization of the precessional mode. If, on the other hand, Eq. (5) is satisfied, then the system responds in an unstable manner somewhat similar to that of conventional MHD modes.

POSSIBLE RESOLUTION OF THEORY AND EXPERIMENT

Given these unfavorable theoretical predictions, how does one explain the large hot particle pressures that are observed in experiment? Recently, Berk and Zhang[39] have speculated that the explanation may be because a minimum-B diamagnetic well was not achieved in the Elmo-type experiments. They showed that the precessional mode can be stable if minimum-B is not achieved. In such a configuration there would be negative dissipation, which can overcome the positive dissipation always present and thereby stabilize a negative energy precessional mode. If the experiments have broader pressure profiles than originally supposed with the same stored energy, the stability of the experiments could be understood. The total stored energy of hot electrons is fairly straightforward to measure, but the radial pressure profile is more difficult to determine. The magnitude of the total magnetic field gradient, $\partial B/\partial r$, is proportional to Δ^{-2}, where Δ is the radial thickness of the hot electron layer. Thus, the determination of whether a mimimum-B configuration has been experimentally achieved depends sensitively on how well the radial profile has been determined.

Attempts have been made to determine the radial pressure profile by inverting Hall probe measurements of magnetic fields, made primarily on the axis of a hot electron annulus. The conclusion of the investigators was that a minimum-B configuration was definitively established.[40,41] However, since this measurement is indirect and the convolution procedure is not unique, it is the opinion of the author of this review that the demonstration of minimum-B formation has not been definitively shown. For example, close fits with broader non-minimum-B profiles were obtained,[41] but were rejected because the total deviations of the reconstruction with the original data exceeded experimental error. However, if more parameters were used to classify the equilibrium, data fits to broad non-minimum-B profiles would have likely been achieved.

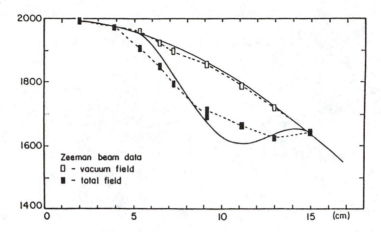

FIGURE 3   Comparison of Hall probe results with Zeeman beam data. Bars are magnetic fields induced from Zeeman data. The upper solid curve is the vacuum field. The lower solid curve is the reconstructed magnetic field inferred from Hall probe measurements. [From Ref. 43]

An alternative direct method of determining the magnetic field with the use of Zeeman splitting techniques has been developed and implemented on the STM mirror experiment.[42,43] These measurements seem to give somewhat broader pressure profiles than would be inferred with Hall probe measurements, as shown in Fig. 3. The bars are the results of the direct Zeeman splitting measurements, and the curve with the slight magnetic field depression is the result of the reconstruction using Hall probe measurements. To date there have been no direct measurements that show the formation of a minimum-B field. However, these results still do not give a definitive answer to the question of whether a minimum-B configuration can be obtained. These experiments were limited to low energy storage, viz., less than half the stored energy that was obtained in the magnetic field reconstruction of the experiments reported in Ref. 40. Unfortunately, the experiments in STM have been terminated, and a definitive answer as to whether a minimum-B configuration can be established in an Elmo-like plasma configuration is still an important open question.

From a theoretical point of view, a plasma disc-like configuration would appear to be more favorable for stability. In this case one forms a radial current profile similar to that of Astron. It has been shown theoretically that the combination of wall stabilization and finite Larmor radius stabilization[33,34] can convert the negative energy precessional mode to a positive energy wave.

To convert the precessional mode to a positive energy wave requires a large enough hot particle beta value. This represents a sort of second stability, which we characterize as robust stability, since all MHD-like perturbations are then positive definite. If the beta of the background plasma is too low, this system is susceptible to destabilization with respect to the negative energy precessional mode; if, however, the background beta exceeds a critical value, the MHD-like modes (including the precessional mode and the Lee-Van Dam constraints) are stabilized. At lower beta there is a fragile stability band present, where MHD-like modes are stable, but there are still negative energy precessional modes possible if a diamagnetic well is achieved. Analysis shows that if one changes equilibrium parameters to pass from the fragile stability regime to the robust stability regime, one must pass an unstable band that is characterized by an MHD-like unstable spectrum.[44] Thus, theory indicates that a hot particle stable diamagnetic well is possible in principle, but that the route to this stable regime is difficult. Fast formation techniques, such as in the Astron experiment or in a field-reversed theta pinch,[45] are one way of achieving the robust stable case. It has been suggested that a laser-produced plasma[46] can produce such a robust stable case. Alternative techniques such as the use of ponderomotive stabilization[47] or the use of transient quadrupole fields,[48] may also be possibilities.

Even with the achievement of robust stability conditions, several stability questions remain to be resolved. Resistivity in the wall can allow a resistive rigid-layer precessional mode,[49] and feed-back techniques need to be designed to stabilize against such a slow growing resistive mode. The diamagnetic current of the energetic particles has associated with it a negative energy mechanism that is similar to the universal drift-wave mechanism. In the relatively high beta systems that we have discussed here, it is possible for the compressional Alfvén wave to have a phase velocity less than diamagnetic drift frequency of the hot component. This leads to a class of compressional drift instabilities, first discussed by Hasegawa[50] and later by Migliuolo[51] and recently in Ref. 44 in the context of hot particle stabilization. Further study of this area is needed.

APPLICATION TO TOKAMAKS

Attempts have also been made to apply the hot particle stabilization principle[37] to establish a bridge to the second stability regime[52] in tokamaks. It was first noted[15] that the unstable ballooning mode gap can be circumvented if hot particles are present, whose response can be treated in the low frequency limit. In a tokamak, magnetic shear aids stability, and consequently the low frequency energy principle can be fulfilled as long as gradient-B reversal is not achieved in the unfavorable curvature region where the hot particles are trapped. Using these results, attempts have been made to design global equilibrium on each field line.[53-55]

Unfortunately, hot particle stability in tokamaks must contend with finite frequency modes near the precessional frequency of the hot particles. For example, the observed fishbone oscillations[56] in tokamaks were explained by Liu, White, and

Rosenbluth[57] as the interaction of the internal kink mode and the precessional oscillation. Detailed studies by several investigators[58-61] have indicated that the finite frequency response of the hot particles due to the interaction of the hot particles' drift resonance with other plasma oscillations, such as shear Alfvén waves[58] and the ion diamagnetic drift wave,[59] may cause additional instability problems with hot particles.

At present, more work is needed to determine if hot particles can be a favorable or a destabilizing effect in tokamak systems. This is quite important in ignition systems where the alpha particles may have an improved role in producing hot particle stabilization or, conversely, in supplying an additional source of instability. In addition, experiments[54] using beams or ECRH have been proposed to test the role of hot particle stabilization in tokamaks. Recent data from the JET experiment[62] with ICRF heating have indicated a suppression of MHD sawtooth instabilities. The stabilization has been interpreted[63,64] as being due to the hot particle mechanism discussed here, with the hot particles arising from the ICRF runaway tail formed during the heating.

CONCLUSION

We have discussed the progress that has been achieved in understanding what role hot particles can play in altering stability criteria in plasma confinement schemes such as those in mirror machines and tokamaks. It is clear that there are plasma regimes where the unstable plasma response predicted from MHD behavior can be stabilized by hot particles. This is apparent from high stored energies that have been achieved in experiments. However, it is not entirely clear whether hot particles will improve the overall stability response in a plasma containing significant pressure in the background plasma, since the combined system tends to introduce other instability sources, such as seen in fishbone oscillations. The possibility of achieving improved containment properties is an important objective and more investigation is needed. It is clear, however, that several subtle instability sources must be controlled in order to attain improved plasma containment.

REFERENCES

1. I.B. Bernstein, E.A. Frieman, M.D. Kruskal, and R.M. Kulsrud, *Proc. Roy. Soc. (London)* **A244**, 17 (1958).
2. D.E Baldwin, *Rev. Mod. Phys.* **49**, 1317 (1977).
3. M.N. Rosenbluth and C.L. Longmire, *Ann. Phys. (N.Y.)* **1**, 120 (1957).
4. M.S. Ioffe, R.I. Sobolev, V.G. Telkovskii, and E.E. Yushmanov, *Zh. Eksp. Teor. Fiz.* **40**, 40 (1961) [*Soviet Phys. - JETP* **13**, 27 (1961)].
5. C.C. Damm, J.H. Foote, A.H. Futch, Jr., A.L. Gardner, F.J. Gordon, A.L. Hunt, and R.F. Post, *Phys. Fluids* **8**, 1472 (1965).

6. A.M.M. Todd, M.S. Chance, J.M. Greene, R.C. Grimm, J.L. Johnson, and J. Manickam, *Phys. Rev. Lett.* **38**, 826 (1977).

7. F. Troyon, R. Gruber, H. Saarenmann, S. Semenzato, and S. Succi, *Plasma Phys. & Contr. Fusion* **26**, 209 (1984).

8. N.C. Christofilis, in *Proceedings of the Second United Nations International Conference on the Peaceful Uses of Atomic Energy* (United Nations, Geneva, 1958), Vol. 32, p. 279.

9. J.J. Bzura, T.J. Fessenden, H.H. Fleischman, D.A. Phelps, A.C. Smith, and D.M. Woodall, *Phys. Rev. Lett.* **29**, 256 (1972).

10. R.A. Dandl, H.O. Eason, P.H. Edmonds, and A.C. England, *Nucl. Fusion* **11**, 411 (1971).

11. R.A. Dandl, H.O. Eason, P.H. Edmonds, A.C. England, G.E. Guest, C.L. Hedrick, J.T. Hogan, and J.C. Sprott, in *Plasma Physics and Controlled Nuclear Fusion Research 1970* (IAEA, Vienna, 1971), Vol. II, p. 607.

12. M. Fujiwara *et al.*, in *Plasma Physics and Controlled Nuclear Fusion Research 1982* (IAEA, Vienna, 1983), Vol. II., p. 197 .

13. R.N. Sudan and M.N. Rosenbluth, *Phys. Fluids* **22**, 282 (1979).

14. R.N. Sudan and P.K. Kaw *Phys. Rev. Lett.* **47**, 575 (1981).

15. M.N. Rosenbluth, S.T. Tsai, J.W. Van Dam, and M.G. Engquist, *Phys. Rev. Lett.* **51**, 1967 (1983).

16. G. Benford, D.L. Book, N.C. Christofilis, T.K. Fowler, V.K. Neil, and L.D. Pearlstein, in *Plasma Physics and Controlled Nuclear Fusion Research 1968* (IAEA, Vienna 1969), Vol. I, p. 981.

17. D.B. Nelson and C.L. Hedrick, *Nucl. Fusion* **19**, 283 (1979).

18. M.D. Kruskal and C.R. Oberman, *Phys. Fluids* **1**, 275 (1958).

19. M.N. Rosenbluth and N. Rostoker *Phys. Fluids* **2**, 23 (1959).

20. H.P. Furth, J. Killeen, and M.N. Rosenbluth, *Phys. Fluids* **6**, 459 (1963).

21. M.N. Rosenbluth, N.A. Krall, and N. Rostoker, *Nucl. Fusion Suppl.* Pt. **1**, 143 (1962).

22. N.A. Krall, *Phys. Fluids* **9**, 820 (1966).

23. H.L. Berk, *Phys. Fluids* **19**, 1255 (1976).

24. I.G. Brown, A.J. Lichtenberg, M.A. Lieberman, and N.C. Wyeth, *Phys. Fluids* **19**, 1203 (1976).

25. S. Hiroe *et al.*, *Phys. Fluids* **27**, 1019 (1984).

26. H.L. Berk, J.W. Van Dam, and M.N. Rosenbluth, *Phys. Fluids* **26**, 201 (1983).

27. A.M. El Nadi, *Phys. Fluids* **25**, 2019 (1982).

28. H.L. Berk, C.Z. Cheng, M.N. Rosenbluth, and J.W. Van Dam, *Phys. Fluids* **25**, 2642 (1982).

29. J.M. Greene and B. Coppi, *Phys. Fluids* **8**, 1745 (1965).

30. H.P. Furth, *Phys. Fluids* **8**, 2020 (1965).

31. J.W. Beal *et al.*, in *Plasma Physics and Controlled Nuclear Fusion Research 1968* (IAEA, Vienna, 1969), Vol. I, p. 967.

32. N.C. Christofilis, R.J. Briggs, R.E. Hester, E.J. Lauer, and P.B. Weiss, Lawrence Livermore Laboratory Report, UCRL-14282 (1965).

33. H.L. Berk, M.N. Rosenbluth, H.V. Wong, and T.M. Antonsen, Jr., *Phys. Fluids* **27**, 2705 (1984).
34. H.L. Berk and H.V. Wong, *Phys. Fluids* **28**, 1881 (1985).
35. J.W. Van Dam and Y.C. Lee, in *Proceedings of the EBT Ring Physics Workshop* (Oak Ridge National Laboratory, Oak Ridge, 1979), CONF-791228, p. 471.
36. D.B. Nelson, *Phys. Fluids* **23**, 1850 (1980).
37. J.W. Van Dam, M.N. Rosenbluth, and Y.C. Lee, *Phys. Fluids* **25**, 1349 (1982).
38. T.M. Antonsen and Y.C. Lee, in *Proceedings of Hot Electron Workshop* (Oak Ridge National Laboratory, Oak Ridge, 1981), CONF-811203, p. 191.
39. H.L. Berk and Y.Z. Zhang, *Phys. Fluids* **30**, 1123 (1987).
40. B.H. Quon, R.A. Dandl, W. DiVergilio, G.E. Guest, L.L. Lao, N.H. Lazar, T.K. Samac, and R.F. Wuerker, *Phys. Fluids* **28**, 1503 (1985).
41. C.L. Hedrick, L.W. Owen, B.H. Quon, and R.A. Dandl, *Phys. Fluids* **30**, 1860 (1987).
42. H. Wickham, N.H. Lazar, and R. Dandl, *Bull. Am. Phys. Soc.* **29**, 1339 (1984).
43. R. Dandl *et al.*, "Determination of Electron Ring Properties in STM," Final Report, AMPC-26-027 (March, 1987).
44. H.L. Berk, H.V. Wong, and K.T. Tsang, *Phys. Fluids* **30**, 2681 (1987).
45. R.F. Gribble *et al.*, in *Plasma Physics and Controlled Nuclear Fusion Research 1974* (IAEA, Vienna 1975), Vol. III, p. 381.
46. F.J. Mayer, H.L. Berk, and D.W. Forslund, *Comments Plasma Phys. Controlled Fusion* **9**, 139 (1985).
47. J.R. Ferron, N. Hershkowitz, R.H. Breun, S.N. Golovato, and R. Goulding, *Phys. Rev. Lett.* **51**, 1955 (1983).
48. R.A. Close, B.K. Kang, A.J. Lichtenberg, M.A. Lieberman, and H. Meuth, *Phys. Fluids* **29**, 1217 (1986).
49. H.H. Fleischmann, K. Kupter, and R.E. Kribel, *Phys. Fluids* **28**, 1917 (1985).
50. H. Hasegawa, *Phys. Rev. Lett.* **27**, 11 (1971).
51. S. Migliuolo, *J. Geophys. Res.* **89**, 11023 (1984).
52. B. Coppi, A. Ferreira, and J.J. Ramos, *Phys. Rev. Lett.* **44**, 990 (1980).
53. H. Naitou, J.W. Van Dam, and D.C. Barnes, *Nucl. Fusion* **27**, 765 (1987).
54. G.A. Navratil and T.C. Marshall, *Comments Plasma Phys. Controlled Fusion* **10**, 185 (1986).
55. R.L. Miller and J.W. Van Dam, *Nucl. Fusion* **27**, 2101 (1987).
56. K. McGuire *et al.*, *Phys. Rev. Lett.* **50**, 891 (1983).
57. L. Chen, R.B. White, and M.N. Rosenbluth, *Phys. Rev. Lett.* **52**, 1122 (1984).
58. C.Z. Cheng, L. Chen, and M.S. Chance, *Ann. Phys. (N.Y.)* **161**, 21 (1984).
59. D.A. Spong, D.J. Sigmar, W.A. Cooper, D.E. Hastings, and K.T. Tsang, *Phys. Fluids* **28**, 2494 (1984).
60. D.P. Stotler and H.L. Berk, *Phys. Fluids* **30**, 1429 (1987).
61. B. Coppi and F. Porcelli, *Phys. Rev. Lett.* **57**, 2272 (1986).
62. J. Jacquinot *et al.*, in *Plasma Physics and Controlled Nuclear Fusion 1986* (IAEA, Vienna, 1987), Vol. I, p. 449.

63. F. Pegoraro, F. Porcelli, and R.J. Hastie, in *Proc. Sherwood Controlled Fusion Theory Conference* (18-20 April 1988, Gatlinburg, Tennessee), paper 2B1.

64. R.B. White, P.H. Rutherford, P. Colestock, and M.N. Bussac, in *Proc. Sherwood Controlled Fusion Theory Conference* (18-20 April 1988, Gatlinburg, Tennessee), paper 2B2.

List of Publications of Marshall N. Rosenbluth

This list of publications covers through the end of 1987. Scientific work published in abstract form has not been included; work published in report form is listed if not published elsewhere.

1949

1. M. N. Rosenbluth, "Electromagnetic Interactions of a Vector Meson with a Scalar Excited State" (Ph.D. dissertation), *Physical Review* **76**, 951-957 (1949).
2. M. N. Rosenbluth, "A Model for Nuclear Capture of μ-Mesons," *Physical Review* **75**, 532 (1949).
3. T. D. Lee, M. N. Rosenbluth, and C. N. Yang, "Interaction of Mesons with Nucleons and Light Particles," *Physical Review* **75**, 905 (1949).

1950

4. M. N. Rosenbluth, "High Energy Elastic Scattering of Electrons on Protons," *Physical Review* **79**, 615–619 (1950); reprinted in *Electron Scattering and Nuclear and Nucleon Structure*, edited by R. Hofstadter (W. A. Benjamin, New York, 1963), pp. 89–93.

1953

5. N. Metropolis, A. W. Rosenbluth, M. N. Rosenbluth, A. H. Teller, and E. Teller, "Equation of State Calculations by Fast Computing Machines," *Journal of Chemical Physics* **21**, 1087–1092 (1953).

1954

6. M. N. Rosenbluth and A. W. Rosenbluth, "Further Results on Monte Carlo Equations of State," *Journal of Chemical Physics* **22**, 881–884 (1954).

1955

7. M. N. Rosenbluth and A. W. Rosenbluth, "Monte Carlo Calculation of the Average Extension of Molecular Chains," *Journal of Chemical Physics* **23**, 356–359 (1955).
8. J. T. Waber and M. N. Rosenbluth, "Mathematical Studies of Galvanic Corrosion. II. Coplaner Electrodes with One Electrode Infinitely Large and with Equal Polarization Parameters," *Journal of the Electrochemical Society* **102**, 344–353 (1955).
9. D. L. Judd, W. M. MacDonald, and M. N. Rosenbluth, "End Leakage Losses from the Mirror Machine," in *Conference on Controlled Thermonuclear Reactions*, Berkeley California, 1955 (US Atomic Energy Commission Report, WASH-289), p. 158.

1956

10. M. N. Rosenbluth, "Stability of the pinch," Los Alamos Scientific Laboratory Report, No. LA-2030 (1956).
11. C. L. Longmire and M. N. Rosenbluth, "Diffusion of Charged Particles Across a Magnetic Field," *Physical Review* **103**, 507–510 (1956).

1957

12. M. N. Rosenbluth, "Dynamics of a Pinched Gas," in *Magnetohydrodynamics*, First Symposium, Palo Alto, California, 1956, edited by R. K. M. Landshoff (Stanford University Press, Stanford, 1957), pp. 57–66.

13. W. M. MacDonald, M. N. Rosenbluth, and W. Chuck, "Relaxation of a System of Particles with Coulomb Interactions," *Physical Review* **107**, 350-353 (1957).

14. M. N. Rosenbluth, W. M. MacDonald, and D. L. Judd, "Fokker-Planck Equation for an Inverse-Square Force," *Physical Review* **107**, 1–6 (1957).

15. A. Andresen, A. W. McReynolds, M. S. Nelkin, M. N. Rosenbluth, and W. L. Whittemore, "Neutron Investigation of Optical Vibration Levels in Zirconium Hydride," *Physical Review* **108**, 1092-1093 (1957).

16. M. N. Rosenbluth and C. L. Longmire, "Stability of Plasmas Confined by Magnetic Fields," *Annals of Physics (New York)* **1**, 120–140 (1957); reprinted in *Magnetohydrodynamic Stability and Thermonuclear Containment*, edited by A. Jeffrey and T. Taniuti (Academic Press, New York, 1966), pp. 109–129.

17. M. N. Rosenbluth, "The Stabilized Pinch," in *Proceedings of the Third International Conference on Ionization Phenomena in Gases*, Venice, Italy, 1957 (Societa Italiana di Fisica, Milan, 1957), p. 903.

18. M. N. Rosenbluth, "Diffusion and Ohmic Heating in Plasma Magnetic Field Systems," General Atomic Report, No. GA-99 (April 15, 1957).

1958

19. M. N. Rosenbluth, "Hydromagnetic Basis for Treatment of Plasmas," in *The Plasma in a Magnetic Field*, Second Symposium on Magnetohydrodynamics, Palo Alto, California, 1957, edited by R. K. M. Landshoff (Stanford University Press, Stanford, 1958), pp. 23–31.

20. M. N. Rosenbluth and A. N. Kaufman, "Plasma Diffusion in a Magnetic Field," *Physical Review* **109**, 1–5 (1958).

21. M. N. Rosenbluth, "Stability and Heating in the Pinch Effect," in *Proceedings of the Second United Nations International Conference on the Peaceful Uses of Atomic Energy. Volume 31. Theoretical and Experimental Aspects of Controlled Nuclear Fusion* (United Nations, Geneva, 1958) pp. 85–92; reprinted in *Plasma Physics and Thermonuclear Research*, Progress in Nuclear Energy, Series XI, edited by C. L. Longmire, J. L. Tuck, and W. B. Thompson (Macmillan Company, New York, 1963), Vol. 1, pp. 463–472; reprinted in *Magnetohydrodynamic Stability and Thermonuclear Containment*, edited by A. Jeffrey and T. Taniuti (Academic Press, New York, 1966), pp. 205–213.

22. M. N. Rosenbluth and N. Rostoker, "Theoretical Structure of Plasma Equations," in *Proceedings of the Second United Nations International Conference*

on the *Peaceful Uses of Atomic Energy. Vol. 31. Theoretical and Experimen-tal Aspects of Controlled Nuclear Fusion* (United Nations, Geneva, 1958), pp. 144–150; reprinted in *Plasma Physics and Thermonuclear Research*, Prog-ress in Nuclear Energy, Series XI, edited by C. L. Longmire, J. L. Tuck, and W. B. Thompson (Macmillan Company, New York, 1963), Vol. 1, pp. 475–490.

23. A. W. McReynolds, M. S. Nelkin, M. N. Rosenbluth, and W. L. Whittemore, "Neutron Thermalization by Chemically-Bound Hydrogen and Carbon," in *Proceedings of the Second United Nations International Conference on the Peaceful Uses of Atomic Energy. Vol. 16. Nuclear Data and Reactor Theory* (United Nations, Geneva, 1958), pp. 297–313.

1959

24. M. N. Rosenbluth and N. Rostoker, "Theoretical Structure of Plasma Equa-tions," *Physics of Fluids* **2**, 23–30 (1959).
25. M. N. Rosenbluth and N. Rostoker, "Test Particles in a Plasma," in *Ionization Phenomena in Gases*, Proceedings of the Fourth International Conference, Up-psala, 17-21 August 1959, edited by N. R. Nilsson (North-Holland Publishing Company, Amsterdam, 1960), Vol. II, pp. 566-576.

1960

26. R. F. Post, R. E. Ellis, F. C. Ford, and M. N. Rosenbluth, "Stable Confinement of a High-Temperature Plasma," *Physical Review Letters* **4**, 166–170 (1960).
27. N. Rostoker and M. N. Rosenbluth, "Test Particles in a Completely Ionized Plasma," *Physics of Fluids* **3**, 1–14 (1960).
28. W. E. Drummond and M. N. Rosenbluth, "Cyclotron Radiation from a Hot Plasma," *Physics of Fluids* **3**, 45–51 (1960). [Errata: *Physics of Fluids* **3**, 491 (1960)].
29. I. B. Bernstein, E. A. Frieman, R. M. Kulsrud, and M. N. Rosenbluth, "Ion Wave Instabilities," *Physics of Fluids* **3**, 136–137 (1960).
30. K. M. Watson, S. A. Bludman, and M. N. Rosenbluth, "Statistical Mechanics of Relativistic Streams. I.," *Physics of Fluids* **3**, 741–747 (1960).
31. S. A. Bludman, K. M. Watson, and M. N. Rosenbluth, "Statistical Mechanics of Relativistic Streams. II," *Physics of Fluids* **3**, 747–757 (1960).
32. M. N. Rosenbluth, "Long-Wavelength Beam Instability," *Physics of Fluids* **3**, 932–936 (1960).
33. M. N. Rosenbluth, "Plasma Physics ('Quantum' and 'Classical')," *Physics Today* **13**(8), 27–30 (1960).
34. M. N. Rosenbluth, chapter editor, "The Problems of Thermonuclear Fusion and High Temperature Plasmas," in *Symposium of Plasma Dynamics*, Woods

Hole, Massachusetts, 1958, edited by F. H. Clauser (Addison-Wesley, Reading, Massachussetts, 1960), pp. 19–53.

35. M. N. Rosenbluth, "Introduction to Plasma Physics," in *International Summer Course in Plasma Physics 1960* (Danish Atomic Energy Commission, Risö, Roskilde, Denmark, 1960) Risö Report, No. 18, pp. 1–11.

36. M. N. Rosenbluth, "Microinstabilities," in *International Summer Course in Plasma Physics 1960* (Danish Atomic Energy Commission, Risö, Roskilde, Denmark 1960), Risö Report, No. 8, pp. 189–200.

37. M. N. Rosenbluth, "Synchrotron Radiation," in *International Summer Course in Plasma Physics 1960* (Danish Atomic Energy Commission, Risö, Roskilde, Denmark, 1960), Risö Report, No. 18, pp. 301–312.

38. M. N. Rosenbluth, "Pinch Dynamics," in *International Summer Course in Plasma Physics 1960* (Danish Atomic Energy Commission, Risö, Roskilde, Denmark, 1960), Risö Report, No. 18, pp. 431–441.

1961

39. M. N. Rosenbluth, "Summary Talk of Ideas Presented at the 1959 International Plasma Physics Institute," *Journal of Nuclear Energy, Part C: Plasma Physics* **2**, 235–238 (1961).

40. J. L. Tuck, M. Kruskal, S. A. Colgate, and M. N. Rosenbluth, "Controlled Nuclear Fusion Reactor," U.S. Patent No. 3016342 (1962).

41. E. Gerjuoy and M. N. Rosenbluth, "Pinch with Rotating Plasma," *Physics of Fluids* **4**, 112–122 (1961).

42. N. A. Krall and M. N. Rosenbluth, "Stability of a Slightly Inhomogeneous Plasma," *Physics of Fluids* **4**, 163–172 (1961).

43. W. E. Drummond and M. N. Rosenbluth, "Comments on Synchrotron Radiation," *Physics of Fluids* **4**, 277–278 (1961).

44. M. N. Rosenbluth and N. Rostoker, "Fokker-Planck Equation for a Plasma with a Constant Magnetic Field," *Journal of Nuclear Energy, Part C: Plasma Physics* **2**, 195–205 (1961).

1962

45. E. A. Frieman, M. L. Goldberger, K. M. Watson, S. Weinberg, and M. N. Rosenbluth, "Two-Stream Instability in Finite Beams," *Physics of Fluids* **5**, 196–209 (1962).

46. M. N. Rosenbluth and N. Rostoker, "Scattering of Electromagnetic Waves by a Nonequilibrium Plasma," *Physics of Fluids* **5**, 776–788 (1962).

47. N. A. Krall and M. N. Rosenbluth, "Trapping Instabilities in a Slightly Inhomogeneous Plasma," *Physics of Fluids* **5**, 1435–1446 (1962).

48. W. E. Drummond and M. N. Rosenbluth, "Anomalous Diffusion Arising from Microinstabilities in a Plasma," *Physics of Fluids* **5**, 1507–1513 (1962).

49. M. N. Rosenbluth, N. A. Krall, and N. Rostoker, "Finite Larmor Radius Stabilization of 'Weakly' Unstable Confined Plasmas," *Nuclear Fusion: 1962 Supplement*, Part 1, pp. 143–150; reprinted in *Magnetohydrodynamic Stability and Thermonuclear Containment*, edited by A. Jeffrey and T. Taniuti (Academic Press, New York, 1966), pp. 131–138.

50. M. N. Rosenbluth, "Controlled Nuclear Fusion Research, September, 1961: Review of Theoretical Results," *Nuclear Fusion: 1962 Supplement*, Part 1 pp. 21–24.

51. M. N. Rosenbluth, "Stability Theory in Plasma Physics," in *Hydrodynamic Instability*, Proceedings of Symposia in Applied Mathematics, Vol. XIII, edited by G. Birkhoff, R. Bellman, and C. C. Lin (American Mathematical Society, Providence, Rhode Island, 1962), pp. 35–40.

1963

52. N. A. Krall and M. N. Rosenbluth, "Low-Frequency Stability of Nonuniform Plasmas," *Physics of Fluids* **6**, 254–265 (1963).

53. W. E. Drummond and M. N. Rosenbluth, "Cyclotron Radiation from a Hot Plasma," *Physics of Fluids* **6**, 276–283 (1963).

54. M. N. Rosenbluth and G. W. Stuart, "Relativistic Virial Theorem," *Physics of Fluids* **6**, 452–453 (1963).

55. H. P. Furth, J. Killeen, and M. N. Rosenbluth, "Finite-Resistivity Instabilities of a Sheet Pinch," *Physics of Fluids* **6**, 459–484 (1963); reprinted in *Magnetohydrodynamic Stability and Thermonuclear Containment*, edited by A. Jeffrey and T. Taniuti (Academic Press, New York, 1966), pp. 77–102.

56. M. N. Rosenbluth, L. D. Pearlstein, and G. W. Stuart, "Collective Electron Instability in a Bounded Region," *Physics of Fluids* **6**, 1289-1298 (1963).

57. A. Simon and M. N. Rosenbluth, "Single-Particle Cyclotron Radiation Near Walls and Sheaths," *Physics of Fluids* **6**, 1566-1573 (1963).

58. M. N. Rosenbluth, "Infinite Conductivity Theory of the Pinch. V. Surface Layer Model in the Limit of No Collisions," in *Plasma Physics and Thermonuclear Research*, Progress in Nuclear Energy, Series XI, edited by C. L. Longmire, J. L. Tuck, and W. B. Thompson (MacMillan Company, New York, 1963) Vol. 2, pp. 271-277 [from Los Alamos Report, No. LA-1850 (dated September 14, 1954) by R. Gerwin, A. W. Rosenbluth, and M. N. Rosenbluth].

59. L. D. Pearlstein, M. N. Rosenbluth, and A. W. Stuart, "Neutralization of Ion Beams," in *Electric Propulsion Development*, Progress in Astronautics and Aeronautics, Vol. 9, edited by E. Stuhlinger (Academic Press, New York, 1963) pp. 379–406.

1964

60. M. N. Rosenbluth and A. Simon, "Necessary and Sufficient Condition for the Stability of Plane Parallel Inviscid Flow," *Physics of Fluids* **7**, 557-558 (1964).
61. H. P. Furth and M. N. Rosenbluth, "Closed Magnetic Vacuum Configurations with Periodic Multipole Stabilization," *Physics of Fluids* **7**, 764-766 (1964).
62. N. A. Krall and M. N. Rosenbluth, "Invariance of the Magnetic Moment for Short Wavelength Perturbations," *Physics of Fluids* **7**, 1094-1095 (1964).
63. M. N. Rosenbluth, "Topics in Microinstabilities," in *Advanced Plasma Theory*, Proceedings of the International School of Physics "Enrico Fermi," Course XXV, Varenna, Italy, 9-12 July 1962, edited by M. N. Rosenbluth (Academic Press, New York, 1964), pp. 137-158.

1965

64. M. N. Rosenbluth and R. F. Post, "High-Frequency Electrostatic Plasma Instability Inherent to 'Loss-Cone' Particle Distributions," *Physics of Fluids* **8**, 547-550 (1965).
65. M. N. Rosenbluth and N. A. Krall, "Demonstration of the Minimum B Stability Theorem for the Universal Instability,"*Physics of Fluids* **8**, 1004-1005 (1965).
66. M. N. Rosenbluth and A. Simon, "Finite Larmor Radius Equations with Nonuniform Electric Fields and Velocities," *Physics of Fluids* **8**, 1300-1322 (1965).
67. N. A. Krall and M. N. Rosenbluth, "Universal Instability in Complex Field Geometries," *Physics of Fluids* **8**, 1488-1503 (1965).
68. J. H. Malmberg and M. N. Rosenbluth, "High Frequency Inductance of a Torus," *Review of Scientific Instruments* **36**, 1886-1887 (1965).
69. D. B. Chang, L. D. Pearlstein, and M. N. Rosenbluth, "On the Interchange Stability of the Van Allen Belt," *Journal of Geophysical Research* **70**, 3085-3097 (1965).
70. M. N. Rosenbluth, "Microinstabilities," in *Plasma Physics* (International Atomic Energy Agency, Vienna, 1965), pp. 485-513.

1966

71. W. B. Kunkel and M. N. Rosenbluth, "Introduction to Plasma Physics," in *Plasma Physics in Theory and Application*, edited by W. B. Kunkel (McGraw-Hill Book Company, New York, 1966), pp. 1-19.
72. H. P. Furth and M. N. Rosenbluth, "Apparatus for Magnetically Confining a Plasma," U.S. Patent No. 3230145 (1966).

73. R. F. Post and M. N. Rosenbluth, "Electrostatic Instabilities in Finite Mirror-Confined Plasmas," *Physics of Fluids* **9**, 730–749 (1966).
74. B. Coppi and M. N. Rosenbluth, "Collisional Interchange Instabilities in Shear and $\int d\ell/B$ Stabilized Systems," in *Plasma Physics and Controlled Nuclear Fusion Research 1965*, Proceedings of the Second Conference, Culham, England (International Atomic Energy Agency, Vienna, 1966), Vol. I, pp. 617–641.
75. D. B. Chang, L. D. Pearlstein, and M. N. Rosenbluth, "Corrections and additions to the paper entitled: On the Interchange Stability of the Van Allen Belt," *Journal of Geophysical Research* **71**, 351–355 (1966).
76. H. P. Furth, J. Killeen, M. N. Rosenbluth, and B. Coppi, "Stabilization by Shear and Negative V''," in *Plasma Physics and Controlled Nuclear Fusion Research 1965*, Proceedings of the Second International Conference, Culham, England (International Atomic Energy Agency, Vienna, 1966), Vol. I, pp. 103–125.
77. B. Coppi, G. Laval, R. Pellat, and M. N. Rosenbluth, "Convective Modes Driven by Density Gradients," *Nuclear Fusion* **6**, 261–267 (1966).
78. L. D. Pearlstein, M. N. Rosenbluth, and D. B. Chang, "High Frequency 'Loss Cone' Flute Instabilities Inherent to Two-Component Plasmas," *Physics of Fluids* **9**, 953–956 (1966).
79. A. Simon and M. N. Rosenbluth, "Flute Instability at Low Density," *Physics of Fluids* **9**, 726–729 (1966).
80. H. L. Berk, M. N. Rosenbluth, and R. N. Sudan, "Plasma Wave Propagation in Hot Inhomogeneous Media," *Physics of Fluids* **9**, 1606-1608 (1966).
81. B. Coppi, H. P. Furth, M. N. Rosenbluth, and R. Z. Sagdeev, "Drift Instability Due to Impurity Ions," *Physical Review Letters* **17**, 377–379 (1966).
82. M. N. Rosenbluth, R. Z. Sagdeev, J. B. Taylor, and G.M. Zaslavski, "Destruction of Magnetic Surfaces by Magnetic Field Irregularities," *Nuclear Fusion* **6**, 297-300 (1966).

1967

83. M. N. Rosenbluth, "Problems in Plasma Microinstability," in *Magneto-Fluid and Plasma Dynamics*, Proceedings of Symposia in Applied Mathematics, Vol. XVIII, edited by H. Grad (American Mathematical Society, Providence, Rhode Island, 1967), pp. 249–256.
84. M. N. Rosenbluth and D. B. Chang, "Finite-β Resistive Instabilities of Magnetospheric Tails," *Journal of Geophysical Research* **72**, 143–158 (1967).
85. B. Coppi, M. N. Rosenbluth, and R. Z. Sagdeev, "Instabilities Due to Temperature Gradients in Complex Magnetic Field Configurations," *Physics of Fluids* **10**, 582–587 (1967).
86. M. N. Rosenbluth and R. K. Varma, "Approximate Equations for Plasmas in Mirror Machines," *Nuclear Fusion* **7**, 33–55 (1967).

87. R. N. Sudan, A. Cavaliere, and M. N. Rosenbluth, "Nonlinear Interaction of Helicons (Whistlers) in Inhomogeneous Media," *Physical Review* **158**, 387–396 (1967).
88. H. L. Berk, C. W. Horton, M. N. Rosenbluth, and R. N. Sudan, "Plasma Wave Reflection in Slowly Varying Media," *Physics of Fluids* **10**, 2003–2016 (1967).
89. M. N. Rosenbluth, R. Bickerton, R. A. Dory, E. A. Frieman, H. P. Furth, W. Kunkel, J. Marshall, K. Symon and S. O. Dean, "Report of Ad Hoc Panel on Low-Beta Toroidal Plasma Research," Atomic Energy Commission Report, No. TID-24228 (1967).

1968

90. B. Coppi, G. Laval, R. Pellat, and M. N. Rosenbluth, "Collisionless Microinstabilities in Configurations with Periodic Magnetic Curvature," *Plasma Physics* **10**, 1–22 (1968).
91. M. N. Rosenbluth and N. A. Krall, "Comments on 'Magnetic Moment Under Short-Wave Electrostatic Perturbations,'" *Physics of Fluids* **11**, 921–922 (1968).
92. B. Coppi, M. N. Rosenbluth, and P. H. Rutherford, "Fluid-Like Electron and Ion Modes in Inhomogeneous Plasmas," *Physical Review Letters* **21**, 1055–1059 (1968).
93. B. Coppi, M. N. Rosenbluth, and S. Yoshikawa, "Localized (Ballooning) Modes in Multipole Configurations," *Physical Review Letters* **20**, 190–192 (1968).
94. B. Coppi, S. Ossakow, and M. N. Rosenbluth, "Low-Density Modes in Multipole Devices," *Plasma Physics* **10**, 571–580 (1968).
95. M. N. Rosenbluth, "Low-Frequency Limit of Interchange Instability," *Physics of Fluids* **11**, 869–872 (1968).
96. H. L. Berk, C. W. Horton, M. N. Rosenbluth, D. E. Baldwin, and R. N. Sudan, "Nonlocal Reflection in Inhomogeneous Media," *Physics of Fluids* **11**, 365–371 (1968).
97. H. P Furth and M. N. Rosenbluth, "Static-Field Acceleration of Electron Rings," in *Symposium on Electron Ring Accelerators*, Lawrence Livermore Laboratory Report, No. UCRL–18103, pp. 210–218 (1968).
98. B. Coppi and M. N. Rosenbluth, "Model for the Earth's Magnetic Tail," European Space Research Organization Report, No. SP-36, pp. 1–3 (1968).

1969

99. M. N. Rosenbluth, J. L. Johnson, J. M. Greene, and K. E. Weimer, "Stability Limitations for Stellerators with Sharp Surfaces," *Physics of Fluids* **12**, 726–728 (1969).

100. H. L. Berk, T. K. Fowler, L. D. Pearlstein, R. F. Post, J. D. Callen, C. W. Horton, and M. N. Rosenbluth, "Criteria for Stabilization of Electrostatic Modes in Mirror-Confined Plasmas," in *Plasma Physics and Controlled Nuclear Fusion Research 1968*, Proceedings of the Third International Conference, Novosibirsk, USSR (International Atomic Energy Agency, Vienna, 1969), Vol. II, pp. 151–164.

101. P. H. Rutherford, M. N. Rosenbluth, W. Horton, E. A. Frieman, and B. Coppi, "Low-Frequency Stability of Axisymmetric Toruses," in *Plasma Physics and Controlled Nuclear Fusion Research 1968*, Proceedings of the Third International Conference, Novosibirsk, USSR (International Atomic Energy Agency, Vienna, 1969), Vol. I, pp. 367–387.

102. H. P. Furth and M. N. Rosenbluth, "Low-Frequency Plasma Loss Mechanisms in MHD-Stabilized Torus," in *Plasma Physics and Controlled Nuclear Fusion Research*, Proceedings of the Third International Conference, Novosibirsk, USSR (International Atomic Energy Agency, Vienna, 1969), Vol. I, pp. 821–845.

103. M. N. Rosenbluth, B. Coppi, and R. N. Sudan, "Non-Linear Interactions of Positive and Negative Energy Modes in Plasmas," in *Plasma Physics and Controlled Nuclear Fusion Research*, Proceedings of the Third International Conference, Novosibirsk, USSR (International Atomic Energy Agency, Vienna, 1969), Vol. I, pp. 771–793.

104. B. Coppi, M. N. Rosenbluth, and R. N. Sudan, "Nonlinear Interactions of Positive and Negative Energy Modes in Rarefied Plasmas (I)," *Annals of Physics (New York)* **55**, 207–247 (1969).

105. M. N. Rosenbluth, B. Coppi, and R. Sudan, "Nonlinear Interactions of Positive and Negative Energy Modes in Rarefied Plasmas (II)," *Annals of Physics (New York)* **55**, 248–270 (1969).

106. A. A. Galeev, R. Z. Sagdeev, H. P. Furth, and M. N. Rosenbluth, "Plasma Diffusion in a Toroidal Stellarator," *Physical Review Letters* **22**, 511–514 (1969).

107. M. N. Rosenbluth and J. B. Taylor, "Plasma Diffusion and Stability in Toroidal Systems," *Physical Review Letters* **23**, 367–370 (1969).

108. M. N. Rosenbluth, "Plasma Physics: General Survey," in *Contemporary Physics: Trieste Symposium, 1968* (International Atomic Energy Agency, Vienna, 1969), Vol. I, pp. 205–220.

109. R. M. Kulsrud, C. Oberman, J. M. Dawson, and M. N. Rosenbluth, "Comments on 'Enhanced Bremsstrahlung from Supraluminous and Subluminous Waves in an Isotropic, Homogeneous Plasma,'" *Physics of Fluids* **12**, 1957–1959 (1969).

110. H. L. Berk, L. D. Pearlstein, J. D. Callen, C. W. Horton, and M. N. Rosenbluth, "Destabilization of Negative-Energy Waves in Inhomogeneous Mirror Geometry," *Physical Review Letters* **22**, 876–879 (1969).

1970

111. F. L. Ribe and M. N. Rosenbluth, "Feedback Stabilization of a High-β, Sharp-Boundaried Plasma Column with Helical Fields," *Physics of Fluids* **13**, 2572–2577 (1970).
112. F. L. Ribe and M. N. Rosenbluth, "Feedback Stabilization of a High β, Sharp-Boundaried Plasma Column with Helical Fields," in *Feedback and Dynamic Control of Plasmas*, AIP Conference Proceedings, No 1, edited by T. K. Chu and H. W. Hendel (American Institute of Physics, New York, 1970), pp. 80–83.
113. S. A. Colgate, E. P. Lee, and M. N. Rosenbluth, "Radio Emission from a Turbulent Plasma," *Astrophysical Journal* **162**, 649–664 (1970).
114. H. P. Furth, M. N. Rosenbluth, P. H. Rutherford, and W. Stodiek, "Thermal Equilibrium and Stability of Tokamak Discharges," *Physics of Fluids* **13**, 3020–3030 (1970).
115. M. N. Rosenbluth, J. L. Johnson, J. M. Greene, and K. E. Weiner, "Reply to Comments of H. Grad and H. Weitzner," *Physics of Fluids* **13**, 1419–1420 (1970).
116. S. Ichimaru and M. N. Rosenbluth, "Relaxation Processes in Plasmas with Magnetic Field. Temperature Relaxations," *Physics of Fluids* **13**, 2778–2789 (1970).
117. G. Laval, E. K. Maschke, R. Pellat, and M. N. Rosenbluth "Limiting β for Tokamak with Elliptical Magnetic Surfaces," International Centre for Theoretical Physics Report, No. IC/70/35, Trieste (1970).
118. R. D. Hazeltine, E. P. Lee, and M. N. Rosenbluth, "Rotation of Tokamak Equilibria" *Physical Review Letters* **25**, 427–430 (1970).
119. S. L. Adler, J. N. Bahcall, C. G. Callan, and M. N. Rosenbluth, "Photon Splitting in a Strong Magnetic Field," *Physical Review Letters* **25**, 1061–1065 (1970).
120. M. N. Rosenbluth, "Synchrotron Radiation in Tokamaks," *Nuclear Fusion* **10**, 340–343 (1970).
121. P. H. Rutherford, L. M. Kovrizhnikh, M. N. Rosenbluth, and F. L. Hinton, "Effect of Longitudinal Electric Field on Toroidal Diffusion," *Physical Review Letters* **25**, 1090–1093 (1970).
122. F. L. Hinton and M. N. Rosenbluth, "Variational Principle for Neoclassical Transport Properties," International Centre for Theoretical Physics Report, No. IC/70/111, Trieste (1970).
123. E. K. Maschke, R. Pellat, M. N. Rosenbluth, and M. Tanaka, "MHD Equilibrium of the Tokamak Multipole," International Centre for Theoretical Physics Report, No. IC/70/125, Trieste (1970).

1971

124. H. P. Furth and M. N. Rosenbluth, "High Energy Ion Accelerator," U.S. Patent No. 3626305 (1971).
125. R. D. Hazeltine, M. N. Rosenbluth, and A. M. Sessler, "Diffraction Radiation by a Line Charge Moving Past a Comb: A Model of Radiation Losses in an Electron Ring Accelerator," *Journal of Mathematical Physics* **12**, 502–514 (1971).
126. S. Ichimaru and M. N. Rosenbluth, "Anomalous Contribution of Low-Frequency, Long-Wavelength Fluctuations to Spatial Plasma Diffusion in a Magnetic Field," in *Plasma Physics and Controlled Nuclear Fusion Research 1970*, Proceedings of the Fourth International Conference, Madison, Wisconsin (International Atomic Energy Agency, Vienna, 1971), Vol. II, pp. 373-379.
127. M. N. Rosenbluth, P. H. Rutherford, J. B. Taylor, E. A. Frieman, and L. M. Kovrizhnikh, "Neoclassical Effects on Plasma Equilibria and Rotation," in *Plasma Physics and Controlled Nuclear Fusion Research 1970*, Proceedings of the Fourth International Conference, Madison, Wisconsin (International Atomic Energy Agency, Vienna, 1971), Vol. I, pp. 495–508.
128. P. H. Rutherford, H. P. Furth, and M. N. Rosenbluth, "Non-Linear Kink and Tearing-Mode Effects in Tokamaks," in *Plasma Physics and Controlled Nuclear Fusion Research 1970*, Proceedings of the Fourth International Conference, Madison, Wisconsin (International Atomic Energy Agency, Vienna, 1971), Vol. II, pp. 553–570.
129. R. D. Hazeltine, E. P. Lee, and M. N. Rosenbluth, "Resistive Plasma Rotation and Shock Formation in Toroidal Geometry," *Physics of Fluids* **14**, 361–370 (1971).
130. C. W. Horton, Jr., J. D. Callen, and M. N. Rosenbluth, "Microinstabilities in Axisymmetric Mirror Machines," *Physics of Fluids* **14**, 2019–2032 (1971).
131. M. N. Rosenbluth and M. L. Sloan, "Finite-β Stabilization of the Collisionless Trapped Particle Instability," *Physics of Fluids* **14**, 1725–1741 (1971).

1972

132. T. O'Neil and M. N. Rosenbluth, "Two-Stream Coupling of Corona and Core in Laser-Heated Pellets," Institute for Defense Analysis Report, No. AD-757790 (July, 1972), 24 pp.
133. M. N. Rosenbluth and S. Ichimaru, "Reply to comments of D. Voslamber," *Physics of Fluids* **15**, 956 (1972).
134. M. N. Rosenbluth, "Application of Computers to Problems of Controlled Thermonuclear Reactors," in *Computing as a Language of Physics* (International Atomic Energy Agency, Vienna, 1972), pp. 157–164.

135. M. N. Rosenbluth, "Monte Carlo Techniques in Statistical Mechanics," in *Computing as a Language of Physics* (International Atomic Energy Agency, Vienna, 1972), pp. 165–169.

136. M. N. Rosenbluth, D. W. Ross, and D. P. Kostomarov, "Stability Regions of Dissipative Trapped-Ion Instability," *Nuclear Fusion* **12**, 3–37 (1972).

137. M. N. Rosenbluth, R. D. Hazeltine, and F. L. Hinton, "Plasma Transport in Toroidal Confinement Systems," *Physics of Fluids* **15**, 116–140 (1972).

138. M. N. Rosenbluth and R. Z. Sagdeev, "Laser Fusion and Parametric Instabilities," *Comments on Plasma Physics and Controlled Fusion* **1**, 129–138 (1972).

139. M. N. Rosenbluth, "Superadiabaticity in Mirror Machines," *Physical Review Letters* **29**, 408–410 (1972).

140. M. N. Rosenbluth, "Parametric Instabilities in Inhomogeneous Media," *Physical Review Letters* **29**, 565–567 (1972).

141. M. N. Rosenbluth and C. S. Liu, "Excitation of Plasma Waves by Two Laser Beams," *Physical Review Letters* **29**, 701–705 (1972).

142. M. N. Rosenbluth and C. S. Liu, "Current-Driven Drift Wave Instability in a Sheared Magnetic Field," *Physics of Fluids* **15**, 1801–1803 (1972).

143. C. S. Liu, M. N. Rosenbluth, and C. W. Horton, Jr., "Electron Temperature-Gradient Instability and Anomalous Skin Effect in Tokamaks," *Physical Review Letters* **29**, 1489–1492 (1972).

144. R. D. Hazeltine and M. N. Rosenbluth, "Effect of Field Asymmetry on Neoclassical Transport in a Tokamak," *Physics of Fluids* **15**, 2211–2217 (1972).

1973

145. F. L. Hinton and M. N. Rosenbluth, "Transport Properties of a Toroidal Plasma at Low-to-Intermediate Collision Frequencies," *Physics of Fluids* **16**, 836–854 (1973).

146. M. N. Rosenbluth, M. Ruderman, F. Dyson, J. N. Bahcall, J. Shaham, and J. Ostriker, "Nuclear Fusion in Accreting Neutron Stars," *Astrophysical Journal* **184**, 907–910 (1973).

147. M. N. Rosenbluth and J. N. Bahcall, "Nonspherical Thermal Instabilities" *Astrophysical Journal* **184**, 9–16 (1973).

148. P. J. Catto, M. N. Rosenbluth, and C. S. Liu, "Parallel Velocity Shear Instabilities in an Inhomogeneous Plasma with a Sheared Magnetic Field," *Physics of Fluids* **16**, 1719–1729 (1973).

149. R. D. Hazeltine, F. L. Hinton, and M. N. Rosenbluth, "Plasma Transport in a Torus of Arbitrary Aspect Ratio," *Physics of Fluids* **16**, 1645–1653 (1973).

150. R. B. White, C. S. Liu, and M. N. Rosenbluth, "Parametric Decay of Obliquely Incident Radiation," *Physical Review Letters* **31**, 520–523 (1973).

151. C. S. Liu, M. N. Rosenbluth, and R. B. White, "Parametric Scattering Instabilities in Inhomogeneous Plasmas," *Physical Review Letters* **31**, 697–700 (1973).

152. M. N. Rosenbluth, R. B. White and C. S. Liu, "Temporal Evolution of a Three-Wave Parametric Instability," *Physical Review Letters* **31**, 1190–1193 (1973).
153. M. N. Rosenbluth, R. Y. Dagazian, and P. H. Rutherford, "Nonlinear Properties of the Internal $m = 1$ Kink Instability in the Cylindrical Tokamak," *Physics of Fluids* **16**, 1894–1902 (1973).
154. A. El Nadi and M. N. Rosenbluth, "Infinite-β Limit of the Drift Instability," *Physics of Fluids* **16**, 2036–2037 (1973).
155. J. N. Bahcall, M. N. Rosenbluth, and R. M. Kulsrud, "Model for X-Ray Sources Based on Magnetic Field Twisting," *Nature: Physical Science* **243**, 27–28 (1973).
156. A. A. Galeev, G. Laval, T. M. O'Neil, M. N. Rosenbluth, and R. Z. Sagdeev, "Parametric Backscattering of a Linear Electromagnetic Wave in a Plasma," *Zh. Eksp. Teor. Fiz. Pisma Red.* **17**, 48–52 (1973) [*JETP Letters* **17**, 35–38 (1973)].
157. M. N. Rosenbluth and R. Z. Sagdeev, "Interaction Between Plasmas and Intense Laser and Electron Beams—Summary of the Activities of the Trieste Working Group 13–31 August 1973," *Nuclear Fusion* **13**, 941–944 (1973).
158. A. A. Galeev, G. Laval, T. O'Neil, M. N. Rosenbluth, and R. Z. Sagdeev, "Interaction between an Intense Electromagnetic Wave and a Plasma," *Zh. Eksp. Teor. Fiz.* **65**, 973–989 (1973) [Soviet Physics–JETP **38**, 482–489 (1974)].

1974

159. C. E. Lee, C. L. Longmire, and M. N. Rosenbluth, "Thomas-Fermi Calculation of Potential Between Atoms," Los Alamos Scientific Laboratory Report, No. LA-5694-MS (July, 1974), 4 pp.
160. M. N. Rosenbluth and R. Z. Sagdeev, "Summary of Activities of the Trieste Working Group on the Interaction Between Plasmas and Intense Laser and Electron Beams," *Comments on Plasma Physics and Controlled Fusion* **2**, 3–10 (1974).
161. M. N. Rosenbluth and F. W. Perkins, "General Plasma Physics II," Princeton Plasma Physics Laboratory Lecture Notes (1973–1974).
162. M. N. Rosenbluth, "Present Status of Controlled Thermonuclear Theory," in *Advances in Plasma Physics*, edited by A. Simon and W. B. Thompson (John Wiley, New York, 1974), Vol. 5, pp. 75–83.
163. R. B. White, P. K. Kaw, D. Pesme, M. N. Rosenbluth, G. Laval, R. Huff, and R. Varma, "Absolute Parametric Instabilities in Inhomogeneous Plasmas," *Nuclear Fusion* **14**, 45–51 (1974).
164. M. N. Rosenbluth, H. P. Furth, and K. M. Case, "Minimization of Conductor Surface Heating by a Pulsed Magnetic Field," *Journal of Applied Physics* **45**, 1097–1099 (1974).

165. P. K. Kaw, R. B. White, D. Pesme, M. N. Rosenbluth, G. Laval, R. Varma, and R. Huff, "Linear Parametric Instabilities in Inhomogeneous Plasmas," *Comments on Plasma Physics and Controlled Fusion* **2**, 11–20 (1974).

166. J. F. Drake, P. K. Kaw, Y. C. Lee, G. Schmidt, C. S. Liu, and M. N. Rosenbluth, "Parametric Instabilities of Electromagnetic Waves in Plasmas," *Physics of Fluids* **17**, 778–785 (1974).

167. P. J. Catto, A. M. El Nadi, C. S. Liu, and M. N. Rosenbluth, "Stability of a Finite-β Inhomogeneous Plasma in a Sheared Magnetic Field," *Nuclear Fusion* **14**, 405–418 (1974).

168. C. S. Liu, M. N. Rosenbluth, and R. B. White, "Raman and Brillouin Scattering of Electromagnetic Waves in Inhomogeneous Plasmas," *Physics of Fluids* **17**, 1211–1219 (1974).

169. H. L. Berk, C. W. Horton, Jr., M. N. Rosenbluth, and P. H. Rutherford, "Microinstability Theory for Toroidal Plasmas Heated by Intense Energetic Ion Beams," in *Symposium on Plasma Heating in Toroidal Devices*, International School of Plasma Physics, Varenna, Italy, 4–17 September 1973 (Editrice Compositori, Bologna, 1974), pp. 182–187.

1975

170. H. L. Berk and M. N. Rosenbluth, "Ion Cyclotron Instability Induced by the Trapped-Ion Mode," *Nuclear Fusion* **15**, 1013–1023 (1975).

171. W. Horton, Jr., D. W. Ross, W. M. Tang, H. L. Berk, E. A. Frieman, R. E. LaQuey, R. V. Lovelace, S. M. Mahajan, M. N. Rosenbluth, and P. H. Rutherford, "Stability Theory of Dissipative Trapped-Electron and Trapped-Ion Modes," in *Plasma Physics and Controlled Nuclear Fusion Research 1974*, Proceedings of the Fifth International Conference, Tokyo (International Atomic Energy Agency, Vienna 1975), Vol. I, pp. 541–548.

172. R. B. White, D. Monticello, M. N. Rosenbluth, H. Strauss, and B. B. Kadomtsev, "Numerical Studies of Non-Linear Evolution of Kink and Tearing Modes in Tokamaks," in *Plasma Physics and Controlled Nuclear Fusion Research 1974*, Proceedings of the Fifth International Conference, Tokyo (International Atomic Energy Agency, Vienna, 1975), Vol. I, pp. 495–503.

173. C. S. Liu, M. N. Rosenbluth, and R. B. White, "Parametric Instabilities in Inhomogeneous Spherical Plasmas," in *Plasma Physics and Controlled Nuclear Fusion Research 1974*, Proceedings of the Fifth International Conference, Tokyo (International Atomic Energy Agency, Vienna, 1975), Vol. II, pp. 515–524.

174. H. L. Berk, H. P. Furth, D. L. Jassby, R. M. Kulsrud, C. S. Liu, M. N. Rosenbluth, P. H. Rutherford, F. H. Tenney, T. Johnson, J. Killeen, A. A. Mirin, M. E. Rensink, and C. W. Horton, Jr., "Two-Energy-Component Toroidal Fusion Devices," in *Plasma Physics and Controlled Nuclear Fusion Research*

1974, Proceedings of the Fifth International Conference, Tokyo (International Atomic Energy Agency, Vienna, 1975), Vol. III, pp. 569–580.

175. M. N. Rosenbluth and P. H. Rutherford, "Excitation of Alfvén Waves by High-Energy Ions in a Tokamak," *Physical Review Letters* **34**, 1428–1431 (1975).

176. H. L. Berk, W. Horton, Jr., M. N. Rosenbluth, and P. H. Rutherford, "Microinstability Theory of Two-Energy-Component Toroidal Systems," *Nuclear Fusion* **15**, 819–844 (1975).

177. M. N. Rosenbluth and P. J. Catto, "An Improved Calculation of Critical Magnetic Shear in an Inhomogeneous Plasma," *Nuclear Fusion* **15**, 573–582 (1975).

178. H. L. Berk and M. N. Rosenbluth, "Ion Cyclotron Instability Induced by the Trapped-Ion Mode," *Nuclear Fusion* **15**, 1013–1023 (1975).

1976

179. A. Simon and M. N. Rosenbluth, "Single-Mode Saturation of the Bump-on-Tail Instability," in *School on Plasma Physics and Controlled Thermonuclear Fusion*, Tbilissi, USSR, September 1976, Report No. COO-3497-16 (CONF-760983-1) (1976).

180. M. N. Rosenbluth and C. S. Liu, "Cross-Field Energy Transport by Plasma Waves," *Physics of Fluids* **19**, 815–818 (1976).

181. C. S. Liu, M. N. Rosenbluth, and W. M. Tang, "Dissipative Universal Instability Due to Trapped Electrons in Toroidal Systems and Anomalous Diffusion," *Physics of Fluids* **19**, 1040–1044 (1976).

182. W. M. Tang, C. S. Liu, M. N. Rosenbluth, P. J. Catto, and J. D. Callen, "Finite-Beta and Resonant-Electron Effects on Trapped-Electron Instabilities," *Nuclear Fusion* **16**, 191–202 (1976).

183. M. N. Rosenbluth, D. A. Monticello, H. R. Strauss, and R. B. White, "Numerical Studies of Nonlinear Evolution of Kink Modes in Tokamaks," *Physics of Fluids* **19**, 1987–1996 (1976).

184. B. I. Cohen, J. A. Krommes, W. M. Tang, and M. N. Rosenbluth, "Non-Linear Saturation of the Dissipative Trapped-Ion Mode by Mode Coupling," *Nuclear Fusion* **16**, 971–992 (1976).

185. C. S. Liu and M. N. Rosenbluth, "Parametric Decay of Electromagnetic Waves into Two Plasmons and its Consequences," *Physics of Fluids* **19**, 967–971 (1976).

186. A. Simon and M. N. Rosenbluth, "Single-Mode Saturation of the Bump-on-Tail Instability: Immobile Ions," *Physics of Fluids* **19**, 1567–1580 (1976).

187. R. N. Sudan and M. N. Rosenbluth, "Stability of Field-Reversed Ion Rings in a Background Plasma," *Physical Review Letters* **36**, 972–975 (1976).

188. B. V. Waddell, M. N. Rosenbluth, D. A. Monticello, and R. B. White, "Non-Linear Growth of the $m = 1$ Tearing Mode," *Nuclear Fusion* **16**, 528–532 (1976).

189. B. Coppi, R. Galvão, R. Pellat, M. N. Rosenbluth, and P. H. Rutherford, "Resistive Internal Kink Modes," *Fizika Plasmy* **2**, 961–966 (1976) [*Soviet Journal of Plasma Physics* **2**, 533–535 (1976)].

1977

190. R. B. White, D. A. Monticello, M. N. Rosenbluth, and B. V. Waddell, "Non-Linear Tearing Modes in Tokamaks," in *Plasma Physics and Controlled Nuclear Fusion Research 1976*, Proceedings of the Sixth International Conference, Berchtesgaden (International Atomic Energy Agency, Vienna, 1977), Vol. I, pp. 569–577.

191. M. N. Rosenbluth, "Summary of Magnetic Confinement (Theory)," in *Plasma Physics and Controlled Nuclear Fusion Research 1976*, Proceedings of the Sixth International Conference, Berchtesgaden (International Atomic Energy Agency, Vienna, 1977), Vol. III, pp. 555–561; included in R. S. Pease, M. N. Rosenbluth, O. N. Krokhin, and K.-H. Schmitter, "Summaries of the Sixth International Conference on Plasma Physics and Controlled Nuclear Fusion Research, Berchtesgaden, 1976," *Nuclear Fusion* **16**, 1047–1062 (1976).

192. R. B. White, D. A. Monticello, M. N. Rosenbluth, and B. V. Waddell, "Saturation of the Tearing Mode," *Physics of Fluids* **20**, 800–805 (1977).

193. W. M.-W. Tang, J. C. Adam, B. I. Cohen, E. A. Frieman, J. A. Krommes, G. Rewoldt, D. W. Ross, M. N. Rosenbluth, P. H. Rutherford, P. J. Catto, K. T. Tsang, and J. D. Callen, "Linear and Non-Linear Theory of Trapped-Particle Instabilities," in *Plasma Physics and Controlled Nuclear Fusion Research 1976*, Proceedings of the Sixth International Conference, Berchtesgaden (International Atomic Energy Agency, Vienna, 1977), Vol. II, pp. 489–497.

194. F. F. Chen, J. M. Dawson, B. D. Fried, H. P. Furth, and M. N. Rosenbluth, "Comments on 'Generalized Criterion for Feasibility of Controlled Fusion and Its Application to Nonideal D-D Systems,'" *Journal of Applied Physics* **48**, 415–417 (1977).

195. H. R. Strauss, D. A. Monticello, M. N. Rosenbluth, and R. B. White, "Nonlinear Helical Perturbations of a Tokamak," *Physics of Fluids* **20**, 390–395 (1977).

196. J. A. Krommes, M. N. Rosenbluth, and W. M. Tang, "Stability of Ripple-Assisted Neutral Beam Injection Against Loss-Cone Modes," *Nuclear Fusion* **17**, 667–680 (1977).

197. L. Chen, R. L. Berger, J. G. Lominadze, M. N. Rosenbluth, and P. H. Rutherford "Nonlinear Saturation of the Dissipative Trapped-Electron Instability," *Physical Review Letters* **39**, 754–757 (1977).

198. R. E. Aamodt, Y.C. Lee, C. S. Liu, and M. N. Rosenbluth, "Nonlinear Dynamics of Drift-Cyclotron Instability," *Physical Review Letters* **39**, 1660–1664 (1977).

199. G. Laval, R. Pellat, D. Pesme, A. Ramani, M. N. Rosenbluth, and E. A. Williams, "Parametric Instabilities in the Presence of Space-Time Random Fluctuations," *Physics of Fluids* **20**, 2049–2057 (1977).
200. R. B. White, D. A. Monticello, and M. N. Rosenbluth, "Simulation of Large Magnetic Islands: A Possible Mechanism for a Major Tokamak Disruption," *Physical Review Letters* **39**, 1618–1621 (1977).
201. B. V. Waddell, G. Laval, and M. N. Rosenbluth, "Reduction of the Growth Rate of the $m = 1$ Resistive Magneto-Hydrodynamic Mode by Finite Gyroradius Effects," Oak Ridge National Laboratory Report, No. ORNL/TM-5968 (1977).

1978

202. B. V. Waddell, M. N. Rosenbluth, D. A. Monticello, R. B. White, and B. Carreras, "Non-Linear Numerical Algorithms for Studying Tearing Modes," in *Theoretical and Computational Plasma Physics* (International Atomic Energy Agency, Vienna, 1978), pp. 79–91.
203. C. G. Callan, R. F. Dashen, R. L. Garwin, R. A. Muller, B. Richter, and M. N. Rosenbluth, "Heavy-Ion-Driven Inertial Fusion," SRI International Report, No. SAN-0115/130-1 (February, 1978), 20 pp.
204. R. B. White, D. A. Monticello, and M. N. Rosenbluth, "Reply to the Comments on 'Simulation of Large Magnetic Islands: A Possible Mechanism for a Major Tokamak Disruption,'" *Physical Review A* **18**, 2735 (1978).
205. D. P. Chernin and M. N. Rosenbluth, "Ion Losses from End-Stoppered Mirror-Trap," *Nuclear Fusion* **18**, 47–62 (1978).
206. M. N. Rosenbluth, "Mode Crossing in a Lagrangian System," *Physics of Fluids* **21**, 297–298 (1978).
207. A. B. Rechester and M. N. Rosenbluth, "Electron Heat Transport in a Tokamak with Destroyed Magnetic Surfaces," *Physical Review Letters* **40**, 38–41 (1978).
208. G. Ara, B. Basu, B. Coppi, G. Laval, M. N. Rosenbluth, and B. V. Waddell, "Magnetic Reconnection and $m = 1$ Oscillations in Current-Carrying Plasmas," *Annals of Physics (New York)* **112**, 443–476 (1978).
209. P. H. Rutherford, L. Chen, and M. N. Rosenbluth, "Stability Limit on Beta in a Tokamak Using the Collisionless Energy Principle," Princeton Plasma Physics Laboratory Report, No. PPPL-1418 (February, 1978).

1979

210. B. Coppi, R. Galvão, R. Pellat, M. N. Rosenbluth, and P. H. Rutherford, "Resistive Internal Kink Modes," in *Plasma Heating in Toroidal Devices*, Proceedings of the Third Symposium, Varenna, Italy, 6–17 September 1976 (Pergamon Press, Oxford, 1979), pp. 199–201.

211. D. A. Monticello, R. B. White, and M. N. Rosenbluth, "Feedback Stabilization of Magnetic Islands in Tokamaks," in *Plasma Physics and Controlled Nuclear Fusion Research 1978*, Proceedings of the Seventh International Conference, Innsbruck (International Atomic Energy Agency, Vienna, 1979), Vol. I., pp. 605–613.

212. N. A. Krall, S. Hamasaki, J. B. McBride, N. T. Gladd, P. H. Ng, H. H. Chen, J. D. Huba, R. C. Davidson, R. E. Aamodt, Y. C. Lee, C. S. Liu, D. R. Nicholson, M. N. Rosenbluth, and D. P. Chernin, "Drift Wave Stability and Transport Theory in Fusion Systems," in *Plasma Physics and Controlled Nuclear Fusion Research 1978*, Proceedings of the Seventh International Conference, Innsbruck (International Atomic Energy Agency, Vienna, 1979), Vol. II, pp. 483–495.

213. M. N. Bussac, H. P. Furth, M. Okabayashi, M. N. Rosenbluth, and A. M. M. Todd, "Low-Aspect-Ratio Limit of the Toroidal Reactor: The Spheromak," in *Plasma Physics and Controlled Nuclear Fusion Research 1978*, Proceedings of the Seventh International Conference, Innsbruck (International Atomic Energy Agency, Vienna, 1979), Vol. III, pp. 249–264.

214. P. J. Catto and M. N. Rosenbluth, "Collisionless Limit of Bumpy Torus Transport," in *Proceedings of the EBT Transport Workshop*, Gaithersburg, MD, May, 1979, Department of Energy, Washington, DC, Office of Fusion Energy Report, No. DOE/ET-0112, CONF-7905112 (1979), 16 pp., no pagination.

215. E. A. Williams, J. R. Albritton, and M. N. Rosenbluth, "Effect of Spatial Turbulence on Parametric Instabilities," *Physics of Fluids* **22**, 139–149 (1979).

216. R. N. Sudan and M. N. Rosenbluth, "Stability of Axisymmetric Field-Reversed Equilibria of Arbitrary Ion Gyroradius," *Physics of Fluids* **22**, 282–293 (1979).

217. M. N. Rosenbluth and M. N. Bussac, "MHD Stability of Spheromak," *Nuclear Fusion* **19**, 489–498 (1979).

218. A. B. Rechester, M. N. Rosenbluth, and R. B. White, "Calculation of the Kolmogorov Entropy for Motion Along a Stochastic Magnetic Field," *Physical Review Letters* **42**, 1247–1250 (1979).

219. H. L. Buchanan, F. W. Chambers, E. P. Lee, S. S. Yu, R. J. Briggs, and M. N. Rosenbluth, "Transport of Intense Particle Beams with Application to Heavy Ion Fusion," presented at the Third International Topical Conference on High Power and Ion Beam Research and Technology, Novosibirsk, USSR, 3–6 July, 1979, Lawrence Livermore National Laboratory Report, No. UCRL-82586 (June, 1979).

220. A. B. Rechester, M. N. Rosenbluth, and R. B. White, "Statistical Description of Stochastic Orbits in a Tokamak," in *Intrinsic Stochasticity in Plasmas*, International Workshop, Cargese, Corsica, France, June, 1979, edited

by G. Laval and D. Gresillon (Ecole Polytechnique, Palaiseau, France, 1979), pp. 239–259.

221. P. J. Catto, M. N. Rosenbluth, and K. T. Tsang, "Resistive Drift-Alfvén Waves in Sheared Magnetic Fields," *Physics of Fluids* **22**, 1284–1288 (1979).

1980

222. N. M. Kroll, P. L. Morton, and M. N. Rosenbluth, "Variable Parameter Free-Electron Laser," in *Free-Electron Generators of Coherent Radiation*, Physics of Quantum Electronics, Vol. 7, Proceedings of the Second ONR Workshop on Free Electron Generators of Coherent Radiation, Telluride, CO, 1979, edited by S. F. Jacobs, H. S. Pilloff, M. Sargent III, M. O. Scully, and R. Spitzer (Addison-Wesley Publishing Co., Reading, Massachusetts, 1980), pp. 89–112.

223. N. M. Kroll, P. L. Morton, and M. N. Rosenbluth, "Enhanced Energy Extraction in Free-Electron Lasers by Means of Adiabatic Decrease of Resonant Energy," in *Free-Electron Generators of Coherent Radiation*, Physics of Quantum Electronics, Vol. 7, Proceedings of the Second ONR Workshop on Free Electron Generators of Coherent Radiation, Telluride, CO, 1979, edited by S. F. Jacobs, H. S. Pilloff, M. Sargent III, M. O. Scully, and R. Spitzer (Addison-Wesley Publishing Co., Reading, Massachusetts, 1980), pp. 113-145.

224. N. M. Kroll and M. N. Rosenbluth, "Sideband Instabilities in Trapped Particle Free-Electron Lasers," in *Free-Electron Generators of Coherent Radiation*, Physics of Quantum Electronics, Vol. 7, Proceedings of the Second ONR Workshop on Free Electron Generators of Coherent Radiation, Telluride, CO, 1979, edited by S. F. Jacobs, H. S. Pilloff, M. Sargent III, M. O. Scully, and R. Spitzer (Addison-Wesley Publishing Co., Reading, Massachusetts, 1980), pp. 147-174.

225. E. P. Lee, H. L. Buchanan, and M. N. Rosenbluth, "Filamentation of a Converging Heavy Ion Beam," in *Heavy Ion Fusion Workshop Proceedings*, Berkeley, CA, October, 1979, Lawrence Livermore National Laboratory Report, No. UCRL-83952 (CONF-7910122-9) (1980).

226. A. B. Rechester, M. N. Rosenbluth, and R. B. White, "Statistical Description of Stochastic Orbits in a Tokamak," *Journal de Physique*, Vol. 41, Colloque C3, Suppl. No. 4, pp. 351–358 (1980).

227. F. L. Hinton and M. N. Rosenbluth, "Convective Amplication of Universal Drift Modes," *Physics of Fluids* **23**, 528–536 (1980).

228. P. J. Catto, R. E. Aamodt, M. N. Rosenbluth, J. A. Byers, and L. D. Pearlstein, "Axis Encircling Ion Gyroinstability," *Physics of Fluids* **23**, 764–770 (1980).

229. P. J. Catto, M. N. Rosenbluth, and K. T. Tsang, "Reply to the Comments by K. Itoh and S. Inoue," *Physics of Fluids* **23**, 848 (1980).

230. E. P. Lee, S. Yu, H. L. Buchanan, F. W. Chambers, and M. N. Rosenbluth, "Filamentation of a Heavy-Ion Beam in a Reactor Vessel," *Physics of Fluids* **23**, 2095–2110 (1980).

1981

231. N. M. Kroll, P. L. Morton, M. N. Rosenbluth, J. N. Eckstein, and J. M. J. Madey, "Theory of the Transverse Gradient Wiggler," *IEEE Journal of Quantum Electronics* **QE-17**, 1496–1507 (1981).
232. D. A. Spong, J. W. Van Dam, H. L. Berk, and M. N. Rosenbluth, "Numerical Solutions of the EBT Radial Eigenmode Problem," in *EBT Stability Theory*, Proceedings of the Workshop, Oak Ridge, Tennessee, 13–4 May 1981, edited by N. A. Uckan (Oak Ridge National Laboratory, Oak Ridge, 1981, CONF-810512), pp. 115-140.
233. J. W. Van Dam, H. L. Berk, M. N. Rosenbluth, and D. A. Spong, "Eigenmode Stability Analysis for a Bumpy Torus," in *EBT Stability Theory*, Proceedings of the Workshop, Oak Ridge, Tennessee, 13-14 May 1981, edited by N. A. Uckan (Oak Ridge National Laboratory, Oak Ridge, 1981, CONF-810512), pp. 97-114.
234. A.B. Rechester, M. N. Rosenbluth, and R.B. White, "Fourier-Space Paths Applied to the Calculation of Diffusion for the Chirikov-Taylor Model," *Physical Review A* **23**, 2664–2672 (1981).
235. N. M. Kroll, P. Morton, and M. N. Rosenbluth, "Free Electron Lasers with Variable Parameter Wigglers," *IEEE Journal of Quantum Electronics* **QE-17**, 1436–1468 (1981).
236. F. L. Hinton, R. D. Hazeltine, D. A. Hitchcock, W. Horton, S. M. Mahajan, D. W. Ross, H. R. Strauss, A. A. Ware, J. W. Wiley, P. J. Catto, and M. N. Rosenbluth, "Relation between Tokamak Temperature Profiles and Localized Instabilities," in *Plasma Physics and Controlled Nuclear Fusion Research 1980*, Proceedings of the Eigth International Conference, Brussels (International Atomic Energy Agency, Vienna, 1981), Vol. I, pp. 365–373.
237. M. N. Rosenbluth and P. H. Rutherford, "Tokamak Plasma Stability," in *Fusion, Vol. 1: Magnetic Confinement, Part A*, edited by E. Teller (Academic Press, New York, 1981), pp. 31–121.
238. R. E. Aamodt, B. I. Cohen, Y. C. Lee, C. S. Liu, D. R. Nicholson, and M. N. Rosenbluth, "Nonlinear Evolution of Drift Cyclotron Modes," *Physics of Fluids* **24**, 55–65 (1981).
239. P. J. Catto and M. N. Rosenbluth, "Trapped Electron Modifications to Tearing Modes in the Low Collision Frequency Limit," *Physics of Fluids* **24**, 243–255 (1981).
240. B. A. Carreras, M. N. Rosenbluth, and H. R. Hicks, "Nonlinear Destabilization of Tearing Modes," *Physical Review Letters* **46**, 1131–1134 (1981).

241. K. Molvig, S. P. Hirshman, A. B. Rechester, R. B. White, M. N. Rosenbluth, J. F. Drake, N. T. Gladd, A. B. Hassam, C. S. Liu, and C. L. Chang, "Theory of Stochastic Magnetic Fluctuations and Anomalous Thermal Conductivity in Tokamaks," in *Plasma Physics and Controlled Nuclear Fusion Research 1980*, Proceedings of the Eighth International Conference, Brussels (International Atomic Energy Agency, Vienna, 1981), Vol. I, pp. 73–81.

242. P. J. Catto, R. D. Hazeltine, S. Hamasaki, H. H. Klein, N. A. Krall, J. B. McBride, J. L. Sperling, C. L. Hedrick, D. B. Batchelor, L. E. Deleanu, R. C. Goldfinger, E. F. Jaeger, D. B. Nelson, D. A. Spong, J. T. Tolliver, N. A. Uckan, J. W. Van Dam, Y. C. Lee, and M. N. Rosenbluth, "Elmo Bumpy Torus (EBT) Transport, Heating, and Stability," in *Plasma Physics and Controlled Nuclear Fusion Research 1980*, Proceedings of the Eighth International Conference, Brussels (International Atomic Energy Agency, Vienna, 1981), Vol. I, pp. 821–830.

243. M. N. Rosenbluth, "Magnetic Trapped-Particle Modes," *Physical Review Letters* **46**, 1525–1528 (1981).

244. P. H. Diamond and M. N. Rosenbluth, "Theory of the Renormalized Dielectric for Electrostatic Drift Wave Turbulence in Tokamaks," *Physics of Fluids* **24**, 1641–1649 (1981).

245. P. H. Diamond and M. N. Rosenbluth, "Theory of Ion Compton Scattering for Drift Wave Turbulence in a Sheared Magnetic Field," Institute for Fusion Studies Report (University of Texas at Austin), No. IFSR-24 (1981).

1982

246. D. A. Spong, H. L. Berk, J. W. Van Dam, and M. N. Rosenbluth, "Radial Structure of Instability Modes in the EBT Hot Electron Annulus," in *Hot Electron Ring Physics*, Proceedings of the Second Workshop, San Diego, CA, December, 1981, edited by N. A. Uckan (Oak Ridge National Laboratory, Oak Ridge, 1982, CONF-811203), Vol. 1, pp. 93–116.

247. H. L. Berk, J. W. Van Dam, M. N. Rosenbluth, and D. A. Spong, "Curvature-Driven Instabilities in a Hot Electron Plasma: Radial Analysis," in *Hot Electron Ring Physics*, Proceedings of the Second Workshop, San Diego, CA, December, 1981, edited by N. A. Uckan (Oak Ridge National Laboratory, Oak Ridge, 1982, CONF-811203), Vol. 1, pp. 83-91.

248. P. H. Diamond, M. N. Rosenbluth, S. P. Hirshman, and J. R. Myra, "Renormalized Theories of Low Frequency Turbulence," in *Fusion Energy–1981* (International Centre for Theoretical Physics, Trieste, 1982, IAEA-SMR-82), pp. 25-28.

249. N. M. Kroll and M. N. Rosenbluth, "Theory of Gain Expanded Free Electron Lasers," *Journal de Physique (Paris)*, Vol. 44, Colloque C1, Supplement No. 2 (Proceedings of the Bendor Free Electron Laser Conference, 26 September– 1 October 1982, Bendor, France), pp. 85-107.

250. M. N. Rosenbluth, H. V. Wong, and B. N. Moore, "Annual Technical Report for Theoretical Studies on Free Electron Lasers," ARA Report I-ARA-82-U-89, NTIS AD-A121673/8 (August, 1982).
251. F. L. Hinton and M. N. Rosenbluth, "Stabilization of Axisymmetric Mirror Plasmas by Energetic Ion Injection," *Nuclear Fusion* **22**, 1547–1557 (1982).
252. M. N. Rosenbluth, "Topics in Plasma Instabilities: Trapped-Particle Modes and MHD," *Physica Scripta* **T2/1**, 104–109 (1982).
253. J. W. Van Dam, M. N. Rosenbluth, and Y. C. Lee, "A Generalized Kinetic Energy Principle," *Physics of Fluids* **25**, 1349–1354 (1982).
254. A. B. Rechester, M. N. Rosenbluth, R. B. White, and C. F. F. Karney, "Statistical Description of the Chirikov-Taylor Model in the Presence of Noise," in *Long Time Prediction in Dynamics*, edited by C. W. Horton, Jr., L. Reichl, and V. G. Szebehely (John Wiley Publishers, New York, 1982), pp. 471–483.

1983

255. J. W. Van Dam, M. N. Rosenbluth, S. T. Tsai, and M. G. Engquist, "Energetic Particles in Tokamaks: Stabilization of Ballooning Modes," in *Advanced Bumpy Torus Concepts*, Proceedings of the Workshop, Rancho Santa Fe, CA, 11–13 July 1983, edited by N. A. Uckan (Oak Ridge National Laboratory, Oak Ridge, 1983, CONF-830758), pp. 167-176.
256. H. L. Berk, J. W. Van Dam, M. N. Rosenbluth, D. A. Spong, and C. Z. Cheng, "EBT Stability Theory," *Nuclear Instruments and Methods in Physics Research* **207**, 267–270 (1983).
257. H. L. Berk, M. N. Rosenbluth, and L. Shohet, "Ballooning Mode Code in Stellarators," in *Proceedings of the Fourth U.S. Stellarator Design Center Workshop* (Oak Ridge National Laboratory, Oak Ridge, 1983, CONF-830428), pp. 198-211.
258. P. H. Diamond, P. L. Similon, P. W. Terry, C. W. Horton, S. M. Mahajan, J. D. Meiss, M. N. Rosenbluth, K. Swartz, T. Tajima, R. D. Hazeltine, and D. W. Ross, "Theory of Two-Point Correlation for Trapped Electrons and the Spectrum of Drift Wave Turbulence," in *Plasma Physics and Controlled Nuclear Fusion Research 1982*, Proceedings of the Ninth International Conference, Baltimore (International Atomic Energy Agency, Vienna, 1983), Vol. I, pp. 259–269.
259. M. N. Rosenbluth, H. V. Wong, and B. N. Moore, "Theoretical Studies on Free Electron Lasers," ARA Report I-ARA-83-U-62, NTIS AD-A136333/2 (November, 1983).
260. H. L. Berk, M. N. Rosenbluth, H. V. Wong, T. M. Antonsen, Jr., D. E. Baldwin, and B. Lane, "Curvature-Driven Trapped-Particle Modes in Tandem Mirrors," in *Plasma Physics and Controlled Nuclear Fusion Research 1982*, Proceedings of the Ninth International Conference, Baltimore (International Atomic Energy Agency, Vienna, 1983), Vol. II, pp. 175–182.

261. H. Abe, H. L. Berk, C. Z. Cheng, M. N. Rosenbluth, J. W. Van Dam, D. A. Spong, N. A. Uckan, T. M. Antonsen, Jr., Y. C. Lee, K. T. Tsang, P. J. Catto, X. S. Lee, K. T. Nguyen, and T. Kammash, "Curvature-Driven Instabilities in the ELMO Bumpy Torus (EBT)," in *Plasma Physics and Controlled Nuclear Fusion Research 1982*, Proceedings of the Ninth International Conference, Baltimore (International Atomic Energy Agency, Vienna, 1983), Vol. III, pp. 427–441.

262. H. L. Berk, M. N. Rosenbluth, H. V. Wong, T. M. Antonsen, and D. E. Baldwin, "Fast Growing Trapped-Particle Modes in Tandem Mirrors," *Fizika Plazmy* **9**, 176–183 (1983) [*Soviet Journal of Plasma Physics* **9**, 108–112 (1983)].

263. H. L. Berk, J. W. Van Dam, M. N. Rosenbluth, and D. A. Spong, "Curvature-Driven Instabilities in a Hot Electron Plasma: Radial Analysis," *Physics of Fluids* **26**, 201–215 (1983).

264. M. N. Rosenbluth, P. J. Catto and X. S. Lee, "Effects of Ion Bounce Resonances on Ballooning-Interchange Modes," *Physics of Fluids* **26**, 216–222 (1983).

265. A. Bhattacharjee, J. E. Sedlak, P. L. Similon, M. N. Rosenbluth, and D. W. Ross, "Drift Waves in a Straight Stellarator," *Physics of Fluids* **26**, 880–882 (1983).

266. H. L. Berk, M. N. Rosenbluth, and J. L. Shohet, "Ballooning Mode Calculations in Stellarators," *Physics of Fluids* **26**, 2616–2620 (1983).

267. H. L. Berk, C. Z. Cheng, M. N. Rosenbluth, and J. W. Van Dam, "Finite Larmor Radius Stability Theory of ELMO Bumpy Torus Plasmas," *Physics of Fluids* **26**, 2642–2651 (1983).

268. D. A. Spong, H. L. Berk, J. W. Van Dam, and M. N. Rosenbluth, "Anisotropy Effects on Curvature-Driven Flute Instabilities in a Hot-Electron Plasma," *Physics of Fluids* **26**, 2652–2656 (1983).

269. M. N. Rosenbluth, S. T. Tsai, J. W. Van Dam, and M. G. Engquist, "Energetic Particle Stabilization of Ballooning Modes in Tokamaks," *Physical Review Letters* **51**, 1967–1970 (1983).

270. T. M. Antonsen, Jr., Y. C. Lee, H. L. Berk, M. N. Rosenbluth, and J. W. Van Dam, "Ballooning Instabilities in Hot Electron Plasmas," *Physics of Fluids* **26**, 3580–3594 (1983).

271. M. N. Rosenbluth, H. V. Wong and B. N. Moore, "Final Technical Report for Theoretical Studies on Free Electron Lasers," Austin Research Associates Report, No. I-ARA-83-U-62 (ARA-502), AD-A136333/2, (November, 1983).

272. M. N. Rosenbluth, H. V. Wong, B. N. Moore, C. A. Brau, S. F. Jacobs, and M. O. Scully, "Free Electron Laser (Oscillator)—Linear Gain and Stable Pulse Propagation," in *Free-Electron Generators of Coherent Radiation*, Proceedings of the Workshop, 1983 (Proc. SPIE 453, International Society for Optical Engineering, Bellingham, Washington, 1984), pp. 25–40.

1984

273. M. N. Rosenbluth, H. V. Wong, B. N. Moore, and G. I. Bourianoff, "Final Theoretical Studies of Two-Dimensional Effects in Free Electron Lasers," ARA Report I-ARA-84-U-121, NTIS AD-A158229/5, December, 1984.
274. H. L. Berk, M. N. Rosenbluth, and J. L. Shohet, "Reply to the comments by D. Correa-Restrepo," *Physics of Fluids* **27**, 1342 (1984).
275. H. L. Berk, M. N. Rosenbluth, H. V. Wong, and T. M. Antonsen, Jr., "Stabilization of an Axisymmetric Tandem Mirror Cell by a Hot Plasma Component," *Physics of Fluids* **27**, 2705–2710 (1984).
276. L. Chen, R. B. White, and M. N. Rosenbluth, "Excitation of Internal Kink Modes by Trapped Energetic Beam Ions," *Physical Review Letters* **52**, 1122–1125 (1984).

1985

277. R. W. Conn, R. C. Davidson, T. K. Fowler, H. P. Furth, J. A. Gilleland, and M. N. Rosenbluth, "A Strategic Framework for the Magnetic Fusion Energy Program," *Journal of Fusion Energy* **4**, 5–10 (1985).
278. Z. G. An, P. H. Diamond, R. D. Hazeltine, J. N. Leboeuf, M. N. Rosenbluth, R. D. Sydora, T. Tajima, B. A. Carreras, L. Garcia, T. C. Hender, H. R. Hicks, J. A. Holmes, V. E. Lynch, and H. R. Strauss, "Role of Multiple Helicity Non-Linear Interaction of Tearing Modes in Dynamo and Anomalous Thermal Transport in Reversed-Field Pinch," in *Plasma Physics and Controlled Nuclear Fusion Research 1984*, Proceedings of the Tenth International Conference, London (International Atomic Energy Agency, Vienna, 1985), Vol. 2, pp. 231–243.
279. L. Chen, R. B. White, C. Z. Cheng, F. Romanelli, J. Weiland, R. Hay, J. W. Van Dam, M. N. Rosenbluth, S. T. Tsai, and D. C. Barnes, "Theory and Simulation of Fishbone-Type Instabilities in Beam-Heated Tokamaks," in *Plasma Physics and Controlled Nuclear Fusion Research 1984*, Proceedings of the Tenth International Conference, London (International Atomic Energy Agency, Vienna, 1985), Vol. II, pp. 59–66.
280. H. L. Berk, C. W. Horton, Jr., M. N. Rosenbluth, H. V. Wong, J. Kesner, B. Lane, T. M. Antonsen, Jr., K. T. Tsang, X. S. Lee, B. Hafizi, J. A. Byers, R. H. Cohen, J. H. Hammer, W. M. Nevins, T. B. Kaiser, L. Lodestro, L. D. Pearlstein, G. R. Smith, H. Ramachandran, and W. M. Tang, "Stabilization of an Axisymmetric Mirror Cell and Trapped Particle Modes," in *Plasma Physics and Controlled Nuclear Fusion Research 1984*, Proceedings of the Tenth International Conference, London (International Atomic Energy Agency, Vienna, 1985), Vol. II, pp. 321–335.

281. H. L. Berk, M. N. Rosenbluth, R. H. Cohen, and W. M. Nevins, "Dissipative Trapped Particle Modes in Tandem Mirrors," *Physics of Fluids* **28**, 2824–2837 (1985).

282. M. N. Rosenbluth, "Two-Dimensional Effects in Free-Electron Lasers," *IEEE Journal of Quantum Electronics* **QE-21**, 966–969 (1985).

283. M. N. Rosenbluth, B. N. Moore, and H. V. Wong, "Steady-State Operation of FEL Oscillator Using Phase Area Displacement Wiggler," *IEEE Journal of Quantum Electronics* **QE-21**, 1026–1033 (1985).

1986

284. S. Eliezer, T. Tajima, and M. N. Rosenbluth, "High Intensity Particle Beams for a Muon-Catalyzed Fusion-Fission Reactor," in *Laser Interactions and Related Plasma Phenomenon*, edited by H. Hora and G. Miley (Plenum Press, New York, 1986), Vol. 7, pp. 613–617.

285. A. Ishida, R. N. Sudan, M. N. Rosenbluth, and M. G. Engquist, "Kink Stability of a Field-Reversed Ion Layer in a Background Plasma," *Physics of Fluids* **29**, 2347–2350 (1986).

286. M. N. Rosenbluth and R. N. Sudan, "Almost Two-Dimensional Strong Turbulence in a Magnetized Plasma," *Physics of Fluids* **29**, 2347–2350 (1986).

287. M. N. Rosenbluth, "The Quest for Magnetically Confined Thermonuclear Fusion," Mortimer and Raymond Sackler Institute of Advanced Studies, Tel-Aviv, Israel, University Report #IAS 828-86 (1986).

1987

288. A. Y. Aydemir, H. L. Berk, V. Mirnov, O. P. Pogutse, and M. N. Rosenbluth, "Linear and Nonlinear Description of Drift Instabilities in a High-Beta Plasma," *Physics of Fluids* **30**, 3083–3092 (1987).

289. J. W. Van Dam, M. N. Rosenbluth, H. L. Berk, N. Dominguez, G. Y. Fu, X. Llobet, D. W. Ross, D. P. Stotler, D. A. Spong, W. Cooper, D. Sigmar, D. E. Hastings, J. J. Ramos, H. Naitou, J. Todoroki, S. T. Tsai, S. G. Guo, and J. W. Shen, "Effects of Energetic Particles on Tokamak Stability," in *Plasma Physics and Controlled Nuclear Fusion Research 1986*, Proceedings of the Eleventh International Conference, Kyoto (International Atomic Energy Agency, Vienna, 1987), Vol. II, pp. 135–145.

290. S. Eliezer, T. Tajima, and M. N. Rosenbluth, "A Muon Catalysed Fusion-Fission Reactor," in *Plasma Physics and Controlled Nuclear Fusion Research 1986*, Proceedings of the Eleventh International Conference, Kyoto (International Atomic Energy Agency, Vienna, 1987), Vol. III, pp. 301–307.

291. M. N. Rosenbluth, "Development of Nuclear Fusion Research," *Kakuyugo Kenkyu* **57**, No. 6, 354–363 (1987).

292. S. Eliezer, T. Tajima, and M. N. Rosenbluth, "Muon Catalysed Fusion-Fission Reactor Driven by a Recirculating Beam," *Nuclear Fusion* **27**, 527–547 (1987).

293. M. N. Rosenbluth, H. L. Berk, I. Doxas, and W. Horton, "Effective Diffusion in Laminar Convective Flows," *Physics of Fluids* **30**, 2636–2647 (1987).

294. X. Llobet, H. L. Berk, and M. N. Rosenbluth, "Finite Pressure Ballooning Mode Stability in Toroidal Equilibriums," *Physics of Fluids* **30**, 2750–2758 (1987).

Index

Credits

Coppi article

Figure 4 adapted from M. Greenwald et al., *Plasma Physics and Controlled Nuclear Fusion Research* 1984, (IAEA, Vienna, 1985), Vol. I, p. 45, Fig. 5. Reprinted by permission.

White article

Figure 2 from M. N. Rosenbluth, D. A. Monticello, H. R. Strauss, R. B. White; *Physics of Fluids* **19** (1976), p. 1987, Fig. 2. Reprinted permission of the American Institute of Physics.

Figure 4 from B. V. Waddell, M. N. Rosenbluth, D. A. Monticello, R. B. White; *Nuclear Fusion* **16** (1976), p. 528, Fig. 2.

Figure 5 courtesy Dr. Wonchull Park.

Figure 7 from D. A. Monticello, R. B. White, M. N. Rosenbluth; *Plasma Physics and Controlled Nuclear Fusion Research* 1978, (IAEA, Vienna, 1979), Vol. 1, p. 605.

Figure 8 from B. Carreras, H. R. Hicks, J. A. Holmes, B. V. Waddell; *Physics of Fluids* **23** (1980), p. 1811, Fig. 6. Reprinted permission of the American Institute of Physics.

Figure 9 reproduced in R. B. White, *Handbook of Plasma Physics, Vol. 1. Basic Plasma Physics I*, ed. A. A. Gallev, R. N. Sudan. ©1983. Fig. 3.5.17, p. 650. Reprinted permission of Elsevier Science Publishers, Physical Sciences and Engineering Division, Amsterdam.

Figure 11 from L. Chen et al., *Plasma Physics and Controlled Nuclear Fusion Research* 1984 (IAEA, Vienna, 1985), Vol. II, p. 59, Fig. 1.

Pearlstein article

Figure 4 from the University of California Lawrence Livermore National Laboratory and the U.S. Department of Energy.

Figure 5 from H. L. Berk, L. D. Pearlstein, J. G. Cordey; *Physics of Fluids* **15** (1972), p. 891, Fig. 5. Reprinted permission of the American Institute of Physics.

Figure 6, 7a, and 7b from *Status of Mirror Fusion Research* 1980, ed. B. I. Cohen, Lawrence Livermore National Laboratory, University of California Report,
No. UCAR-10049-80-Rev. 1 (1980) Figs. 2.4–22 and 2.4–25(a) & (b).

Figure 8 adapted from B. E. Kaneev *Nuclear Fusion* **19** (1979), p. 347, Figs. 9 and
13.

Figures 10a, 10b, 11a, 11b, and 12 from *Status of Mirror Fusion Research* 1980,
ed. B. I. Cohen, Lawrence Livermore National Laboratory, University of California
Report, No. UCAR-10049-80-Rev. 1 (1980); Figs. 2.4–20, 2.4–21, 2.4–14, and 2.4–
18.

Figure 13 from T. A. Casper and G. R. Smith, *Physical Review Letters* **48** (1982),
p. 1015, Fig. 3. Reprinted permission of the American Physical Society.

Figure 14 from T. Simonen et al., *Mirror-Based and Field-Reversed Approaches
to Magnetic Fusion*, ed. R. F. Post et al. (EUR-8961-EN, International School of
Plasma Physics–Monotypia Franchi, Varenna, Italy, 1983). Vol. 1, p. 187, Fig. 12.

Figures 16, 17, 18, and 19 from G. R. Smith, W. M. Nevins, W. M. Sharp; *Physics of
Fluids* **27** (1984), p. 2120; Figs. 3(16), 8(17 and 18), and 9(19). Reprinted permission
of the American Institute of Physics.

Robinson article

Figure 2 from D. C. Robinson. *Nuclear Fusion* **18** (1978), p. 939, Figure 10.

Figure 4 from D. C. Robinson and M. G. Rusbridge, *Physics of Fluids* **14** (1971),
p. 2499, Fig. 11. Reprinted permission of the American Institute of Physics.

Figure 5 from T. C. Hender and D. C. Robinson, *Plasma Physics and Controlled
Nuclear Fusion Research* 1982, (IAEA, Vienna, 1983), Vol. III, p. 417, Fig. 4.

Table 1 from a patent (originally classified) by G.P. Thompson and Blackman
(patent number unknown).

Table 2 from M. N. Rosenbluth, in *Proceedings of the Second United Nations International Conference on the Peaceful Uses of Atomic Energy*, Vol. 31. *Theoretical
and Experimental Aspects of Controlled Nuclear Fusion* (United Nations, Geneva,
1956), p. 85, Table 1.

Dawson article

Figures 4 and 5 from T. Katsouleas et al., *Laser Acceleration of Particles*. AIP
Conference Proceedings No. 130, C. Joshi and T. Katsouleas (eds.) p. 63, Figs. 4
and 3. Reprinted permission of the American Institute of Physics.

Kroll article

Figure 2 from N. M. Kroll, P. L. Morton, M. N. Rosenbluth; *IEEE Journal of
Quantum Electronics* QE-17, p. 1436, Fig. 7. ©1981 IEEE.

Figure 3 from N. M. Kroll, P. L. Morton, M. N. Rosenbluth; "Free-Electron Generators of Coherent Radiation," in *Physics of Quantum Electronics*, Vol. 7, edited by Jacobs/Pilloff/Sargent/ Scully/Spitzer. ©1980, Addison-Wesley Publishing Company, Inc., Reading, MA. Reprinted with permission.

Figures 4 and 5 from N. M. Kroll, P. L. Morton, M. N. Rosenbluth; *IEEE Journal of Quantum Electronics* QE-17, p. 1436, Figs. 12 and 15. ©1981 IEEE.

Figure 6 from E. T. Scharlemann, A. M. Sessler, J. Wurtele; *Physical Review Letters* **54** (1985), p. 1925, Fig. 1. Reprinted permission of the American Physical Society.

Figure 7 from M. N. Rosenbluth, H. V. Wong, B. N. Moore; unpublished ARA Report No. I-ARA-83-U-62 (NTIS No. AD-136333) November, 1983.

Figures 8, 9, and 10 from R. W. Warren et al., *Journal of Quantum Electronics* QE-19, p. 391, Figs. 15, 12, and 14. ©1983 IEEE.

Figure 12 adapted from T. J. Orzechowski et al., *Physical Review Letters* **57** (1986), p. 2172, Figs. 1 and 3. Reprinted permission of the American Physical Society.

Kennel article

Figures 7 and 8 from A. Zachary, C. E. Max, J. Arons, and B. I. Cohen, manuscript in preparation; published in preliminary form as part of the Ph.D theses of A. Zachary, Lawrence Livermore National Laboratory Report, No. UCRL-53793 (1987). Reprinted by permission.

Furth article

Figure 3 from S. Yoshikawa, *Nuclear Fusion* **13** (1973), p. 433, Fig. 18.

Figure 4 from J. Sinnis, M. Okabayashi, J. Schmidt, S. Yoshikawa; *Physical Review Letters* **29** (1972), p. 1214, Fig. 2. Reprinted permission of the American Physical Society.

Figures 5 and 6 from S. Ejima and M. Okabayashi. *Physics of Fluids* **18** (1975), p. 904, Figs. 5 and 6. Reprinted permission of the American Institute of Physics.

Figures 8 and 9 from M. G. Bell et al., *Plasma Physics and Controlled Fusion* **28**, p. 1329, Figs. 4 and 9. ©1986 Pergamon Press.

Figure 11 adapted from R. Hawryluk et al., *Plasma Physics and Controlled Nuclear Fusion Research* 1986, (IAEA, Vienna, 1987), Vol. 1, p. 51,, Fig. 7.

Rutherford article

Figure 1 from "The Tokamak," by H. P. Furth, in *Fusion*, ed. E. Teller; Vol. 1, p. 123, Fig. 44. Academic Press, 1981. Reprinted by permission.

Figure 3 from A. H. Glasser, H. P. Furth, P. H. Rutherford; *Physical Review Letters* **38** (1977), p. 234, Fig. 3(a). Reprinted permission of the American Physical Society.

Figure 4 from P. H. Rutherford, *Physics of Fluids* **16** (1973), p. 1903, Fig. 1. Reprinted permission of the American Institute of Physics.

Figure 5 from R. Carreras, H. R. Hicks, J. A. Holmes, B. V. Waddell; *Physics of Fluids* **23** (1980), p. 1811, Fig. 6. Reprinted permission of the American Institute of Physics.

Figure 8 from K. M. McGuire et al., *Plasma Physics and Controlled Nuclear Fusion Research* 1986 (IAEA, Vienna, 1987), Vol I, p. 421, Fig. 1.

Figure 10 from H. Soltwisch, W. Stodiek, J. Manickam, J. Schluter; *Plasma Physics and Controlled Nuclear Fusion Research* 1986, (IAEA, Vienna, 1987), Vol. 1, p. 263, Fig. 4.

Figure 15 from D. W. Ignat, P. H. Rutherford, and H. Hsuan, *Course and Workshop on Application of rf Waves to Tokamak Devices*, Varenna, Italy, 1985 (CEC EUR 10333 EN, Brussels, 1986), p. 525.

Figure 16 from P. H. Rutherford, in *Course and Workshop on Basic Physical Processes of Toroidal Fusion Plasmas*, Varenna, Italy, 1985 (CEC EUR 10418 EN, Brussels, 1986), p. 531.

Sudan article

Figures 3 and 5 from R. N. Sudan and D. Pfirsch, *Physics of Fluids* **28** (1985), p. 1702, Figs. 1 (adapted) and 10(c). Reprinted permission of the American Institute of Physics.

Figure 7 from R. N. Sudan and D. Pfirsch, *Physics of Fluids* **29** (1986), p. 2347, Fig. 1. Reprinted permission of the American Institute of Physics.

Krall article

Table 1 adapted from P. C. Liewer, *Nuclear Fusion* **25** (1985), p. 543, Table III.

Figure 3 from N. A. Krall, *Physics of Fluids* **30** (1987), p. 878, Fig. 1. Reprinted permission of the American Institute of Physics.